DATE			

BAKER & TAYLOR

BATTERY HAZARDS AND ACCIDENT PREVENTION

BATTERY HAZARDS AND ACCIDENT PREVENTION

Samuel C. Levy
Sandia National Laboratories
Albuquerque, New Mexico

and

Per Bro
Southwest Electrochemical Company
Santa Fe, New Mexico

PLENUM PRESS • NEW YORK AND LONDON

Library of Congress Cataloging-in-Publication Data

Levy, Samuel C.
    Battery hazards and accident prevention / Samuel C. Levy and Per
Bro.
        p.   cm.
    Includes bibliographical references and index.
    ISBN 0-306-44758-4
    1. Electric batteries--Safety measures. 2. Electric batteries-
-Accidents.   I. Bro, P. (Per) II. Title.
TK2941.L43   1994
621.31'242'0289--dc20   94-30814
                                                              CIP

ISBN 0-306-44758-4

©1994 Plenum Press, New York
A Division of Plenum Publishing Corporation
233 Spring Street, New York, N.Y. 10013

Preface

This book is about how to avoid the accidents and injuries that may occur when batteries are abused or mishandled. It is the first book to deal specifically with this subject in a reasonably comprehensive manner accessible to readers ranging from regular consumers to technical specialists. Batteries and battery processes are described in sufficient detail to enable readers to understand why and how batteries cause accidents and what can be done to prevent them. Each year in the United States alone, thousands of individuals are injured by battery accidents, some of which are severely disabling. The tragedy is that such accidents need not occur.

The book is intended to satisfy the needs of a varied group of readers: battery users in general, battery engineers, and designers of battery-operated equipment and consumer electronics. Since the book is a reference source of information on batteries and battery chemicals, we believe it may also be useful to those studying the environment as well as to medical personnel called upon to treat battery injuries. There are no prerequisites for an understanding of the text other than an interest in batteries and their safe usage.

Nontechnical readers may skip the more technical segments of the book, such as the chapters of Part II, with no loss of understanding of the essential characteristics of batteries and their hazards. They may wish to begin with Chapter 2 and Part V, both of which deal with accidents and accident prevention, before dipping into Chapters 1, 7, and 8 for information on batteries in general and on specific battery systems. If they are using lithium batteries, Part IV should be considered mandatory reading. Battery engineers will find that they know some of the material in the book already. They may want to skim lightly over Chapter 1 and Part III and to concentrate their reading on Parts II, IV, and V. The book should prove useful as a resource on batteries for designers of battery-operated equipment and consumer electronic devices.

Designers play an essential role in ensuring the safety of battery-operated equipment and electronic devices for general consumer use. In order to accomplish that objective, they need to understand battery processes, the causes of battery hazards, and the possible effects of electronic and electrical circuitry on the development of hazardous conditions in batteries. Designers of new and advanced equipment may find Part IV on lithium batteries of particular value. It contains essential information on the advanced power sources likely to power consumer electronic devices of the future. Parts III and IV contain much of interest to environmentalists. There, we discuss the chemical composition of batteries and the toxicity of battery materials. We do not discuss the effects of battery materials on ecosystems, nor the disposal or recycling of batteries. These topics are treated in considerable detail in existing literature to which we refer in the text.

We need to say a few words about the different manner in which we have organized the material on aqueous battery systems and that on nonaqueous lithium battery systems. Aqueous systems represent an essentially mature technology, and they have many common features. This enabled us to organize the existing information within a simply structured framework. Lithium batteries, on the other hand, are still in development and display far greater chemical and physical variations than do aqueous batteries. Some lithium systems have found important applications, but the majority of them have yet to make a significant commercial impact. It is too early to say which among the many systems will become important in the general battery market. In addition, there is little commonality between the different lithium systems. We chose, therefore, to discuss each lithium system separately and in some detail to provide readers with information that may allow them to assess the relative merits of the various systems from an operational and safety point of view. Thus, lithium batteries occupy a disproportionate fraction of text relative to their commercial importance. We think this approach is justified both by the promise of lithium batteries for future applications and by their greater hazard potential compared with that of aqueous electrolyte battery systems.

We benefited greatly from our discussions with friends and colleagues and by their advice during the preparation of the book. George Schwartz was very helpful with medical advice, Paul Krehl provided useful information on the practical aspects of lithium battery safety, and D. B. Adolf, J. W. Braithwaite, R. G. Buchheit, Jr., S. N. Burchett, W. R. Cieslak, J. M. Freese, and R. A. Guidotti of the Sandia National Laboratories were all very helpful with information on material properties relevant to battery safety. We are greatly indebted to them all and thank them for their contributions.

This work was partially supported by the United States Department of Energy under contract DE-ACO4-94 AL 8500.

Contents

PART I. INTRODUCTION TO BATTERIES AND SAFETY

1. Batteries and Battery Processes 3

    1.1. Battery Structures 6
    1.2. Active Materials 8
    1.3. Elementary Processes 12
        1.3.1. Primary Aqueous Electrolyte Cells 13
        1.3.2. Primary Lithium Cells 14
        1.3.3. Rechargeable Nickel/Cadmium Cells 17
        1.3.4. Rechargeable Lead–Acid Cells 18
        1.3.5. Primary Multicell Batteries 19
        1.3.6. Rechargeable Multicell Batteries 20
    References 21

2. The Nature of Battery Hazards and Accidents 23

    2.1. Battery Hazards 23
    2.2. Accident Reports 35
    References 38

PART II. FUNDAMENTAL ASPECTS OF BATTERY SAFETY

3. Battery Leakage 43

    3.1. Gas Generation in Batteries 43
        3.1.1. Spontaneous Gas Generation, Internal Drivers 44
        3.1.2. Electrical Gas Generation, External Drivers ... 48
        3.1.3. Gas-Generated Cell Pressures 55

3.2. Other Internal Leakage Drivers 58
3.3. Leakage Paths 60
    3.3.1. Crimp Seals 61
    3.3.2. Hermetic Seals 65
    3.3.3. Corrosion-Induced Leakage Paths 69
3.4. Externally Induced Leakage 71
3.5. Leakage Rates 76
    3.5.1. Electrolyte Leakage 77
    3.5.2. Gas Leakage 80
References 84

4. Ruptures 87

4.1. Mechanical Stress and Pressure Tolerance of Cells 88
4.2. Safety Vents 95
References 97

5. Explosions 99

5.1. The Cause of Battery Explosions 101
5.2. The Explosive Process 104
References 111

6. Thermal Runaway 113

6.1. High Discharge Rates 115
6.2. Short Circuits 117
6.3. Charging and Overcharging 125
Bibliography 131

PART III. AQUEOUS ELECTROLYTE BATTERIES

7. Primary Batteries 135

7.1. Zinc/Carbon Batteries (Leclanché Batteries) 136
7.2. Zinc/Chloride Batteries (Heavy-Duty Zinc/Carbon
    Batteries) 138
7.3. Alkaline Manganese Batteries 139
7.4. Silver Oxide Batteries 140
7.5. Mercuric Oxide Batteries 142
7.6. Cadmium/Mercuric Oxide Batteries 144
7.7. Zinc/Air Batteries 145

7.8. Primary Battery Materials and Toxicity 147
    7.8.1. Cadmium 148
    7.8.2. Manganese 148
    7.8.3. Mercury 149
    7.8.4. Silver 150
    7.8.5. Zinc 151
7.9. Aqueous Electrolyte Toxicity 151
    7.9.1. Ammoniacal Electrolytes 152
    7.9.2. Alkaline Electrolytes 153
    7.9.3. Zinc Chloride Electrolytes 155
References 156

8. Rechargeable Batteries 157

8.1. Lead–Acid Batteries, Vented and Sealed 158
8.2. Nickel/Cadmium Batteries, Vented and Sealed 162
8.3. Nickel/Metal Hydride Batteries 165
8.4. Nickel/Zinc Batteries 168
8.5. Silver/Zinc Batteries 169
8.6. Manganese/Zinc Batteries 173
8.7. Zinc/Air Batteries 174
8.8. Rechargeable Battery Materials and Toxicity 177
    8.8.1. Antimony 177
    8.8.2. Arsenic 178
    8.8.3. Cadmium 179
    8.8.4. Cobalt 179
    8.8.5. Lead 180
    8.8.6. Manganese 181
    8.8.7. Nickel 181
    8.8.8. Silver 182
    8.8.9. Zinc 182
    8.8.10. Metal Hydrides 183
8.9. Aqueous Electrolyte Toxicity 184
    8.9.1. Alkaline Electrolytes 184
    8.9.2. Sulfuric Acid 184
References 186

PART IV. LITHIUM BATTERIES

9. Solid Cathode Lithium Systems 189

9.1. Lithium/Manganese Dioxide 189
    9.1.1. General Safety Considerations 190
    9.1.2. Chemistry 190

    9.1.3.   Safety-Related Design Features   191
    9.1.4.   Abuse Tests   192
  9.2.   Lithium/Poly(Carbonmonofluoride)   193
    9.2.1.   Safety-Related Designs   194
    9.2.2.   Safety/Abuse Tests   194
  9.3.   Lithium/Copper Oxide   195
    9.3.1.   Safety-Related Design Features   196
    9.3.2.   Safety/Abuse Tests   196
  9.4.   Lithium/Copper Oxyphosphate   197
    9.4.1.   Safety Features   198
  9.5.   Lithium/Copper Sulfide   198
    9.5.1.   Chemistry   198
    9.5.2.   Design Features   199
    9.5.3.   Safety/Abuse Tests   199
  9.6.   Lithium/Vanadium Pentoxide   200
    9.6.1.   Chemistry   200
    9.6.2.   Abuse Conditions   201
  9.7.   Lithium/Chromium Oxide   201
    9.7.1.   Safety/Abuse Tests   201
  9.8.   Lithium/Iron Sulfide   202
    9.8.1.   Design Features   202
    9.8.2.   Safety/Abuse Tests   203
  9.9.   Lithium/Silver Vanadium Oxide   203
    9.9.1.   Safety Tests   203
  9.10.   Lithium/Iodine   204
  9.11.   Other Systems   205
    9.11.1.   Lithium/Silver Chromate   205
    9.11.2.   Lithium/Silver–Bismuth Chromate   205
    9.11.3.   Lithium/Bismuth Oxychromate   206
    9.11.4.   Lithium/Bismuth Oxide   206
    9.11.5.   Lithium/Lead Iodide   206
    9.11.6.   Lithium/Bismuth–Lead Oxide   206
    9.11.7.   Lithium/Cobalt Polysulfide   206
  References   206

10.   Lithium/Thionyl Chloride Batteries   211

  10.1.   Documented Safety Incidents   211
  10.2.   Hazardous Reactions   212
    10.2.1.   Normal Discharge   212
    10.2.2.   Forced Overdischarge into Reversal   213
    10.2.3.   Charging   213
    10.2.4.   Other Studies   214
    10.2.5.   Increased Pressure   214

	10.3.	Cell Designs	215
		10.3.1.	Configuration	215
		10.3.2.	Balance	216
		10.3.3.	Vents	219
		10.3.4.	Separators	220
		10.3.5.	Other Safety Innovations	220
	10.4.	Thermal Studies.................................	222
	10.5.	Catalysts.......................................	223
	10.6.	Modeling	223
	10.7.	Safety/Abuse Tests..............................	224
		10.7.1.	Wound Cells	224
		10.7.2.	Bobbin Cells	225
		10.7.3.	Prismatic Cells	225
		10.7.4.	Flat Circular Cells	225
		10.7.5.	Small and Large Cells	225
		10.7.6.	Alloy Anode Cells......................	226
	References	226

11. **Lithium/Sulfur Dioxide Batteries** 233

	11.1.	Hazardous Reactions.............................	233
		11.1.1.	Normal Discharge......................	233
		11.1.2.	Forced Overdischarge	235
		11.1.3.	Charging	236
		11.1.4.	Reaction Enhancement	236
	11.2.	Cell Designs	237
		11.2.1.	Balance	237
		11.2.2.	Vents	238
		11.2.3.	Other Safety-Related Features	238
	11.3.	Thermal Studies.................................	239
	11.4.	Modeling	241
	11.5.	Safety/Abuse Tests..............................	241
	11.6.	Multicell Battery Packs	242
		11.6.1.	Battery Design.........................	242
		11.6.2.	SO_2 Getter	243
	References	243

12. **Other Soluble Depolarizer Lithium Systems** 249

	12.1.	Lithium/Bromine Complex.......................	249
		12.1.1.	Chemistry	249
		12.1.2.	Safety/Abuse Tests	250
		12.1.3.	Design	251

12.2. Lithium/Sulfuryl Chloride 251
       12.2.1. Chemistry 251
       12.2.2. Abuse Tests 252
12.3. Lithium/Chlorine in Sulfuryl Chloride 252
       12.3.1. Safety Design Features 252
       12.3.2. Abuse Tests 253
       References 254

13. **Rechargeable Lithium Systems** 257

13.1. Lithium/Molybdenum Disulfide 257
       13.1.1. General/Design 257
       13.1.2. Safety/Abuse Tests 258
13.2. Lithium Titanium Disulfide 259
       13.2.1. Safety-Related Chemistry 260
       13.2.2. Design 261
       13.2.3. Overcharge Protection 261
       13.2.4. Safety/Abuse Tests 261
13.3. Lithium/Manganese Dioxide 262
       13.3.1. Design 262
       13.3.2. Chemistry 262
       13.3.3. Abuse Tests 263
13.4. Lithium/Sulfur Dioxide 263
       13.4.1. Chemistry 264
       13.4.2. Safety/Abuse Tests 264
13.5. Lithium/Cobalt Oxide 265
       13.5.1. Design Features 265
       13.5.2. Gas Evolution 266
       13.5.3. Abuse Tests 266
13.6. Lithium/Niobium Triselenide 266
       13.6.1. Safety/Abuse Tests 267
13.7. Lithium/Vanadium Oxide 267
13.8. Lithium/Chromium Oxide 268
13.9. Lithium/Niobium Pentoxide 268
13.10. Lithium Ion Systems 268
        13.10.1. Lithium Ion/Cobalt Oxide 269
        13.10.2. Other Systems 269
       References 270

14. **High-Temperature Systems** 273

14.1. Sodium/Sulfur 273
       14.1.1. Chemistry 273
       14.1.2. Cell Design 274

          14.1.3. Multicell Battery Design   275
          14.1.4. Safety/Abuse Tests   276
          14.1.5. Safety Status   278
     14.2. Lithium/Iron Sulfide   278
          14.2.1. Design   279
          14.2.2. Safety Tests   279
     14.3. Lithium/Iron Disulfide "Thermal Batteries"   280
          14.3.1. Design   280
          14.3.2. Safety/Abuse Tests   281
     References   281

PART V. ACCIDENT PREVENTION

15. Safety Evaluation   287

     15.1. Types of Safety/Abuse Tests   287
          15.1.1. Electrical Tests   287
          15.1.2. Environmental Tests   289
     15.2. Test Protocols for Lithium Batteries   291
          15.2.1. U.S. Navy   291
          15.2.2. Underwriters Laboratory   291
     15.3. Test Protocols for Nonlithium Batteries   293
     15.4. Safety Testing during Design   294
          15.4.1. Accelerating Rate Calorimetry   295
          15.4.2. Cell Pressure Measurements   296
          15.4.3. Other Considerations   296
     15.5. Safety Procedures   296
          15.5.1. Hot Cells   297
          15.5.2. Fires   297
     References   298

16. Aqueous Electrolyte Batteries   299

     16.1. Designing for Safety   299
     16.2. Battery Operating Procedures   306
     16.3. Regulatory Aspects   309
     16.4. Battery Disposal   311

17. Ambient Temperature Lithium Systems   313

     17.1. Primary Cells   314
          17.1.1. Designing Cells for Safety   314
          17.1.2. Safe Handling of Lithium Cells and Batteries . .   316

17.1.3. Multicell Batteries   317
17.1.4. Fire Safety   319
17.2. Rechargeable Cells   319
17.2.1. Chemical Considerations   320
17.2.2. Overcharge Protection   320
17.2.3. Short-Circuit Prevention   321
17.2.4. Lithium Dendrite Suppression   321
17.2.5. Multicell Batteries   322
References   322

18. Storage, Transportation, and Disposal of Lithium Cells   325

18.1. Storage   325
18.2. Transportation   326
18.2.1. Cells Containing Less than 0.5 g of Lithium . . .   326
18.2.2. Cells Containing More than 0.5 g of Lithium . .   326
18.2.3. Special Exemptions   328
18.2.4. Discharged Cells   328
18.2.5. Shipment outside the United States   329
18.3. Disposal   329
18.3.1. Treatments   329
18.3.2. Technologies   330
References   332

Definitions and Abbreviations   333

Appendix: Specialty Batteries   337

1. Nickel/Hydrogen Cells   337
2. Silver/Hydrogen Cells   338
3. Redox Flow Batteries   338
3.1. Iron–Chromium Redox System   339
3.2. Zinc/Ferricyanide System   339
4. Liquid Ammonia Battery   339
5. Aluminum–Air System   340
6. Zinc/Bromine Battery   341
7. Zinc/Chlorine Battery   342
References   343

Index   345

I

INTRODUCTION TO BATTERIES AND SAFETY

1

Batteries and Battery Processes

The storage of electricity in small or large packages depends upon the existence of chemical reactions that occur when electrons are transferred from one chemical species to another. In batteries, electrons are stored in electron-rich substances (donors), and when the batteries are discharged, the electrons flow from the donor, through the external circuit where they give up much of their energy, and return to the battery to be stored as low-energy electrons in chemical substances that are electron acceptors. This process is sketched schematically in Fig. 1.1. Since a net negative charge cannot accumulate on the acceptor without stopping the further flow of electrons, the charge must be neutralized by an influx of positively charged species (ions) generated at the electron donor. This flow of ions completes the electrical circuit, and current continues to flow until there are no more electrons available from the donor or until the acceptor becomes saturated with electrons. The chemical process itself consists of the creation of charged chemical species at the electron donor and their incorporation into the structure of the electron acceptor.

Many chemical reactions occur by virtue of such electron transfer processes (redox reactions), but few lend themselves to the efficient storage and generation of electricity. The most suitable systems are those in which the donor has a strong tendency to give up electrons, and the acceptor a strong tendency to hold onto electrons. In addition, the chemical species must have a high capacity, on a weight or volume basis, to store electrons. It is also required that the donor and acceptor be chemically stable in contact with any other material present in batteries, particularly when in contact with the ionically conducting medium, the electrolyte, whether it be a liquid or a solid. This is a problem with most of the high-energy-density battery systems and with high-voltage systems. Lithium batteries, in particular, tend to be chemically unstable in this regard, and special precautions are needed to prevent the batteries from losing their capacities by such parasitic reactions at an unacceptably high rate. Lithium is a strong electron donor and reacts spontaneously with many substances, for example, water and many of the organic solvents that are used in lithium batteries.

The tendency of donors to give up electrons and of acceptors to hold on to electrons is expressed by their individual voltages. The difference between the two voltages of any given donor–acceptor pair, the cell voltage, is a measure of the electrical force available to drive the electrons through an external load. That voltage, after some adjustments, multiplied by the rate of flow of the electrons (generally expressed in amperes) gives the power delivered by the battery to the load. The practical voltages and energy densities of some of the more common primary battery systems are given below.

Battery system	Single-cell voltage	Energy density (Wh/lb)
Zinc/air	1.1	130
Zinc/silver oxide	1.5	60
Alkaline manganese	1.3	42
Zinc/carbon	1.1	30
Lithium/thionyl chloride	3.5	150
Lithium/sulfur dioxide	2.6	120
Lithium/vanadium oxide	2.8	120
Lithium/carbon mono fluoride	2.6	100
Lithium/copper sulfide	1.8	90
Lithium/copper oxide	1.5	75
Lithium/manganese oxide	2.8	70

The realizable energy densities and voltages depend very much on the size of the batteries, the discharge rates, the operating temperatures, and the particular battery designs.

Given that substances exist that can donate and accept electrons, it is reasonable to inquire if, once a substance D has given up its electrons to an acceptor A, it may be possible to reverse the process and pump electrons back from A to D. Indeed, it is possible to do so for some systems, but energy is required to operate the process. The best known electrochemical systems that can be operated efficiently in this manner are the lead/sulfuric acid/lead dioxide and the nickel oxyhydroxide/potassium hydroxide/cadmium systems. Batteries that have the capability of being charged and discharged repeatedly are referred to as secondary or rechargeable batteries. Batteries that cannot be operated in this manner and that are discarded once their electrical charge has been spent are referred to as primary batteries. In primary cells, the electron donors are referred to as anodes and the electron acceptors as cathodes. In rechargeable cells, it is common practice to refer to the charged electron donors

Load

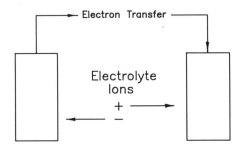

Negative Electrode Positive Electrode
Electron Donor   Electron Acceptor
Anode   Cathode
Oxidation   Reduction

Figure 1.1. Cell discharge process.

as the negative electrodes and the charged electron acceptors as the positive electrodes, regardless of the state of charge of the electrodes. Although some primary battery systems may be partially recharged, they are not designed for such service, and recharging them may create hazardous conditions. Primary batteries should never be recharged. Secondary batteries for consumer applications can be recharged several hundred times, and specially designed Ni/Cd batteries can be recharged several thousand times. The practical voltages and energy densities of the more common rechargeable batteries are given below.

Battery system	Single-cell voltage	Energy density (Wh/lb)
Zinc/silver oxide	1.5	40
Nickel/hydrogen	1.2	25
Nickel/cadmium	1.1	18
Lead–acid	1.8	15

As in the case of primary batteries, the realizable voltages and energy densities depend on the size of the batteries, their design, the operating temperatures, the charge/discharge rates, the depth of discharge, and the past number of charge/discharge cycles.

1.1. BATTERY STRUCTURES

Batteries are available in a wide range of sizes that include cylindrical and prismatic shapes. The cylindrical shapes dominate in the consumer market. Batteries are designated as cylindrical when their heights exceed their diameters and as button cells or coin cells when their diameters exceed their heights. The prismatic shapes dominate in the area of large batteries; the best known prismatic battery is the standard automotive starter battery. The shapes and sizes generally conform to well-defined international standards to accommodate the needs of device manufacturers as well as the needs of users who should not have to depend upon the availability of battery designs unique to any particular battery maker. A great deal of information on batteries, their designs, and their performance characteristics is available in the standard battery handbook[1] as well as in manuals available from battery manufacturers.

The majority of batteries sold in the consumer market are sold as packages containing one or more individual cells, rather than as multicell batteries, and most battery-operated devices have compartments that hold one or more individual cells. Exceptions are the standard 6-V and 9-V primary batteries that are prepackaged by the manufacturers into single multicell units and a number of other multicell batteries designed for particular devices and applications that require higher operating voltages than those available from single cells. The number of cells connected in series to give high voltages depends upon the voltages of the unit cells. Multicell alkaline manganese batteries have nominal voltages that are multiples of 1.5 V, and multicell lithium batteries have nominal voltages that are multiples of about 3 V. Thus, the number of lithium cells in a given high-voltage battery is only half the number of alkaline manganese cells required for the same battery. A cell can never deliver a fractional part of its characteristic voltage unless coupled to a potentiometer.

Some common battery structures are shown schematically in Figs. 1.2–1.4. The simplest button cells consist of wafers of the two electrode materials separated by a porous insulator, and the cells are filled with an appropriate electrolyte. In the absence of a void space within the cells, the cells might rupture at elevated temperatures due to thermal expansion of the cell materials. Two different internal structures are employed in cylindrical cells, each serving a different function. Pellet and bobbin type structures are employed in low- to moderate-rate applications, and spirally wound structures are employed in high-rate applications. The cells may be crimp-sealed using an inert polymeric gasket that also serves as an insulator between the negative and positive terminals of a battery, or the cells may be sealed hermetically with glass-to-metal or ceramic-to-metal

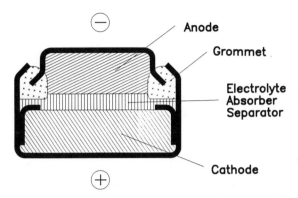

Figure 1.2. Typical button cell.

seals. The type of seal employed depends upon the chemical characteristics of the battery systems and the applications. The majority of alkaline cells are crimp-sealed, and lithium cells are generally hermetically sealed. Large multicell batteries are commonly constructed from prismatic cells with flat-plate electrodes as shown schematically in Fig. 1.5. The individual cell compartments are separated from each other to prevent electrolytic short circuits.

An important consideration in the design of a battery is the possible occurrence of excessive internal cell pressures caused by gas evolution due to corrosion reactions, charging processes, or cell reversals (more detailed descriptions are given in Part 2). To prevent any adverse effects that might result from excessive cell pressures, designers incorporate pressure relief valves or rupture disks in batteries when necessary, or they employ gas recombination systems when it is feasible to do so.

The various types of lithium batteries are also available in many sizes and shapes, generally conforming to the standards applicable to aqueous electrolyte cells, except that cells designed for special applications such as medical implant devices may have nonstandard shapes. The interior configurations of the lithium cells are similar to those of aqueous cells and include pellet type electrodes in button and coin cells, plate type electrodes in large cells, and bobbin type or spirally wound electrodes in cylindrical cells. Because of the extreme reactivity of the components of lithium cells, most of them are provided with hermetic seals to prevent the ingress of air and moisture and egress of electrolyte. Some solid cathode lithium cells are made with crimp seals. When properly made, crimp seals can survive for long times under normal user conditions without creating any leakage or other problems.

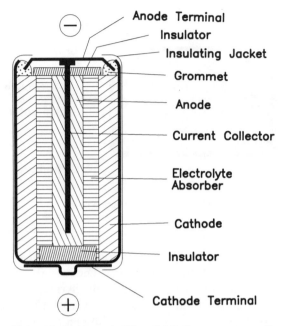

Figure 1.3. Inside-out bobbin cell (alkaline manganese cell).

1.2. ACTIVE MATERIALS

Most of the materials used in batteries are innocuous, but some are toxic or corrosive, or both, and may cause harm if released to the environment. It is desirable that users know what batteries contain and that they have some information about the hazards they may be exposed to if the battery contents should escape in some manner so that they may take appropriate steps to protect themselves before accidents occur. We summarize briefly the principal components in the most commonly used batteries. More detailed descriptions of aqueous electrolyte batteries are given in Part 3, and lithium batteries are described in Part 4. The associated hazards will be reviewed briefly in Chapter 2 and will be discussed in more detail in Part 2.

Zinc/Carbon (Zn/C) Batteries (Regular and Heavy Duty). Zinc/carbon batteries (Leclanché batteries) are used extensively in very many consumer applications that do not require good voltage regulation; that is, the load voltages of the batteries decrease during discharge. On a total-volume basis, these batteries are probably used more frequently than any others on a world-wide basis, although they are being displaced to an increasing extent by alkaline manganese batteries.

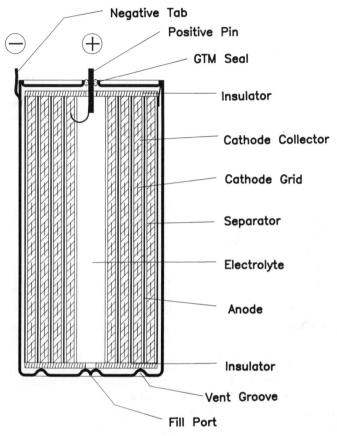

Figure 1.4. Hermetically sealed Li/SO_2 cell with spirally wound electrodes.

The anodes consist of zinc metal and may contain some cadmium or lead in low concentrations. The cathodes contain mostly manganese oxides of variable purity and carbon; some heavy metals may be present as impurities. The electrolytes are concentrated ammonium chloride solutions and/or zinc chloride solutions in water and may contain a starchy paste. Some chromate and other corrosion inhibitors may also be present.

Zinc/Mercuric Oxide (Zn/HgO) Cells. Although the distribution of zinc/mercuric oxide batteries is quite limited and their use has been curtailed in recent years, some may still find their way into consumer applications as small button cells. They are a significant source of mercury pollution.

The anodes are fabricated from zinc metal and may contain measurable concentrations of mercury. The cathodes are made from essentially pure mer-

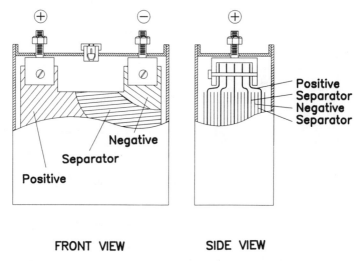

FRONT VIEW   SIDE VIEW

Figure 1.5. Prismatic vented cell with flat electrodes.

curic oxide, to which a binder may be added. The electrolyte is a concentrated solution of potassium hydroxide in water and contains dissolved zinc oxides.

Cadmium/Mercuric Oxide (Cd/HgO) Cells. Cadmium/mercuric oxide cells are not in common use in consumer applications, and their use has been curtailed in recent years because of their cadmium and mercury content. The cells are similar to the zinc/mercuric oxide cells and contain the same materials, except that the zinc is replaced with cadmium metal, a toxic substance.

Silver/Zinc (Zn/Ag₂O) Cells. The silver/zinc cells are used extensively in many consumer applications in the form of small button cells.

The zinc anodes may contain some mercury, although the use of mercury as a corrosion inhibitor is being phased out by battery manufacturers. The mercury may be replaced with low concentrations of other unspecified corrosion inhibitors. The positive electrodes contain primarily silver oxide and a small amount of binder. The electrolyte is a concentrated solution of potassium hydroxide in water, and it contains some dissolved zinc oxides.

Zinc/Air (Zn/O₂) Cells. Because of their high specific capacities and energy densities, the zinc/air cells are finding use in many applications in the consumer market, especially in applications that do not require long active battery life. Both small and medium size cells are available.

The anodes are made from zinc metal, similarly to the anodes used in other alkaline cells, and may contain unspecified corrosion inhibitors in low concentrations. The positive electrodes comprise Teflon-bonded structures containing some carbon and manganese oxide; some heavy metals may also be present as catalysts. The electrolyte is a concentrated solution of potassium hydroxide in water; it may contain other alkali hydroxides as well.

Alkaline/Manganese (Zn/MnO_2) Cells. Alkaline manganese batteries are used in a great variety of consumer applications because of their good voltage regulation and high current-carrying capabilities. Together with zinc/carbon batteries, they constitute by far the majority of batteries employed in portable consumer devices.

The anodes are made from high-purity zinc and may contain small amounts of unspecified corrosion inhibitors to replace the mercury formerly used for this purpose. The cathodes contain high-purity manganese oxide and may contain no more than a trace of heavy metals; carbon may be present in low concentrations. The electrolyte is a concentrated solution of potassium hydroxide in water, and it contains dissolved zinc oxides.

Lithium/Manganese (Li/MnO_2) Cells. Primary lithium batteries have been available for several years and are entering into a growing number of applications, both in the consumer market and in specialty markets. They may be expected to increase in importance during the coming years, especially when rechargeable lithium batteries become available for portable consumer electronics applications.

The anodes of primary lithium batteries are made of pure lithium metal, and the most commonly available lithium battery for consumer applications contains cathodes made from high-purity manganese oxide. The cathodes contain bonding agents such as Teflon or similar materials and may contain small amounts of carbon. Several other cathode materials are used in lithium batteries, and they are discussed more fully in Part 4. Various electrolytes are used in lithium batteries. They comprise solvents such as propylene carbonate, dimethoxyethane, tetrahydrofuran, and others, and salts such as lithium perchlorate, lithium hexafluoroarsenate, lithium tetrafluoroborate, and others. A detailed discussion of lithium batteries is presented in Part 4.

Lead–Acid (Pb/PbO_2) Batteries. Lead–acid batteries are used very widely for traction, starting, lighting, standby power, and a variety of industrial applications as both medium-size and large-size batteries. Smaller lead–acid batteries are used in several portable consumer devices.

The negative electrodes are made of metallic lead and may contain significant amounts of antimony and/or calcium and possibly other metals in low concentrations. The positive electrodes are made of lead dioxide and contain various lead salts whose composition depends upon the state of charge of the batteries. The electrolyte is a concentrated solution of sulfuric acid in water. Its concentration depends upon the state of charge of the battery, being higher in the fully charged state than in the discharged state. In the discharged state, the active masses of the electrodes comprise mostly various lead sulfates.

Nickel/Cadmium (NiOOH/Cd) Batteries. Nickel/cadmium batteries are used in an increasing variety of portable consumer applications, such as power tools and communication equipment, and in many industrial applications. The smaller nickel/cadmium batteries are used principally in consumer devices, and the larger batteries in industrial equipment.

The negative electrodes contain metallic cadmium as the electroactive component, and other materials may be present in low concentrations as conditioners. The positive electrodes contain hydrated nickel oxides as the active material, embedded in a porous nickel matrix; other materials may be present in low concentrations as conditioners. The electrolyte is a concentrated solution of potassium hydroxide in water.

Nickel/Metal Hydride (NiOOH/MH) Batteries. Because of the recognized toxicity of cadmium and its probable dispersion in the environment from uncontrolled landfills that contain discarded nickel/cadmium batteries, the nickel/metal hydride battery system has been developed as an environmentally acceptable replacement for nickel/cadmium batteries. The metal hydride batteries are expected to increase in importance during the coming years. The composition of the nickel/metal hydride batteries is essentially the same as that of the nickel/cadmium batteries, except that the cadmium is replaced with a metal hydride made from hydrogen and a mixture of various metals (see Section 8.3).

1.3. ELEMENTARY PROCESSES

An appreciation of what may happen with and to batteries under various circumstances requires a knowledge of their characteristics and the processes that occur or may occur in them. We present a brief discussion of these processes for the principal types of batteries. Single-cell and multicell batteries will be discussed separately since the operation of multicell batteries involves processes and events that are due entirely to the interactions between their constituent cells.

1.3.1. Primary Aqueous Electrolyte Cells

Zinc/carbon cells are one of the more important types of primary aqueous electrolyte cells. Although designated as zinc/carbon cells, the cathode-active material is manganese dioxide mixed with carbon. The electrolyte is a concentrated aqueous solution of ammonium chloride or zinc chloride. The discharge of the cells leads to the formation of a hydrated manganese oxide, or oxyhydroxide, and to various zinc salts whose composition depends upon the electrolyte used and the discharge conditions; generally, the salts formed are hydrated zinc oxide, zinc hydroxychlorides, or zinc oxychlorides. The zinc/carbon cells are probably the most innocuous cells available in the consumer market, and they have a very good safety record. Their electrolytes are corrosive, however, and the processes that occur in these cells are sufficiently similar to those that occur in alkaline electrolyte cells that they may be discussed together. A main distinction is that the electrolytes in zinc/carbon cells are not as aggressive as the electrolytes of alkaline cells.

The primary alkaline cells of interest are based on the five electrochemical couples:

1. Zinc/manganese dioxide
2. Zinc/silver oxide
3. Zinc/air
4. Zinc/mercuric oxide
5. Cadmium/mercuric oxide

The first two are the most important. Because of environmental considerations in relation to mercury and cadmium toxicity, the batteries with mercuric oxide cathodes are being phased out of production and may be expected to be of decreasing importance during the coming years for consumer applications. For the same reason, the manufacturers have already or are in the process of eliminating mercury completely from the other alkaline cells. Until recently, mercury was the universal inhibitor used in alkaline cells to reduce the evolution of hydrogen due to the corrosion of zinc in the alkaline electrolyte.

During normal discharge of alkaline cells, the zinc anodes are converted to hydrated zinc oxide that has a low solubility in the electrolyte. The manganese oxide is converted to a hydrated lower oxide, and other oxides to their base metals. In zinc/air cells, the metal oxides have been replaced by a catalytic current collector, and, during discharge, the oxygen from the air is converted to hydroxyl ions by reaction with the incoming electrons and the electrolyte. From a safety point of view, the most important process in alkaline and zinc/carbon cells is the corrosion of zinc that leads to the formation of hydrogen gas. Another hazard is the electrolyte. Potassium hydroxide is an effective destroyer of living tissue, and exposure to this electrolyte may cause permanent tissue damage. The evolution of hydrogen in alkaline cells may be represented by

$$Zn + 2H_2O \rightarrow Zn(OH)_2 + H_2$$

Depending on the temperature and the efficacy of the corrosion inhibitors, the rates of gas evolution may vary over a considerable range. The rate of hydrogen evolution and the consequent pressure buildup at elevated temperatures may be sufficient to rupture a cell unless it is designed to allow the excessive pressure to be relieved through specially designed vent structures or by the escape of gas through the crimp seal. All the alkaline and zinc/carbon cells are crimp-sealed. An additional consideration is that highly elevated temperatures may cause the cells to rupture by virtue of the increased vapor pressure of the electrolyte at high temperatures. This is no concern at normal operating temperatures.

Attempts to recharge primary batteries may reverse the normal discharge processes to a limited extent, but the cells are essentially resistant to recharge. Recharging of primary cells is invariably associated with a new set of electrochemical processes, generally, the evolution of hydrogen at the negative electrode, or metal deposition, and oxygen evolution at the positive electrode. In the case of metal deposition, short circuits are likely to form that may destroy the cells. In the case of gas evolution, quite apart from the possibility that the high pressures generated by the gases may rupture the cells, the mixture of hydrogen and oxygen is highly explosive and may detonate if ignited by sparks or by catalytic reaction sites that may be present in the cells or near the cells if the gas has accumulated in a confined space outside the cells.

1.3.2. Primary Lithium Cells

We limit this section to a brief introduction to lithium batteries as a detailed discussion will be presented in Part IV.

Three types of ambient temperature lithium batteries have been developed, differing in the nature of their electrolytes and the cathode-active materials. The solid cathode, organic electrolyte types of cells resemble the aqueous electrolyte types most closely. Typical examples are the Li/MnO_2 and $Li/(CF_x)_n$ cells. A very different type comprises cells in which the cathode-active materials are present in liquid form with or without added organic solvents. Some examples are the Li/SO_2 and $Li/SOCl_2$ cells. The third type includes the solid-state batteries, in which the anodes, cathodes, and electrolytes are all solids or semisolid materials. An example is the Li/I_2 (complex) system used for medical implant devices. The electrolyte is essentially lithium iodide, and the cathode is a plasticized organic complex of iodine. Solid-state cells have low current-carrying capabilities and are used in few applications at the present time.

The discharge reactions of the various lithium batteries depend upon the chemical compositions of their cathodes, but they all involve the oxidation of lithium at the anode with the formation of lithium ions, the transfer of lithium ions to the cathode, and the incorporation of the lithium ions into the cathode material in various ways. Two examples of typical overall reactions are

$$2Li + CuO \rightarrow Li_2O + Cu$$

$$Li + MnO_2 \rightarrow LiMnO_2$$

If for any reason a primary lithium battery is charged, the discharge reactions may be reversed to some extent. Lithium is formed on the negative electrode as a dendritic or mossy deposit that is likely to short-circuit the cell. This is a very reactive form of lithium and may trigger spontaneous exothermic reactions, causing a thermal runaway and associated hazards. Reoxidation of the cathode material may occur on the positive electrode on charging, but solvent oxidation is more likely, a process that may create hazards. Primary lithium batteries are not designed for recharging, and, as in the case of primary aqueous electrolyte batteries, charging is hazardous and should not be done.

Reversal is a process that occurs when a cell is forced into discharge beyond its available capacity. Reversal of primary lithium batteries is similar to charging except that the roles of the electrodes are interchanged. Oxidation occurs at the negative electrode, and reduction at the positive electrode. The hazards created by reversal are the same as described above for charging.

Lithium battery systems are very different from aqueous electrolyte systems in that a greater variety of hazardous chemical reactions may occur in lithium systems, both inside the batteries and outside the batteries if the battery contents are exposed to the atmosphere, and we review some of these reactions very briefly. Lithium is a strong electron donor, and, when in contact with an organic solvent with a moderate to high dielectric constant, such as is used to render the electrolyte conductive, it will react chemically with the solvent. Once formed, however, the reaction products inhibit further reaction to such an extent that the lithium/solvent system becomes sufficiently stable under normal user conditions to render it suitable for power-generating purposes. At elevated temperatures that exceed those of most applications, the protective action may no longer be effective, and the cells may deteriorate at unacceptably high rates or they may enter into a thermal runaway condition, causing cell ruptures or explosions. In the case of liquid cathode systems, the reaction between lithium and the reactive liquid is much more aggressive, but, again, a protective film is formed on the lithium that inhibits any further significant loss of active materials. These systems are only pseudostable, and, if exposed to high temperatures, they are more likely to rupture or explode than are the solid cathode lithium batteries. For this reason, the liquid cathode cells are generally provided with vents and/or other safety devices to prevent excessive overpressures and thermal runaways (these subjects are discussed in Part II). Liquid cathode lithium cells and batteries are not available for general use in consumer applications because of their potential hazards.

The stability of organic solvents in contact with solid cathode materials depends upon the oxidizing power of the cathode materials, that is, on their voltage. A high cell voltage indicates a strongly oxidizing cathode. The higher the voltage, the greater is the tendency of the cathode to decompose the solvent in a spontaneous heat-producing reaction, essentially a combustion reaction. For the cells available in the consumer market, principally Li/MnO_2 cells, the compositions of the organic solvent and the cathode have been controlled to the point where this potential instability is of little concern, at least for small cells and batteries at normal operating temperatures. Another potential safety concern relates to the electrolyte itself. An electrolyte salt such as lithium perchlorate is a strong oxidizing agent, and it may react with organic solvents with explosive rapidity at normal or slightly elevated temperatures unless catalytic impurities (moisture is an effective catalyst for this reaction) have been removed or unless suitable inhibitors are employed to block the reaction. There is the additional possibility that some of the electrolyte salts may be toxic either in their native form or after reaction with other materials inside or outside the cells. Lithium hexafluoroarsenate is an example of a salt that is used in some lithium batteries. The toxicity of the arsenic does not depend very much on the manner in which it is combined with other chemical elements. Safety considerations favor the use of less reactive and less toxic electrolyte salts.

If for any reason a lithium cell ruptures or explodes, the subsequent events depend upon the chemical composition of the system, the conditions leading up to the rupture or explosion, and the ambient conditions. Lithium itself is extremely reactive and may ignite in contact with air or water, causing a high-temperature metal fire that may ignite nearby flammable materials. If, in addition, an organic solvent is present, it may ignite as well and burn or give rise to a fuel/air explosion. The latter is likely to occur if ignition takes place during the period of time when the air/solvent mixture passes through its explosive composition range. The chances of this occurring depend on many factors related to the properties of the solvent, the amount of solvent involved, the ejection process, and the manner in which the solvent is distributed and mixed with the air. If an air/solvent explosion occurs, it may be more violent than any process associated with the air oxidation of lithium. Events of the kinds described above are extremely unlikely with the lithium cells available for consumer applications, but the possibility that they may happen has limited the adoption of the larger size lithium batteries for general consumer applications.

Solid-state lithium batteries are intrinsically safer than any of the other types of lithium batteries. Apart from the very slow corrosion reactions that proceed at finite rates under normal operating conditions, for example, the reaction between lithium and iodine in lithium/iodine cells to form additional lithium iodide electrolyte, there are few processes in solid-state cells that give rise to safety concerns. If the cell temperatures increase to the point where the cells rupture or explode, as is almost certain to occur if the temperature reaches the melting point of lithium (181°C), the ejection of the cell contents may cause lithium fires, and nearby personnel may be exposed to the toxic effects of the battery materials. The severity of any adverse effects depends upon the amounts of materials involved, but since most of the solid-state cells used today are quite small, it may be possible to limit the severity of any accident.

1.3.3. Rechargeable Nickel/Cadmium Cells

In the charged state the positive electrode contains mostly nickel oxy-hydroxide as active material, and the negative electrode cadmium metal. Discharge involves an overall reaction that may be represented by

$$Cd + 2NiOOH + 2H_2O \rightarrow Cd(OH)_2 + 2Ni(OH)_2$$

The reverse reaction occurs on recharge. The active materials and their reaction products are practically insoluble in the potassium hydroxide electrolyte. The reaction shows that water is consumed during discharge and regenerated during recharge. In contrast to primary cells, where current reversal of discharged cells does not lead to a significant reconstitution of the charged electrodes, the electrochemical properties of the nickel/cadmium electrodes and the cell designs lend themselves to the efficient regeneration of the active materials and their structures.

The reconstitution of the electrodes by charging is not a simple process. It requires an appreciable force to do so at practically useful rates; that is, a voltage in excess of the open-circuit voltage is required to convert all the discharged materials back into charged materials. A characteristic of electrochemical systems is that the nature of the chemical transformations that occur is controlled by the voltage applied to a cell. Each chemical reaction has a characteristic voltage above which it occurs if it is an oxidation and below which it occurs if it is a reduction. The rates at which the reactions occur depend upon the applied voltage above or below the open-circuit voltages, upon the design of the cells, upon the relative amounts of converted and unconverted materials in the electrodes, upon the electrolyte concentration, and upon the temperature. As a consequence, when nickel/cadmium cells are recharged at high rates, or even at low rates near the end of charge, higher voltages must be applied to achieve a complete conversion of discharged material to charged material. The applied voltages can easily reach values sufficient to decompose the water in the electrolyte to hydrogen and oxygen. The control of the charging voltages and/or the charging rates is an important consideration for the safe operation of any rechargeable battery. Since different systems have different voltage characteristics, different charging circuits are required for different battery systems. Charging units are not interchangeable between different battery systems. A disregard of this rule may lead to battery damage and personal injuries.

The overall rate of gas evolution on charging and on overcharging depends upon many factors, as does the ratio between the rates of hydrogen evolution at the negative electrode and oxygen evolution at the positive electrode. Since hydrogen and oxygen form an explosive gas mixture, nickel/cadmium cells are designed to prevent or minimize the accumulation of this explosive mixture in one of two ways. Either the cells are designed to vent the gases from the cells (vented cells) or they are designed to remove the oxygen by letting it react with the cadmium metal in the negative electrode to form cadmium oxide or hydroxide (sealed cells). Unless gas recombination occurs, vented nickel/cadmium cells will lose water and may slowly dry out due to poorly controlled charging and overcharging. Poorly vented cells or battery compartments may experience gas accumulations sufficient to create explosive hazards.

1.3.4. Rechargeable Lead–Acid Cells

Lead–acid batteries are used in many applications, and the general consumer is probably most familiar with the automotive starter batteries and some of their hazards, particularly their tendency to leak and corrode nearby metals and to generate gas on charging.

The discharge of lead–acid batteries involves essentially the conversion of the negative lead electrode and the positive lead dioxide electrode to lead sulfate, a process that may be represented by the overall reaction

$$Pb + PbO_2 + 2H_2SO_4 \rightarrow 2PbSO_4 + 2H_2O$$

The reverse of this reaction occurs on charging. It may be noted that water is generated on discharge and sulfuric acid on charge. All the solid materials are practically insoluble in the sulfuric acid electrolyte.

An interesting and important characteristic of lead–acid batteries is the slowness with which water is decomposed in these cells. Water decomposes to hydrogen and oxygen at voltages above about 1.23 V, yet lead–acid cells that have voltages of about 2 V do not decompose water at significant rates under normal conditions. The reason for the relative stability of water in lead–acid batteries is that the rates of gas evolution are kinetically inhibited by the lead and lead dioxide electrodes. If for any reason catalytic impurities were to be present in the cells, they would deteriorate very rapidly with the profuse spontaneous generation of gas owing to the high cell voltage. The decomposition of the electrolyte cannot be entirely avoided, however, and both hydrogen and oxygen may be expected to accumulate in lead–acid cells even when they are not in use. Since these gases form an explosive mixture that may be ignited by sparks, it is desirable to prevent their accumulation either by venting the cells or by providing a mechanism for the safe catalytic recombination of hydrogen and oxygen as soon as they are formed in sealed cells. In the case of vented lead–acid batteries, it is necessary to prevent hydrogen accumulation in compartments outside the batteries by adequate ventilation. Since the venting of gases from lead–acid batteries implies a gradual loss of water, it is necessary to replenish the water from time to time. The slow parasitic reactions that occur in lead–acid cells lead to the gradual loss of battery capacity, and full charge can be maintained only by a continuous low rate of charging, under properly controlled conditions, unless the batteries are recharged on a regular basis at frequent intervals.

As in the case of all rechargeable batteries, excess voltages are required to convert the discharged materials to fully charged materials. This means that gas evolution is bound to occur during the recharge of lead–acid batteries at rates that depend upon the rates of charge, the cell design, the state of charge, temperature, and some other factors related to the past history of the battery. For safety reasons, there is considerable merit associated with the use of sealed lead–acid batteries. By virtue of their construction and composition, the excessive accumulation of explosive gas mixtures is much reduced both inside and outside the cells. Furthermore, the chances of the accidental spillage of corrosive sulfuric acid are much reduced.

1.3.5. Primary Multicell Batteries

Although most batteries are sold as single cells rather than as multicell units, in most applications the cells are employed in multicell configurations. The more common nominal voltages of multicell devices are 6 V, 9 V, 12 V, and 24 V. The coupling of cells in series configurations to reach these voltages creates the possibility that some new processes may occur that merit consideration from a safety point of view. If all the cells in a series stack have exactly the same capacity, they will all reach end of life at the same time since the same current flows through all the cells, and there would be no problem. Consider what may happen if one cell in a series stack has a low capacity, as might happen if a partially discharged cell is used in a series stack of fresh cells. When the inferior cell becomes exhausted, the voltage of the battery decreases and the battery may fail to operate the device, a likely event with series strings that contain very few cells. If the device continues to draw current, even if the voltage has decreased, the continued discharge forces current to flow through the exhausted cell, where it forces electrochemical reactions to occur, principally the evolution of hydrogen and/or oxygen in aqueous electrolyte systems, or other reactions, depending on the composition of the systems. Reactions other than hydrogen and oxygen evolution will occur in nonaqueous systems. This process is referred to as voltage reversal since the cell voltage is driven in a negative direction compared with the natural cell voltage. Voltage reversals can create situations that may develop into ruptures or explosions and can cause equipment and/or personal injuries unless precautions are taken to prevent their occurrence. Reversals will be discussed in more detail in Part II.

1.3.6. Rechargeable Multicell Batteries

Rechargeable cells are generally employed in series-connected configurations to generate the voltages required to operate various devices. This is particularly true for motor-operated devices and emergency lighting applications. In contrast to primary batteries, rechargeable multicell batteries are generally sold as preassembled multicell configurations to eliminate the possibility that users may connect together rechargeable cells with the wrong polarity, an invitation to disaster. The discharge of a series string of rechargeable cells may introduce the same kind of problems that can occur with a series string of primary cells, that is, voltage reversals. The repeated charge and discharge of a series string of rechargeable cells is likely to aggravate any capacity imbalance that may exist between the cells. A mitigating factor is that rechargeable batteries are rarely discharged to their full capacities. Examples are batteries that are used for automotive starter and emergency lighting applications. Only a small fraction of their available capacity is used during any one discharge. For traction applications, on the other hand, batteries are discharged quite deeply, and each discharge cycle may utilize most of the available capacity. Unless the batteries are deeply discharged, there is available a considerable reserve capacity, and voltage reversals are not likely to occur.

A process that is the converse of voltage reversal may take place when a series string of rechargeable cells is charged. Even though the same capacities are removed from all the cells in a series string during discharge, the cells may differ in their ability to handle the charging current with equal ease. Some cells may require a higher voltage across their terminals than other cells in the string for the same current to pass through the cells. Since attempts to achieve a full recharge necessarily mean that voltages must be applied to a battery that exceed its open-circuit voltage, some cells may experience excessive charging voltages that induce side reactions such as gas evolution in aqueous electrolyte cells and other reactions in nonaqueous cells, with the possibility of hazardous consequences that will be discussed in the following chapters. If for any reason a cell in a series string of rechargeable cells has been short-circuited, for example, as a result of overcharging of a rechargeable lithium battery, this process can easily escalate the seriousness of the problem even if the battery is charged with a voltage-controlled device. Then, all the remaining functional cells will be exposed to excessive charging voltages, and undesirable side reactions are likely to occur.

REFERENCES

1. D. Linden (ed.), *Handbook of Batteries and Fuel Cells,* McGraw-Hill, New York (1984).

2

The Nature of Battery Hazards
and Accidents

Batteries are self-contained chemical reactors capable of transforming chemical energy into electrical energy on demand. The chemicals used in batteries are corrosive and toxic and may cause personal injuries or equipment damage if they escape from any battery. Most of the batteries available in regular commerce are quite abuse resistant and effectively contain the corrosive and toxic substances under normal user conditions. The escape of the battery contents, if it occurs, is generally caused by inadvertent or deliberate abuse or some form of mishandling. Battery defects introduced by poorly controlled manufacturing operations may also lead to containment problems.

The majority of batteries, both rechargeable batteries (secondary batteries) and nonrechargeable batteries (primary batteries), are chemically unstable. Slow chemical reactions occur in batteries whether they are in use or not, and energy is released by these reactions. If for any reason batteries are exposed to conditions that accelerate the rate of these reactions to a significant extent, the energy may be released so fast that the batteries rupture or explode. The disposal of batteries in fires or incinerators is certain to cause ruptures or explosions.

2.1. BATTERY HAZARDS

Given the nature of the chemical substances used in batteries and the processes that occur on charge, on discharge, and on open circuit, various hazards may be created. These hazards are of a chemical or mechanical nature; electrical hazards are of little importance because of the low voltage of most batteries in common use. An exception is the generation of sparks from batteries, in particular, automobile batteries, that may ignite nearby flammable gases or liquids. The mechanical hazards are generally caused by chemical or electrochemical reactions that trigger ruptures or explosions.

During the life of a battery, different groups of individuals may be at risk if accidents occur:

- Manufacturing workers
- Personnel involved with storage, transportation, installation, and charging of batteries
- Battery users
- Waste handlers and disposal personnel
- Battery recyclers
- Individuals exposed to leachates from landfills or flue gases from incinerators

In addition, there is the question of possible adverse effects on ecosystems due to improper disposal of toxic battery materials. This subject will be mentioned only briefly since it has been treated extensively in the literature.[1] Nor will we discuss the hazards associated with manufacturing, recycling, and disposal of batteries and wastes from battery manufacturing operations. Personnel involved with these operations are generally familiar with battery hazards and the required safety procedures. Our principal concern will be the hazards associated with battery usage in general. Over the years, manufacturers have learned to design and build robust batteries that are quite abuse resistant, and accidents are rare with most of the batteries in common consumer use. The new generation of high-energy-density lithium batteries have a greater hazard potential than conventional batteries, and they are less forgiving of handling mistakes and abuse. Lithium batteries are entering into special applications in increasing numbers and into the consumer market to a more limited extent. The small lithium batteries available to consumers have a smaller potential for causing damage than the larger lithium batteries employed in special applications. Since the lithium batteries are relative newcomers to the battery family and since their properties are not well known to the general battery user, we describe their characteristics and potential hazards in considerable detail in Part IV.

The hazards created by the improper use and handling of batteries may be classified according to their effects on users and equipment as follows:

- Physical hazards
- Chemical hazards and injuries
- Equipment damage
- Environmental impact

We discuss each of these categories briefly below and the underlying processes in more detail in Part II.

Physical Hazards. The physical hazards include structural damage and personal trauma caused primarily by battery explosions. Explosions may be triggered in various ways (see Part II), and the severity of their effects is related to the characteristics of the explosions themselves, which range from relatively benign battery ruptures to violent detonations, particularly in the case of large lithium batteries, where fuel/air explosions may occur. In the former case, most of the battery components may remain inside the battery, but any accumulated gases and most of the electrolyte are likely to escape. Some flying debris may be created by ruptures. In the case of detonations proper, most of the battery components are likely to be scattered with considerable kinetic energy, sufficient to cause personal injuries and equipment damage. Considerable destructive forces may also be associated with the shock waves and overpressures generated by battery explosions. The primary factors determining the severity of any battery explosion are the total energy involved and the speed with which the energy is released, both of which depend upon the chemical nature of the battery system, the size of the battery, and the mechanical properties of the containing structures. We give three examples below to illustrate some of the events that may occur.

Primary alkaline cells have energy densities of about 40 Wh/lb derived from the active battery materials. Since the active materials are solids that react slowly unless thoroughly mixed, this energy cannot be released very rapidly. These cells are most likely to rupture, if at all, because of hydrogen overpressures in the cells or because of excessive vapor pressures caused by high cell temperatures when, for example, the cells are exposed to fires. If a considerable amount of hydrogen is formed due to the corrosion of zinc electrodes by reaction with an alkaline electrolyte, and if this hydrogen suddenly escapes from an alkaline cell in the presence of a source of ignition, a violent hydrogen explosion may occur that could cause equipment damage and personal injuries. Since most alkaline cells are designed to permit the slow escape of any hydrogen formed in the cells, such an event is extremely unlikely.

Lead–acid batteries have energy densities of about 15 Wh/lb derived from well-separated solid reactants. As in the case of alkaline cells, this energy is not likely to be released very rapidly under any circumstances. The physical hazards from lead–acid batteries are associated primarily with the gases evolved during charging, hydrogen and oxygen, depending on the charging mode and the battery design. This is a very explosive gas mixture, and, if it is ignited in any way, whether inside or outside the battery, considerable damage is likely to occur.

Primary lithium/sulfur dioxide batteries have energy densities of the order of 120 Wh/lb. Since the oxidizing agent sulfur dioxide is present in liquid and vapor form, the possibility exists that a very rapid energy-releasing reaction may occur in these cells, especially at elevated temperatures. If it occurs, it is likely to result in explosions or violent ruptures. To prevent such an eventuality, the cells are provided with vents to permit liquids and gases to escape before any elevated temperatures can trigger an explosion. Interestingly, the electrolyte in these cells contains an organic solvent which, if mixed with air in the presence of an ignition source, can cause violent fuel/air explosions for various fuel/air mixture ratios. The explosive potential of lithium/sulfur dioxide D-size cells in air has been estimated to be equivalent to about 40 grams of TNT.[2] This much energy can cause measurable physical damage and personal injuries.

Explosions and ruptures of small batteries—button cells and cylindrical cells—generally cause limited damage. Large batteries, on the other hand, may cause considerable damage, particularly rechargeable batteries and large to moderately large primary lithium batteries. The types of physical damage likely to occur in battery explosions depend very much on the configurations involved. The partial or complete destruction of battery compartments and adjacent circuitry may be possible in the case of portable electronic devices, but these are rare events. In the case of large stationary batteries and large lithium batteries, if they are abused to the point where explosions occur, considerable damage may be inflicted on nearby equipment and structures. However, such events are unlikely to occur since operators of large battery installations are well aware of the required safe operating procedures, and such installations are not available to the general consumer. The largest battery the general consumer is likely to handle is the standard automotive starter battery, at least until electric vehicles become available. Strict adherence to the safety procedures recommended by the manufacturers should minimize the occurrence of accidents with any battery.

Physical injuries to personnel are rare with ordinary batteries, but, if explosions do occur, injuries may be sustained. Their severity depends upon the proximity of an individual to the center of the explosion, the magnitude of the explosion, and the amount of debris formed, including the possibility of an electrolytic shower. Battery explosions are generally confined to battery compartments that absorb some of the explosive power, except in the case of batteries that are not located within battery compartments, such as lead–acid batteries. Explosions of the most commonly used consumer batteries may cause physical injuries to any exposed body part, especially face and hands. Eyes are particularly vulnerable to both physical and chemical injuries, and safety goggles should be worn by anyone who comes near a battery whenever there is the slightest possibility that it may explode, rupture, or spill electrolyte during handling operations.

Fires may result from the abuse of batteries. Two questions arise in connection with batteries and fires. The most important question for battery users is what may happen when batteries are exposed to fires or very high temperatures. A less important question relates to the possibility of battery fires *per se*. Any kind of battery will rupture or explode when its temperature becomes high enough, such as in any kind of fire, and the battery contents will be expelled in random directions from the fire zone unless channeled in some way or contained. The battery materials may also react with other materials present in the fire before entering the environment. For anyone exposed to batteries being overheated by any means, the more immediate concern is that of being hit by flying debris from battery explosions and of being exposed to electrolyte expelled from ruptured or exploding cells. The consequences of exposure to electrolytes and the remedial actions to be taken will be discussed below. The cardinal rule of never disposing of batteries in fires is not always observed by users.

The abuse of lithium batteries may lead to fires of two kinds: metal fires due to the ignition of lithium and fuel/air fires due to the ignition of flammable solvents escaping from lithium batteries. Such fires can occur only if the batteries rupture or explode and if an adequate ignition source is present, generated either by the battery itself or some other agent. The likelihood of fires is very low with the lithium batteries used in consumer applications. Aqueous electrolyte batteries do not contain any flammable materials and may cause fires or explosions only by the excessive accumulation and ignition of hydrogen/oxygen or hydrogen/air mixtures generated by zinc corrosion in primary cells and by charging and/or corrosion of rechargeable batteries. Another cause of fires attributable to batteries of both the aqueous and nonaqueous electrolyte types is faulty electrical circuitry that may lead to battery short-circuiting. This is most prevalent in battery-operated toys, and the high currents drawn by faulty circuits may generate sufficient heat in battery-operated devices to ignite them or nearby flammable objects.

If lithium batteries were to ignite, the magnitude and severity of any resulting fire would be proportional to the amounts of flammable materials present. In the unlikely event that a small consumer-type lithium battery were to ignite, the recommended procedure would be to let the fire burn to completion and to prevent the fire from spreading by the liberal application of water to nearby flammable objects. If dry sand is available, the lithium fire may be smothered by being covered by sand. As always, the essence of effective fire fighting is speed of application of the appropriate remedies. Attempts to extinguish lithium fires with water run into the problem that lithium reacts with water to generate hydrogen that will feed the fire. Special extinguishers are required to quench lithium fires, but they are not likely to be available in locations other than those where lithium fires may be expected or where appreciable quantities of lithium are stored or used. The use of halocarbon extinguishers is strongly discouraged since they may generate toxic fumes.

Chemical Hazards and Injuries. Batteries contain materials that are toxic and/or corrosive, and any time these materials escape from a battery they may cause personal injuries or equipment damage. The severity of any injury or damage depends upon the chemical nature of the materials involved and the manner in which they are dispersed from a battery. The most benign situation is one where a leak permits electrolyte to seep slowly from a battery. In this case, salt encrustations form at the sites of the leaks, generally, alkaline deposits at the crimp seals of alkaline cells and acidic deposits at the terminals or fill ports of lead–acid batteries. The worst case scenario is one where a battery explodes violently and the battery components are scattered over the explosion site with damage extending beyond damage to the battery itself. In between are cases of less violent explosions and ruptures, ventings, and electrolyte spillages due to leaks. Experience shows that personal injuries are unlikely as long as batteries are handled properly. Basically, individuals at risk are those who abuse or otherwise mishandle batteries. The injuries that may occur will be discussed briefly with reference to the nature of the materials involved.

Most of the batteries in general use do not contain volatile substances, and inhalation of vapors from such batteries is extremely unlikely under normal circumstances. Exceptions occur when fine mists of electrolytes form in violent ruptures or explosions. Then, there is a possibility that exposed individuals may inhale fine droplets of sulfuric acid or potassium hydroxide electrolytes. Although the amounts of inhaled substances may be reduced somewhat by breathing through a moist cloth, the inhalation of any electrolyte creates a serious situation that requires immediate medical attention. It should be noted that inhalation of finely divided solids may occur as well, as part of any mist exposure, and it may increase the severity of any tissue damage.

Lithium batteries present a different problem in regard to vapor inhalation. They contain volatile organic solvents that may be inhaled if the electrolytes escape from the batteries, or they contain strongly oxidizing and toxic inorganic solvents that are very volatile. Because of their volatility, the latter group of electrolyte solvents may be readily inhaled if they escape from batteries. They do, however, have distinct odors and create an immediate reaction that signals that inhalation has occurred. Since they may cause severe physiological damage, immediate medical attention should be sought if inhalation has occurred. Although ruptures and explosions of lithium batteries are rare, they do happen, and affected areas should be evacuated and ventilated to clear the atmosphere.

Body contact with battery electrolytes by means other than inhalation can also be destructive, particularly in the case of sulfuric acid and potassium hydroxide electrolytes. The ingestion of any electrolyte or eye exposures require immediate medical attention. It cannot be overemphasized that immediate flushing of any exposed part with copious amounts of water, especially flushing of the eyes, is the most effective first aid generally available at the time of exposure and should be practiced as the best immediately available means of reducing the severity of any tissue damage. Casual exposure to sulfuric acid is probably the most commonly occurring event because of the widespread use of lead–acid batteries in automobiles and the need for users to open the batteries when replenishing water lost from the batteries as a result of charging. Maintenance-free lead–acid batteries do not have this problem. Furthermore, when lead–acid batteries are charged, the evolution of gas in open batteries may create a fine mist that may expose individuals to contact with sulfuric acid. The more concentrated the acid, that is, the more fully charged the battery, the more aggressive are the chemical burns caused by the acid. If the acid comes in contact with clothing, the clothing will be destroyed more or less rapidly. Whenever clothing and body parts have come in contact with sulfuric acid, the clothing should be removed and the exposed body parts washed as soon as possible with copious amounts of water. Whenever there is the slightest possibility that the eyes may be exposed to sulfuric acid, safety goggles should be worn by every person at risk. Chemical burns caused by battery electrolytes may lead to permanent tissue damage. Any spilled sulfuric acid may be neutralized with sodium bicarbonate or lime. The former is a generally available household item. Rubber gloves and eye protection should be worn by personnel involved with the cleanup of acid spills, and acid-contaminated materials should be stored in acid-resistant receptacles with lids, for example, plastic containers. Care should be exercised whenever sulfuric acid is neutralized since neutralization generates a great deal of heat that may cause boiling and the formation of acid mists.

Potassium hydroxide electrolytes are used in most alkaline cells, and their effects are more insidious than those of sulfuric acid. Whereas the acid makes its presence felt by a burning sensation very soon after exposure, not so potassium hydroxide. Exposure to potassium hydroxide solutions makes the skin feel slippery, but there is no immediate sensation of pain that would alert an exposed person to wash away any alkali in contact with the skin. This makes potassium hydroxide more dangerous since failure to take immediate remedial action may give the alkaline solution time to penetrate underlying tissues, where it is likely to cause irreparable damage. It is of the greatest importance that if there is the slightest suspicion that exposure to potassium hydroxide has occurred, the affected clothing should be removed and the affected body parts should be flushed immediately with copious amounts of water. It is imperative that exposed eyes be flushed immediately and thoroughly with lots of water and medical attention sought without delay on an emergency basis. If there is any possibility that potassium hydroxide electrolyte may be spilled, personnel at risk of exposure should wear protective clothing and safety goggles that give full eye protection. Any spilled potassium hydroxide electrolyte may be neutralized with household vinegar. Rubber gloves and eye protection should be worn by personnel handling such spills and handling contaminated materials. Contaminated materials should be disposed of in plastic containers with lids.

Since solid battery materials are likely to be contaminated with electrolytes, exposure to battery solids may cause chemical burns in either of the categories described above. Another type of injury may result from the inadvertent ingestion of battery solids. The most likely path by which battery solids may be ingested is via foods contaminated by utensils that have been exposed to battery solids, or contaminated persons may handle food without cleaning themselves properly before doing so. Anyone handling battery materials, either as a result of cleanups of ruptured batteries, for purposes of battery maintenance, or as a result of deliberate disassembly of batteries, needs to practise cleanliness to avoid the possibility of ingesting even traces of toxic and corrosive battery materials.

The emergence of lithium batteries as a consumer item, although relatively insignificant to date on a quantitative basis, raises the possibility of exposures to metallic lithium in the rare event that such batteries may rupture or explode. Lithium reacts rapidly with air and water and may ignite spontaneously if exposed to either or both. Since the reaction generates lithium oxide or lithium hydroxide, which is chemically equivalent to potassium hydroxide, contact with lithium or its reaction products produces the same types of injuries as potassium hydroxide. In addition, contact with reacting lithium may cause thermal burns since a great deal of heat is generated by the lithium reactions. Remedial actions should be taken immediately after exposure to lithium or lithium hydroxide, similar to those required after exposures to potassium hydroxide.

A particular form of ingestion involves young infants and babies, who may swallow small batteries. This can occur only with batteries that are readily accessible to infants, whose propensity for putting things in their mouth is well known. The small button cells are the ones most likely to be swallowed. They are strong mechanically and generally pass through the digestive tract without rupturing. The likelihood of any material escaping from the cells and causing chemical harm is extremely small, but medical attention should be sought whenever there is a suspicion that a child may have swallowed a battery.

The presence of a battery in the digestive tract and its progress through that tract can be readily detected and monitored, and remedial action taken whenever required. The best preventive measure is to store both fresh and discarded batteries where they are inaccessible to young children and infants. Since batteries may be removed from some toys with relative ease by children, parents have expressed concern that such toys are unsafe.

Some Toxicity Considerations. We present a brief discussion of some of the factors that affect toxicity. Readers interested in comprehensive information on toxicity, diagnostics, and treatment of toxic injuries may find it useful to consult the texts cited in Refs. 3–5. The assessment of the toxicity hazards of any material requires considerations that extend beyond the *intrinsic toxicity* of the material *per se.* From an operational point of view, the *risk* is a more useful concept. Risk is defined as the probability that an individual, or individuals, may suffer harm or injury, immediate or delayed, as a result of exposure to toxic materials under more or less well defined conditions. Conversely, *tolerance* is defined as the probability that no harm or injury will result from exposure under the same conditions. The quantification of toxicity is a difficult matter. Some chemical species may be essential in low concentrations for the healthy functioning of the human body, but harmful in higher concentrations. Furthermore, the systematic evaluation of the toxicity of many substances is limited to animal experiments (ethical problems in both human and animal experiments!), and extrapolations to the effects on humans in terms of the commonly used dosage per unit body weight may not always represent the correct method of scaling even if the pathophysiologies are similar.

Among the many factors that affect risk, in addition to the intrinsic toxicity of a material, some of the more important ones are:

1. The physical state of the material (gas, vapor, liquid, solid, spray, mist, aerosol, dust, fume, particle size)
2. Chemical environment *in toto* (presence of other chemical agents or factors with synergistic or antagonistic effects)
3. The time–concentration profile of exposure (spatial and temporal distribution at the site of exposure of toxic agents and factors that affect toxicity and intensity of exposure)
4. The exposure pathways (external contact, ingestion, inhalation. The toxic effects of airborne materials are affected by particle size and breathing patterns, whether nasal or oral. Oral breathing is normally associated with strenuous physical activity.)
5. The age of exposed individuals (Children and elderly persons tend to be more vulnerable to toxic effects than others.)

6. The general health of exposed individuals (poor diet and poor physical condition may predispose individuals to toxic injury or harm.)

Quantitative information on tolerance levels is available primarily for exposures to airborne intoxicants in industrial environments, that is, tolerance to inhalation of toxicants, and is expressed by the maximum tolerable exposures based on an 8-hour working day [time-weighted average exposure (TWA)] and the maximum tolerable short-term exposures [threshold limit values (TLV)]. TWA data are not directly useful for the assessment of the toxic hazards that general battery users may be exposed to, but they do provide an indication of the relative toxicity of various battery materials. Ingestion of toxicants is less frequent than inhalation, but there are no comparable general measures available that apply to ingestion of toxicants. The maximum tolerable concentrations of toxicants in drinking water provide some guide to their ingestive toxicity, but these values are of little practical value for our purposes.

The risk of injury is also affected by the extent to which exposed individuals may wear or have immediate access to protective equipment. An important factor in reducing risk is an awareness on the part of potentially exposed individuals of the hazardous properties of the materials they may be exposed to and how and why they may be exposed, as well as their knowledge of first aid procedures in case of exposures. Another consideration is the time of appearance and the nature of any symptoms that may be associated with exposures, whether they are known to exposed individuals, and whether they are immediate, delayed, or transitory. Immediate and appropriate remedial action in response to toxic exposures can be very effective in limiting the extent of both short-term and permanent injuries that may result from exposures to toxic materials. Useful discussions of risk assessments and situations that require immediate action are given in the medical reference texts cited in Refs. 3–5.

Battery users are exposed to risk of toxic injury only if the contents of a battery escape and enter their environment in some manner (see Part II). The possible types of exposure include:

(a) Physical contact with toxic solids as a result of handling ruptured cells and their components
(b) Physical contact with electrolyte if batteries leak, rupture, or explode
(c) Ingestion of toxic materials due to improper cleanliness after handling contaminated parts
(d) Ingestion of small cells, generally button or coin cells
(e) Inhalation of airborne particulate materials formed as a result of violent cell ruptures, punctures, or explosions

Equipment Damage. Structural damage is rarely associated with battery accidents but may occur as a result of explosions. It is generally limited to battery compartments in the case of small batteries and to nearby structures as well in the case of large batteries. Damage to equipment is primarily of a corrosive nature. The severity of any corrosive attack depends upon the nature of the battery materials and the composition of the exposed materials. Most of the plastic materials used in close proximity to batteries are resistant to attack by sulfuric acid, potassium hydroxide, and other electrolytes and may be returned to service after cleaning unless damaged physically by exposure to leaking electrolytes. Corrosion-resistant metals employed in battery construction and in battery-operated devices are not affected adversely by exposures to battery electrolytes. Electrical equipment, however, contains a variety of components that are easily attacked and eventually destroyed by sulfuric acid, potassium hydroxide, and electrolyte salts. Although potassium hydroxide is slowly neutralized by atmospheric carbon dioxide, the residual salts retain their corrosive power. In the case of lithium batteries, any electrolyte that may escape from a battery carries with it salts that promote the corrosion of susceptible metals, especially if the salts are deposited at the junctions of different metals. Electronic circuitry can be protected from corrosion by battery electrolytes to a considerable extent by acid- and alkali-resistant polymer coatings. Contamination and damage to equipment by battery solids are of little concern, except that continued contact with strongly oxidizing materials such as cathode solids in humid environments or in the presence of electrolytes may promote the corrosion of exposed metals. Thorough cleaning and drying of exposed parts is generally sufficient to prevent corrosive damage, unless the parts have been exposed for a length of time sufficient to cause damage before cleaning can be started.

Environmental Impact. Recent years have seen a growing awareness of batteries as a potential source of environmental pollution.[6] The constituents of particular concern are lead, cadmium, mercury, and their chemical compounds. The toxicity of these materials is well recognized and has led to the establishment of maximum allowable concentrations in drinking water and in the air in working environments[7,8]:

Drinking water:

Lead	0.050 mg/liter
Cadmium	0.010 mg/liter
Mercury	0.002 mg/liter

Air:

Lead	0.05 mg/m^3
Cadmium	0.01 mg/m^3 (CdO dust)
Mercury	0.05 mg/m^3

The given limits for air exposures are defined to be:

- Lead: time-weighted average exposure
- Cadmium and mercury: upper limit for short exposures

The major environmental problem is associated with batteries used in portable consumer devices because they are most often discarded as part of general household waste. The larger lead–acid and nickel/cadmium batteries are more often than not returned to the vendors and recycled for the recovery of their metal values. It is estimated that more than 95% of the lead used in lead–acid batteries in the United States (1989) is recycled.[9] The recycling of lead–acid batteries has been practiced for a long time, motivated primarily by economic considerations.

Mercury presents a different picture. It has been used extensively in small primary batteries since the 1960s, and attempts to collect and recycle spent batteries containing mercury have not been very successful. For this reason, and because of the recognized environmental and health problems created by spent batteries containing mercury, battery manufacturers have made determined and successful attempts to discontinue the use of mercury in batteries. The primary batteries manufactured today for the general consumer are essentially free of mercury. This does not mean that there is no longer a mercury problem due to batteries. Landfills may contain batteries discarded during years gone by, and there is a distinct possibility that some of the mercury from old batteries in uncontrolled landfills may enter groundwater supplies by slow leaching processes.

Recently, batteries have received some attention as a source of cadmium pollution, primarily because the amount of cadmium entering the environment from this source has become significant as a result of the growing use of Ni/Cd batteries in a variety of new consumer devices. The larger Ni/Cd batteries have always been recycled quite efficiently because of their intrinsic metal values. The potential seriousness of the cadmium problem has led battery manufacturers to look for alternate materials to replace cadmium as the negative electrode in these batteries. It appears that nickel/metal hydride batteries may provide an acceptable alternative for many applications in the consumer market. In addition, one or more of the newer rechargeable lithium batteries may replace Ni/Cd batteries in various consumer applications.

Considerable progress has been made in reducing the influx of toxic battery materials to the environment, but further improvements are needed.

It is unlikely that municipal solid wastes will ever be free of discarded batteries. So far, no satisfactory method has been found to extract spent batteries from general solid waste. There is a growing interest in the use of incineration to reduce the volume of solid wastes prior to their final disposal or reprocessing. To the extent that incinerator feedstocks contain discarded batteries, the toxic materials in the batteries will enter the incinerator flues or ashes. It is highly desirable that toxic materials in the flues and ashes be rendered harmless. This is possible with some of the organic solvents and polymers used in batteries that are rendered harmless by combustion. Scrubbing of the flue gases may be necessary to remove toxic materials, whatever their source, and the ashes may be reprocessed to recover their metal values or stored in controlled landfills.

The hazards to waste collectors and landfill operators due to batteries are generally greater than the hazards to battery users. Discarded batteries are likely to be damaged during collection and transport, and the operators may be exposed inadvertently to electrolyte from damaged batteries, possibly to fires if lithium batteries are present in the waste, and, in rare cases, to battery explosions and toxic fumes. Insofar as incineration is concerned, batteries will rupture or explode when incinerated, but the force of such explosions is not likely to cause any physical damage in well-designed incinerators. It is important that waste collectors, landfill operators, and incinerator operators be aware of potential battery hazards and that they follow recommended safety procedures when handling wastes containing batteries.

2.2. ACCIDENT REPORTS

It is difficult to obtain accurate and complete information on the battery accidents that have occurred and on the types of batteries involved. The information that is available[10] indicates that the number of accidents is quite small relative to the number of batteries used, probably much less than one accident for every million batteries used. Such a low accident ratio represents a good safety record but provides small comfort to anyone injured in a battery accident. A survey of accident reports[10] shows that there are two major causes of battery accidents:

1. Improper handling on the part of users
2. Malfunctioning of electrical circuitry in battery-operated devices, especially toys

An understanding of battery hazards is essential for the development of safe handling procedures, and we begin our discussion with an examination of the hazards that have been observed. Subsequently, in Part II, we discuss the processes that occur in batteries that may lead to unsafe battery behavior.

Our examination of the reported battery accidents is limited to the information available from the National Injury Information Clearinghouse for the period January 1983–December 1992.[10] Much more information is available in files maintained by battery manufacturers, but that information is not available to the public. Because of the incomplete data base for our discussion, some caution should be exercised in the assessment of the relative frequency of the various types of accidents. Furthermore, many of the accident reports are fragmentary and give incomplete information on the types of batteries involved and the immediate cause of the accidents. For the purpose of our analysis, we have lumped all the different types of batteries together, whether they be small or large, and whether they be primary or secondary batteries. The types of accidents and their relative frequencies were found to be:

Explosions and violent ruptures	38% of accidents
Overheating	27% of accidents
Leaks and mild ruptures	23% of accidents
Flames and fires	12% of accidents

The accidents listed in the reports were related to batteries in some way. For example, one type of accident was reported as a battery accident that caused a fire when in fact the fire was not caused by a battery. It was blamed on the battery because the smoke detector failed to operate; no batteries had been installed in the smoke detector. Another example is the following: "6 year old girl and mother received cat bites following a cat attack believed to have been caused by high-pitched sound from doll that runs on batteries." We have excluded all such reports from our summary. Most of the reported explosions occurred with automotive lead–acid batteries, and many of them were triggered by sparks from the batteries or by flames near the batteries, for example, as the result of the use of cigarette lighters to look at the batteries in the dark. Many of the lead–acid battery explosions occurred during charging, including explosions triggered by faulty jump-starting procedures. Some of the reported explosions occurred when primary batteries were charged or when the wrong types of chargers were used to charge rechargeable batteries. The majority of the accidents reported as overheating occurred with primary batteries in toys. Although the reports were unclear as to the causes of overheating, it would appear that faulty circuits were the dominant cause of the accidents. Many of the overheating cases also involved battery leakage. If such leakage caused no personal injuries, we classified the accidents as overheating. If personal injuries occurred, we classified the accidents as leakages. Therefore, the reported percentage of accidents that involved leakage, 23%, is too low as an estimate of the fraction of batteries that actually leaked. When overheating caused ignition of nearby flammable materials, we put the accident in the flames and fires category. We excluded accidental fires caused by battery sparks that ignited nearby gasoline, primarily sparks from automotive lead–acid batteries.

Some of the reports indicated a certain amount of ignorance on the part of battery users in regard to the proper handling and installation of batteries. An example is: "Consumer feels battery powered toy boat is a fire hazard after batteries became hot to touch and battery compartment had started to melt. Batteries are to be inserted with positive ends together and negative ends together."

Reports from the National Electronic Injury Surveillance System (NEISS)[11] give useful information on the types of injuries associated with battery accidents, even though batteries may not have been the primary cause of the accidents. The list of accidents is based on reports from a sample of hospital emergency rooms and was used by NEISS to estimate the probable number of battery-associated injuries in the United States for the year 1991. Since it is known that injuries from battery accidents occur that are not treated in hospital emergency rooms and not reported to NEISS, the estimates are very likely too low, but they are the best estimates available. The estimated total number of injuries was found to be 12,560, and the injuries were distributed as follows in terms of the affected body parts:

Eyes	40.9%
Internal	19.2%
Head, face, mouth	15.7%
Hands, fingers	7.5%
Legs, ankles, feet, toes	6.1%
Body	5.2%
Ears	3.4%
Arms, shoulders	2.0%

The reports indicated that about 70% of the injuries affected the body parts normally exposed when people handle batteries.

The types of injuries were categorized as follows:

Chemical burns	35.8%
Ingestion	19.2%
Contusion, abrasion	12.0%
Skin injuries	10.3%
Foreign body	8.5%
Laceration	5.2%
Poisoning	2.1%
Others	6.9%

Clearly, the greatest number of injuries was caused by electrolyte spills. The severity of the injuries varied from relatively mild injuries that affected small areas, to serious fractures, and, in the more severe cases, to injuries that required amputation. Deaths were attributed to some of the accidents. Of the total number of cases, 28.5% may be classified as mild injuries, 45% as moderately severe injuries, and 26.2% as severe injuries. It is difficult to determine the types of batteries involved in the various categories of injuries, but the prevalence of lead–acid battery explosions suggests that they may be the major contributor to the more serious injuries. The ingestion of small batteries ranked quite high as a type of accident. In most of the cases, however, the batteries passed through the digestive tract with no injury and without any need for surgical intervention.[12]

An interesting statistic relates to the age distribution of the injured persons:

Age group	Estimated accidents per million members of age group
0–4	214
5–14	56
15–24	62
25–44	42
45–64	28
65 and up	13

Males were found to be four times as likely to be involved in battery-related accidents as females.

The nature of the reported accidents and injuries and the number of injuries indicate that there is a need for battery users to understand more fully the factors and processes that cause and contribute to battery leakage, ruptures, explosions, and overheating. Then, with the knowledge they possess, they may be able to take the necessary precautions when handling and using batteries to avoid accidents and injuries.

REFERENCES

1. P. Bro and S. C. Levy, Batteries and the environment, in *Environmentally Oriented Electrochemistry* (C. A. C. Sequeira, ed.), Elsevier, Amsterdam (1994).
2. T. Våland and S. Eriksen, *J. Power Sources* **7**, 365 (1982).
3. M. J. Ellenhorn and D. G. Barceloux, *Medical Toxicology. Diagnosis and Treatment of Human Poisoning,* Elsevier, New York (1988).
4. F. W. Oehme (ed.), *Toxicity of Heavy Metals in the Environment,* Marcel Dekker, New York (1979).
5. J. B. Sullivan, Jr. and G. R. Krieger (eds.), *Hazardous Materials Toxicology. Clinical Principles of Environmental Health,* Williams and Wilkins, Baltimore (1992).
6. S. P. Wolsky (ed.), *Proceedings International Seminars on Battery Waste Management (ISBWM),* Ansum Enterprises, Boca Raton, Florida (published annually since 1989).
7. N. I. Sax and R. J. Lewis, Jr., *Dangerous Properties of Industrial Materials,* Van Nostrand Reinhold, New York (1989).

 8. United States Code of Federal Regulations, Primary Drinking Water Regulations, 40 CFR 141.11.
 9. B. M. Barnett and S. P. Wolsky, The battery waste problem and alternatives to small sealed rechargeable lead acid and NiCd batteries, in *Proceedings of the 2nd International Seminar on Battery Waste Management (ISBWM),* Ansum Enterprises, Boca Raton, Florida (1990).
10. National Injury Information Clearinghouse (USA), Reported Incident File, Code 0884, Batteries, January 1983–December 1992, Washington, D.C.
11. National Electronic Injury Surveillance System, U.S. Consumer Product Safety Commission, Directorate for Epidemiology, National Injury Information Clearinghouse, Code 0884, Batteries, 1991, Washington, D.C.
12. T. L. Litovitz, Button battery ingestions, *J. Am. Med. Assoc.* **249,** 2495 (1983).

FUNDAMENTAL ASPECTS OF BATTERY SAFETY

Battery hazards may occur whenever the contents of a battery escape and enter the immediate or remote environment. The hazards created depend upon the chemical nature of the battery and upon the rate at which the contents escape. The prerequisite for any escape is the creation of a force sufficient to violate the integrity of the battery, that is, the creation or presence of an escape path and the creation of a driving force to expel the contents from a battery. The rate of escape is determined by the driving force and the conductance of the escape path. An understanding of the processes that lead to the formation of escape paths and of the forces that drive leakages is a prerequisite for a realistic assessment of battery hazards and for the design of safe batteries.

A battery is an electrochemical reactor whose behavior is governed by the thermodynamics and kinetics of its components and by the mechanical properties of the battery structure, which determine its response to the stresses created by the electrochemical reaction system. These responses may be grouped according to the potential or actual damage associated with the escape process, and we consider four groups of events based on their relative violence:

1. Leaks
2. Ruptures
3. Explosions
4. Thermal runaway

The first group comprises slow escape processes in which no measurable mechanical force is transmitted to the environment. Ruptures are characterized by the sudden opening of a battery and the expulsion of some or all of its contents to the environment. Some mechanical force may be transmitted to the environment by flying debris from the battery, and mild shock waves or loud reports may accompany the ruptures. Battery explosions are created by the rapid release of excessive chemical energy inside batteries sufficient to blow them apart. Pressure waves are created that may cause considerable structural and/or bodily damage, depending on the size of the battery. The kinetic energy of flying debris from the battery is sufficient to cause damage as well. In the case of batteries that contain flammable gases or solvents, ruptures may develop into secondary explosions if the flammable fuel/air mixtures are ignited while within their explosive composition range. Thermal runaways are uncontrolled increases in the temperature of a battery driven by exothermic processes inside the battery. They are likely to result in leaks, ruptures, or explosions, unless checked by some means.

3

Battery Leakage

The escape of electrolyte or gas from a battery requires a driving force and an escape path. Apart from batteries with engineered vent structures, batteries are designed to contain moderate pressures to prevent the release of gases and electrolytes. When leakages do occur, they may be attributed to the existence or generation of leakage paths due to defects, excessive driving forces, or the deliberate or inadvertent abuse of the battery. The principal driving force is internal cell pressure, whether we consider slow leakages, ruptures, or explosions. In this chapter, we focus attention on processes that lead to the slow generation of cell pressures, the factors responsible for the creation of leakage paths, and some parametric estimates of leakage rates. A brief discussion of other forces that may also cause leakage is included.

3.1. GAS GENERATION IN BATTERIES

The development of internal cell pressures is most often caused by gas evolution, and it provides a driving force for gas and electrolyte leakage. In the case of pressurization by hydrogen, fires and/or explosions may occur as well. Since most batteries are designed to vent or to contain moderate pressures, our concern is the possibility that excessive pressures may be generated, sufficient to cause intolerable leaks, ruptures, or explosions. Pressurization by gas evolution may be driven by:

1. Internal factors: spontaneous physicochemical processes
2. External factors: electrically energized processes

These processes may occur singly or in combinations.

We have already referred to the rate of pressure increase, i.e., the rate of gas evolution, as a useful parameter to categorize the events that may occur as a result of excessive cell pressures. The slow development of excessive pressures is generally terminated by venting or ruptures, whereas fast pressure-generating processes may cause violent ruptures or explosions. We begin our discussion with the relatively slow pressure-building processes. Processes that lead to ruptures and explosions will be discussed in Chapters 4 and 5, respectively. Furthermore, we limit the discussion in this chapter to basic principles and illustrate them with reference to the characteristics of some of the more commonly used batteries. The principles are of a general nature and may be applied to any battery system, but the quantitative characteristics may differ considerably from one system to another.

3.1.1. Spontaneous Gas Generation, Internal Drivers

Gas evolution is controlled by the thermodynamic and kinetic properties of the system under consideration. The primary driver is the free energy change of a gas-evolving reaction; the more negative the free energy change, the greater is the tendency for gas to be evolved. An equivalent criterion is the voltage of the gas-producing reaction; the more positive the voltage, the greater is the tendency for gas evolution. We examine three systems of considerable practical importance:

1. Zinc electrodes in alkaline cells
2. Lead dioxide electrodes in lead–acid batteries, lead electrodes in lead–acid batteries
3. Nickel oxyhydroxide electrodes in Ni/Cd batteries, cadmium electrodes in Ni/Cd batteries

The associated, spontaneous, gas-generating reactions and their reaction voltages are given below. Note that the standard-state voltages are given. The voltages applicable to operational cells are different, as they depend upon the exact electrolyte compositions and the temperatures of the cells.

$$Zn + 2OH^- \rightarrow ZnO_2^{2-} + H_2 \qquad 0.39 \text{ V}$$

$$PbO_2 + H_2SO_4 \rightarrow PbSO_4 + H_2O + \tfrac{1}{2}O_2 \qquad 0.23 \text{ V}$$

$$Pb + H_2SO_4 \rightarrow PbSO_4 + H_2 \qquad 0.36 \text{ V}$$

$$2NiOOH + H_2O \rightarrow 2Ni(OH)_2 + \tfrac{1}{2}O_2 \qquad 0.09 \text{ V}$$

$$Cd + 2H_2O \rightarrow Cd(OH)_2 + H_2 \qquad -0.01 \text{ V}$$

There is a spontaneous tendency for hydrogen to be evolved in all cells with zinc electrodes, in both primary and secondary cells, and in all lead–acid batteries. Oxygen may be expected to form spontaneously in charged lead–acid and Ni/Cd batteries, but the potential for oxygen evolution is much smaller in Ni/Cd batteries than in lead–acid batteries. Cadmium electrodes are quite stable in alkaline electrolytes. Since all of the reactions with positive reaction voltages occur independently of any external forces, they may be expected to generate gas continuously in a battery whether it is in use or not. On open-circuit stand, gas evolution leads to the gradual loss of battery capacity, and the battery self-discharges, the more so the higher the temperature. The practical utility of zinc electrodes and lead–acid batteries depends upon our ability to reduce the rates of gas evolution to acceptable levels. In the case of zinc electrodes, inhibitors have been found that are reasonably effective in reducing the rate of hydrogen evolution, and in the case of lead–acid batteries, the kinetics of the gas evolution reactions are intrinsically slow, provided certain catalytic agents are excluded from the cells.

The thermodynamic principles apply equally well to lithium batteries. In their case, aqueous electrolytes cannot be used because no way has been found to inhibit adequately the spontaneous reaction of lithium with water. Lithium reacts violently with water to generate hydrogen. It is essential that water be excluded from lithium batteries. Instead, organic solvents, inorganic solvents, or solid electrolytes are used that may or may not react with lithium to cause battery degradation. Although many of the selected materials do react with lithium, the reaction rates are sufficiently slow to be acceptable for a variety of practical applications. Because of their unique characteristics, lithium batteries will be discussed separately in Part IV.

The rates of gas evolution and the total amount of gas accumulated in a cell due to the spontaneous chemical reactions depend upon many factors related to the design of the cell, its chemical composition, state of charge, temperature, and its past history, particularly its temperature history. It is not feasible, therefore, to do a comprehensive phenomenological analysis of the gas evolution kinetics. Instead, an empirical approach is employed. In the following, we illustrate the empirical approach with a simulated set of gas evolution data. The simulation represents a fairly realistic set of conditions for alkaline cells with zinc electrodes. The methods we present are applicable to any battery system, but the numerical values will differ for different systems, cell sizes, and cell designs. .

The Battery: A 10-Ah cylindrical alkaline cell with zinc electrodes. The capacity of the cell is anode limited.

- Void volume: $V_0 = 2$ cm^3, assumed constant (not generally true)
- Hydrogen solubility in electrolyte: negligible.
- Hydrogen recombination rate: negligible.

Remark: The solubility of hydrogen in 40% KOH is approximately 10^{-3} cm^3 (STP)/atm per cubic centimeter of electrolyte between 20°C and 75°C.[1]

Experimental Data: The simulated experimental data are shown in Fig. 3.1. They comprise the cumulative amounts of hydrogen, reported as hydrogen volume at 0°C and 1 atm, generated in the cells as a function of time and temperature. The simulation was based on capacity losses of:

1. 2% per year at 25°C
2. 5% per year at 45°C
3. 15% per year at 65°C

The experimental data may be analyzed by conventional chemical kinetic methods[2] to give an analytical form that fits the data. We obtained the following rate expression:

$$\dot{N} = A \exp(-\beta t), \quad \text{with} \quad A = A(T), \quad \beta = \beta(T) \quad (3.1)$$

where N is the amount of gas formed, t is time, and T is temperature, and a dot above a symbol signifies its time derivative. The experimental data may also be fitted to a linear rate expression, but a statistical analysis shows that

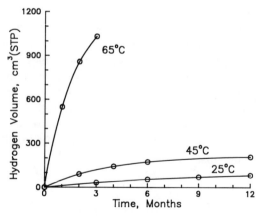

Figure 3.1. Cumulative hydrogen generation due to zinc corrosion (simulated data).

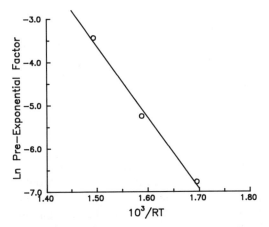

Figure 3.2. Activation energy plot of the pre-exponential factor in Eq. (3.1).

the exponential form gives a better fit. The relevant method of analysis and some associated considerations have been discussed in Ref. 3. The parameters in the rate expression were estimated from the experimental data, and their temperature dependence is shown in Fig. 3.2 and 3.3. They could be approximated by the exponential forms:

$$A = 1.97 \times 10^9 \exp(-16,600/RT) \tag{3.2}$$

$$\beta = 8.93 \times 10^4 \exp(-8000/RT) \tag{3.3}$$

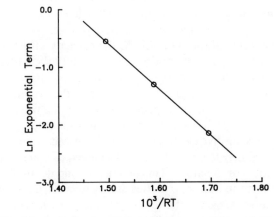

Figure 3.3. Activation energy plot of the exponential term in Eq. (3.1).

The activation energies of the two parameters were found to be 16.6 kcal/mol and 8.0 kcal/mol, respectively. Given these empirical correlations, the rates of gas evolution may be estimated for any given temperature or any given temperature history of interest, assuming that there are no changes in the mechanism of gas evolution at temperatures beyond the range of the experimental data.

3.1.2. Electrical Gas Generation, External Drivers

Electrochemical reactions occur in batteries whenever current is flowing. In discharging cells, these are the normal discharge reactions. Conditions exist, however, when current is forced through a cell from an external source under conditions that lead to unwanted reactions. The external source may be other cells or electrical charging devices, the latter being the most commonly encountered external source.

Cell Reversal. Cell reversal occurs when discharge current is forced through a cell after its capacity has been exhausted. It is a process confined primarily to high-voltage, multicell battery stacks, either primary or secondary battery stacks. Although it may occur in low-voltage, series-connected stacks, it is less likely in such batteries because of the voltage cutoff requirements. When one cell becomes exhausted, the voltage normally drops below useful service levels and the batteries are disconnected from their loads. The cell reversal process is illustrated in Fig. 3.4, where we show the voltage delivered by a stack of series-connected cells. One of the cells has a lower capacity than all the other cells. The capacity deficit of the defective cell has been accentuated for illustrative purposes. The battery voltage is the sum of the individual cell voltages. Current will flow and the battery will deliver useful service as long as the battery voltage remains above the cutoff voltage, point B, for the application under consideration. At point A, the capacity of the defective cell has been exhausted, but current continues to flow until the battery is disconnected from the load. This means that oxidation will continue at the negative electrode and reduction at the positive electrode of the defective cell. In the case of aqueous systems that are capacity limited by the negative electrode, this means that oxygen will be evolved, or the current collector will be oxidized. If the defective cell is positive electrode limited, hydrogen evolution will occur. Reversals are quite likely to occur if partly discharged cells are combined with fresh cells in a series combination. This practice should definitely be avoided. Other reactions occur in lithium batteries during reversal, as will be discussed in Part IV. Since rechargeable batteries are rarely discharged to their full capacities, cell reversals are less likely to occur in rechargeable batteries than

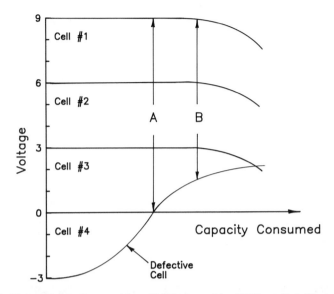

Figure 3.4. Illustration of cell reversal for a 12-V battery with a 7.5-V cutoff. A, Start of reversal; B, cutoff voltage. Load voltage = top curve − Bottom curve.

in primary batteries, except for rechargeable batteries that have been cycled deeply or many times.

The rate of gas evolution in a reversed cell is directly proportional to the reversal current, assuming that no reactions occur other than gas evolution. The amount of gas may be estimated from the equivalence between electrical charge and chemical conversion, according to which 209 cm^3 (STP) of oxygen and 418 cm^3 (STP) of hydrogen are evolved for each ampere-hour of charge that passes through a cell. Given the battery voltage, V volts, and the load resistance, R ohms, the total amount of gas is obtained by integrating V/R over the duration of the reversal. The amounts of gas evolved in the case of a constant-current reversal are shown in Fig. 3.5. Either hydrogen or oxygen, or both, may be evolved, depending on which is the limiting electrode. If the reversal is carried beyond the point where both of the electrodes have discharged their full capacities, an explosive mixture of hydrogen and oxygen will be produced in the cell, a potential bomb in the case of sealed cells.

Battery Charging. An important source of gas in rechargeable aqueous electrolyte batteries is the decomposition of the electrolyte due to excessive charging voltages. This is a very general process, and we describe it with reference to the lead–acid system. All aqueous systems exhibit the same qual-

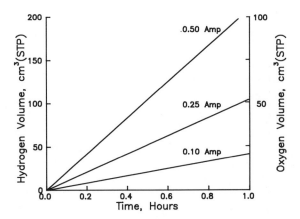

Figure 3.5. Cumulative gas evolution in cell reversal. The reversal currents are indicated on the figure.

itative features with respect to electrolyte decomposition on overcharge, but they may differ considerably in their quantitative characteristics. The same general principles apply to nonaqueous batteries, but other decomposition reactions occur—reductions and/or oxidations—whose nature depends upon the system compositions.

The basic principles are illustrated in Fig. 3.6 with data from the lead–acid battery system for concreteness. The figure shows the general features of

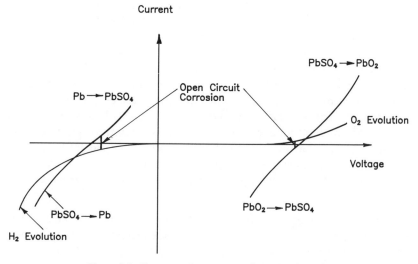

Figure 3.6. Current–voltage patterns in lead–acid cells.

the current–voltage characteristics of the various cell reactions that may occur (simplified reaction schemes). All the reactions are voltage driven, including the corrosion reactions. Consider first the electrochemical decomposition of water. If the voltage between two electrodes immersed in an aqueous electrolyte exceeds the decomposition voltage of water, 1.23 V in terms of the standard-state voltage, water will be decomposed to hydrogen and oxygen, an explosive gas mixture. The actual decomposition voltage will differ from the standard value since it depends upon the composition of the electrolyte and the temperature. To the right of the water decomposition range, we find a curve showing how the rate of oxygen evolution, i.e., the equivalent current, depends upon the voltage of the positive electrode. To the left of this range, we find the corresponding curve for the rate of hydrogen evolution. The actual features of the current–voltage diagram have been exaggerated for reasons of clarity. In the same diagram we also show the current–voltage characteristics of the positive lead dioxide electrode, both for the reduction and for the oxidation processes:

$$\text{Discharge (reduction):} \quad PbO_2 \rightarrow PbSO_4$$

$$\text{Charge (oxidation):} \quad PbSO_4 \rightarrow PbO_2$$

Similarly, for the negative electrode:

$$\text{Discharge (oxidation):} \quad Pb \rightarrow PbSO_4$$

$$\text{Charge (reduction):} \quad PbSO_4 \rightarrow Pb$$

It may be seen that the voltage of the lead dioxide electrode is more positive than the voltage required to oxidize water to oxygen, and the voltage of the lead electrode is more negative than the voltage required to reduce water to hydrogen. Accordingly, hydrogen and oxygen will be produced spontaneously in lead–acid batteries. When no net current is flowing through the battery, the equivalent currents of oxygen evolution and lead dioxide reduction must be equal, but of opposite signs, as shown in the figure. The same holds true for hydrogen evolution and lead oxidation on the negative electrode. During discharge, the positive lead dioxide voltage decreases and the negative lead voltage increases in a positive direction, as a result of which the rates of oxygen and hydrogen evolution decrease to insignificant levels. The contrary happens on battery charging. Then, the voltage of the positive electrode becomes more positive and that of the negative electrode becomes more negative. Two processes occur simultaneously at the positive electrode:

$$PbSO_4 \rightarrow PbO_2$$

$$H_2O \rightarrow O_2$$

both of which consume a positive current, the sum of which equals the current flowing through the cell. Similarly, two processes occur simultaneously at the negative electrode:

$$PbSO_4 \rightarrow Pb$$

$$H_2O \rightarrow H_2$$

both of which consume a negative current, the sum of which equals the total current at the positive electrode, but with the opposite sign. Although the positive and negative currents must be numerically equal, it is not required that hydrogen and oxygen be produced at the same rates. This provides designers an opportunity to control the relative rates of hydrogen and oxygen evolution by modifying the kinetic activity of the electrodes for the gas evolution reactions by the use of inhibitors and other means.

As the positive and negative electrodes approach full charge, they contain an ever decreasing amount of lead sulfate, and ever greater voltages are required to convert the residual sulfate to active material. Consequently, the current–voltage curves become noticeably flatter (Fig. 3.7). A more positive voltage is required on the positive electrode and a more negative voltage on the negative electrode to maintain the same charging current. Because of the increasing voltage, more and more of the charging current is diverted into the decomposition of the electrolyte. It is clearly desirable to employ charging regimes that limit the excessive voltages.

A charging regime of considerable importance is the float charging regime. Once a battery has been fully charged, it will gradually lose its capacity as a result of the spontaneous corrosion reactions discussed above. This capacity loss can be prevented by maintaining a low charging rate on the battery to balance the corrosion reactions. A concomitant of float charging is the continuous evolution of hydrogen and oxygen, as illustrated in Fig. 3.8 for lead–acid batteries. Whether the charging is conducted in a controlled current or a controlled voltage mode, it is constrained by the requirement that the same total current must pass through the two electrodes, causing oxidation at the positive electrode and reduction at the negative electrode; that is, the positive and negative currents must be numerically equal as indicated in Fig. 3.8. In controlled current charging, the voltages applied to each of the two electrodes are given by the total polarization curves and the applied currents, as shown. Conversely, in controlled voltage charging, the voltages applied to each of the

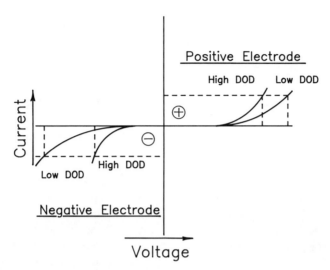

Figure 3.7. Current–voltage patterns in lead–acid cells on charge at start (high DOD) and end (low DOD) of charge.

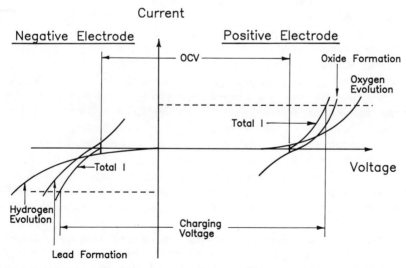

Figure 3.8. Illustration of charging process (exaggerated polarization). OCV, Open-circuit voltage. (Refer to text for discussion.)

two electrodes adjust themselves to equalize the total reducing and oxidizing currents. In both cases, the rates of hydrogen and oxygen evolution are given by the currents corresponding to the intersections of their polarization curves with the voltages applied to the two electrodes. Clearly, the rates of gas evolution may be changed by adjusting the charging regime and by manipulating the electrode kinetics by chemical and physical means, but gas evolution cannot be avoided on float charge. A consequence of float charging is the generation of an explosive mixture of hydrogen and oxygen from the decomposition of the electrolyte and the gradual loss of water from the cells in the case of vented cells.

It is a common practice to charge series-connected cells as a single unit, rather than charging each cell separately. The latter approach may be employed in special applications requiring high reliability and a long cycle life. The principles that apply to the charging of single cells apply to the charging of a series-connected stack of cells as well, but now the charging voltages applied to the individual cells cannot be controlled. Only the charging current or the total charging voltage can be controlled. If, therefore, the cells have different impedances, as is generally the case, different voltages will be applied to the various cells. As a consequence, some may produce more gas than others, and as the total number of charging cycles increases, some cells are likely to deteriorate more rapidly than others, with the development of potential safety problems.

Apart from the deterioration of battery performance associated with gas production, well-defined hazards may be created by the accumulation of hydrogen and oxygen in a battery or in any confined space. Hydrogen and oxygen form explosive mixtures that, if ignited, may explode with considerable violence, sufficient to cause both physical and bodily damage (see Chapter 5). Even if no oxygen is produced, mixtures of hydrogen and air are similarly explosive, and ignition may be triggered by small sparks or hot objects. The rate of hydrogen production during battery charging is directly proportional to the charging current multiplied by the charging efficiency of the negative electrode. It depends upon the state of charge of the battery, the charging regime, the temperature of the battery, and the design characteristics of the battery itself. Generally, the lower the temperature, the lower is the charging efficiency. On float charge, most of the charging current is consumed in gas generation. From the electrochemical equivalence relations, we find the following maximum rates of gas generation as a function of the charging current:

Hydrogen: 0.42 cm^3 (STP) per hour for each milliampere of charging current

Oxygen: 0.21 cm^3 (STP) per hour for each milliampere of charging current

It is clearly desirable to design batteries and to employ charging regimes that minimize the rates of gas production or other adverse processes in batteries. Various means are available to reduce the net rates of gas formation, the most important of which is the control of the charging mode. Gas production can be substantially decreased by decreasing the charging current as charging proceeds. A useful rule of thumb is that the charging current in amperes should not exceed the residual capacity to reach full charge in ampere-hours. Many other factors are involved in the control of the charging process to limit gas production, and interested readers may be referred to the excellent discussions in Refs. 4 and 5. Furthermore, design features may be incorporated in batteries to reduce gas production or the accumulation of gas in batteries. Some of these features will be discussed with reference to the lead–acid battery system. By virtue of the location of the reaction voltages shown in Fig. 3.6, we observe that oxygen may be expected to react spontaneously with lead:

$$Pb + \tfrac{1}{2}O_2 + H_2SO_4 \rightarrow PbSO_4 + H_2O$$

and hydrogen may be expected to react spontaneously with lead dioxide:

$$PbO_2 + H_2 + H_2SO_4 \rightarrow PbSO_4 + 2H_2O$$

Both reactions do in fact occur, the first more readily than the second. If the amount of electrolyte in a battery is decreased to allow easy access of oxygen and hydrogen to the negative and positive electrodes, respectively, and if the battery is sealed to prevent the escape of the gases, these reactions can be effective in scavenging the gases produced on low-rate float charge and the gases produced by corrosion reactions. As an added safety feature, sealed batteries may be provided with pressure-actuated vents designed to operate at moderate pressures. In the case of unsealed batteries, spillproof vents are normally provided to permit the escape of any gases produced in the cells.

Similar considerations apply to nickel/cadmium, nickel/zinc, nickel/metal hydride, and other rechargeable aqueous electrolyte batteries, both in regard to gas generation due to corrosion and charging and in regard to means of preventing overpressures due to gas accumulation. The only significant differences relate to the potentials of the various reactions and their rate characteristics. Although the same principles apply to lithium batteries, the reactions that occur in such batteries are quite different from those that occur in aqueous systems. Lithium batteries will be discussed in Part 4.

3.1.3. Gas-Generated Cell Pressures

Given a favorable free energy change for gas evolution, the next important question concerns the pressures generated by the gas. The rates of pressure increase are related to the rates of gas evolution and the rates of gas leakage, and the total cell pressures are directly proportional to the total amounts of residual gas in a cell. The cell pressures may be measured experimentally as a function of the state of a cell in specially designed cells, or they may be estimated from the experimentally determined gas evolution rates. The first method gives the desired information directly, and it needs no further elaboration. We focus attention on the second method and illustrate it with the data presented in Section 3.1.1 for a primary cell. For simplicity, we employ all the assumptions stated in that section. This means that the estimated pressures will be greater than the actual values. Subsequently, we explore the effects of gas leakage on the estimated pressures. Quite analogous methods apply to secondary batteries, but other terms need to be included to allow for gas generation due to overcharge and gas recombination. Both of these effects are design- and system-specific characteristics, and their incorporation in the equations used for the pressure estimates should present no problems.

The cell pressure may be estimated directly from the experimental data using the ideal gas law, $PV = nRT$, or it may be estimated from the gas evolution rate correlations for a variety of conditions. The results are presented in Fig. 3.9 in terms of the gauge pressures at 25°C after cell exposures for the stated periods of time at the various temperatures. Considering that the simulated gassing rates are close to actually observed values and that the bursting pressures of cylindrical cells may be of the order of 1000 psi (gauge), the results should give rise to concern. They suggest that hermetically sealed alkaline cells with no gas recombination are likely to rupture, sooner or later.

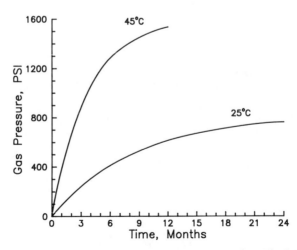

Figure 3.9. Pressure development in D cell based on simulated data from Fig. 3.1 (no gas leaks).

From a safety point of view, it is imperative that means be provided to permit the evolved hydrogen to be removed either by chemical recombination or by the creation of a leakage path. The latter approach is preferred and can be implemented by using slow-leakage-rate crimp seals or by incorporating gas diffusion elements in the cell covers. The diffusion elements would comprise thin polymer membranes with a high hydrogen permeability. Based on the gas permeabilities of some common polymers,[6] we estimate the following conductances for hydrogen in a 0.5-mm-thick circular membrane with a diameter of 2 mm (note that the diffusion membranes need to resist rupture due to internal pressures):

Nylon	0.002 cm^3 (STP)/day $\cdot$ atm
Polypropylene	0.010 cm^3 (STP)/day $\cdot$ atm
Teflon	0.025 cm^3 (STP)/day $\cdot$ atm

The effect of gas leakage rates on cell pressure, P, can be estimated once the relationship between cell pressure and leakage rates is known. For most cases of interest, the rate of escape of a gas is proportional to the cell pressure:

$$\dot{n} = \alpha P \tag{3.4}$$

From the ideal gas law:

$$\dot{n} = V\dot{P}/RT \tag{3.5}$$

A material balance equating the rate of change of gas pressure in the cell with the difference between the rate of gas generation and the rate of gas leakage gives

$$\dot{P} + \alpha RTP/V = (ART/V)\exp(-\beta t) \tag{3.6}$$

where β is a rate constant of gas evolution, α is the conductance of the leakage path, and A is a kinetic constant.

Thus, we obtain

$$P = \frac{ART}{\alpha RT - \beta V}[\exp(-\beta t) - \exp(-\alpha RTt/V)] \tag{3.7}$$

This equation was used to estimate the pressure buildup in the sample cell at 45°C, using a conductance of 0.02 cm^3 (STP)/day $\cdot$ atm (Fig. 3.10). We find that this value suffices to prevent excessive cell pressures. It may be noted that this conductance corresponds to the estimated conductance of

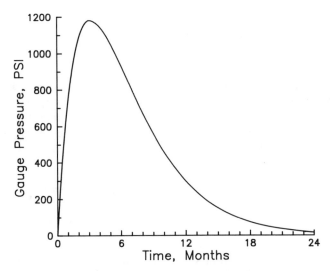

Figure 3.10. Effect of gas leakage on cell pressure. (Refer to text for details.)

hydrogen in Teflon membranes with the dimensions given above. In other words, the incorporation of controlled leakage paths into alkaline cells with zinc electrodes can prevent cell ruptures. Similar estimates can be made for any other set of conditions of interest. The above required rate of escape of hydrogen is too low to create explosive hazards provided the escaping gas is vented from the battery compartment. If the hydrogen is permitted to accumulate in a confined space, an explosive hazard may be created.

3.2. OTHER INTERNAL LEAKAGE DRIVERS

The discharge of primary batteries and the cycling of secondary batteries are accompanied by volume changes of the active materials and of the electrolytes. These volume changes must be allowed for by providing sufficient void space in the cells to accommodate any expansion. An allowance must also be included for the thermal expansion of the cell components and the cell case on the assumption that the batteries will be exposed to temperatures both below and above their normal operating ranges. If an insufficient void volume is available, the forces generated by thermal expansion are due primarily to the electrolyte. Battery solids expand less than do battery electrolytes. As an illustration, we may consider an aqueous electrolyte with a compressibility of the order of 4×10^{-5} per megabar. Organic and inorganic solvent electrolytes have higher compressibilities; that is, they give lower hydrostatic pressures for a given volume expansion compared with aqueous electrolytes. A net volume increase of 1% (difference between the expansion of the cell contents and the cell case) beyond the point of zero void volume would generate a pressure of about 3700 psi, assuming no deformation of the case. Since cell cases yield at significantly lower pressures, this amount of volume expansion would lead to the deformation of the cell, primarily by deflections of the cell base and cover for cylindrical cells and of the flat walls of prismatic cells. Battery leakage or rupture would be a likely result. This is a well-recognized potential problem, and batteries are designed to accommodate any volume changes that may occur during normal battery usage. It is a rare event for a battery to develop leaks or to rupture by this pathway, and, if it occurs, it is most likely due to poorly controlled manufacturing operations.

Capillary forces provide another driver for battery leakage that must be considered for crimp-sealed cells. It can operate only if there is an available leakage path in the crimp seal. Its magnitude depends upon the surface energies of the constituents of the metal–electrolyte–polymer–gas interfacial system and upon the potential of the metal. The driving force may be expressed in terms of the wettability of the system. Since the wettability can be controlled by the application of hydrophobic coatings, capillarity is not generally a problem with aqueous electrolyte batteries. In the case of crimp-sealed, organic electrolyte lithium cells, it is more difficult to control the wettability, and the operation of capillary forces can be counteracted only by the use of high-quality seals with no leakage paths.

A force has been observed in alkaline cells that promotes electrolyte leakage to an extent that exceeds leakage due to both ordinary capillary forces and electrocapillary forces. This type of leakage occurs preferentially at the negatively polarized metal of crimp-sealed alkaline cells. The process, electrochemical creepage, is attributed to the reduction of water and/or oxygen at the negatively polarized metal in a region ahead of the electrolyte meniscus, where it generates hydroxyl ions in a very thin liquid film on the metal.[7-9] It operates both at metals polarized at the cadmium potential (oxygen reduction) and at metals polarized at the zinc potential (water and oxygen reduction). The process is not observed if water and oxygen are blocked from access to the metal/electrolyte interface. This indicates that leakage by this mechanism can be prevented by the use of hydrophobic coatings on the outer crimp seal area and by preventing the formation of leakage paths in the crimp seal. Although no cases of electrochemical creepage have been reported for crimp-sealed lithium cells, the nature of the process suggests that it may operate in these systems as well if any leakage paths are present in the crimp seals.

3.3. LEAKAGE PATHS

The slow escape of electrolyte or gas from a battery is likely to occur via defects introduced during manufacturing or via paths created during the life of a battery. Leakage paths may be classified according to their locations in the cells or according to the processes responsible for their formation:

1. Battery closures (crimp seals and hermetic seals)
2. Corrosion cracks or pinholes
3. Abusive treatments of various kinds

Since the last category comprises events distinctly different from the others, they will be discussed separately in Section 3.4. We do not discuss leakage paths deliberately incorporated into battery designs, for example, vents in unsealed lead–acid batteries. Electrolyte leakage may certainly occur in such batteries, but only as a result of careless handling or operations.

Battery closures provide potential paths for the escape of gases and electrolytes and a possible means of relieving overpressure to prevent explosions. Two types of closures are used in batteries:

1. Polymeric compression seals, generally crimp seals
2. Hermetic glass-to-metal or glass-to-ceramic seals

The polymeric compression seals rely on the use of plastic or elastomeric gaskets of suitable shapes to form closures whose performance depends upon a variety of factors to be discussed below. Although high-quality, relatively leakproof polymer seals can be made, they are not truly hermetic, and their use is confined to batteries where hermeticity is not an absolute requirement. Crimp seals are used in most aqueous electrolyte systems, except for some special purpose alkaline cells that may employ hermetic ceramic-to-metal seals. Lithium cells may use either crimp seals or hermetic seals, depending on the nature of their electrolytes, as discussed in Section 3.3.2 below. Since hermetic seals are more expensive than crimp seals, the latter type is preferred wherever feasible.

Corrosion may be expected to occur whenever metals come in contact with aggressive electrolytes of the types used in batteries. Some batteries may be housed in nonmetallic containers to avoid corrosion (e.g., automotive lead–acid batteries), but the majority of batteries in general usage are housed in metallic containers. The selection of suitable metals or alloys and the use of appropriate processing methods can essentially eliminate the formation of leakage paths due to corrosion. Some of the conditions that may lead to the formation of leakage paths due to corrosion will be discussed in Section 3.3.3.

3.3.1. Crimp Seals

The most commonly employed polymeric compression seal is the crimp seal of cylindrical cells, and we focus attention on that type of seal. Considerations similar to those below apply to other types of polymeric compression seals. Crimp seal leakage occurs quite rarely, and it represents more of a nuisance than a serious hazard. Typically, any alkaline electrolyte that may escape from a cell does so quite slowly, and it is gradually neutralized by atmospheric carbon dioxide. This leaves a mildly alkaline encrustation on the cells that may be removed by wiping with a wet cloth. Care should be taken to avoid skin contact with the salt deposit. The accumulation of such salt deposits over a longer period of time may lead to a corrosive attack on nearby metals and electrical circuitry, but it does not seriously impair the performance of the batteries.

An appreciation of the causes of the formation of leakage paths in crimp seals may be gained by a brief discussion of the seal geometry, the seal-forming operations, the properties of the polymeric material, and the nature of the interfaces between the gasket and the bounding metals. These topics will be discussed with reference to the schematic representation of a typical crimp seal in Fig. 3.11, where we show the pre-crimp and post-crimp configurations of a coin cell and the potential leakage paths. The seal geometry is designed to provide the maximum feasible contact area between the gasket and the metal, consistent with the maintenance of an adequate stress between the

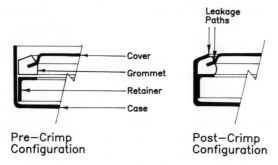

Figure 3.11. Typical crimp seal (electrodes not shown).

metal and the polymer. Since cell closures occupy a significant fraction of the total cell volume, particularly for coin cells, a contrary requirement is that a minimal volume be allocated to the seal structure.

Depending on the seal geometry and the nature of the polymer, the gasket and the cell cover may be fabricated and inserted separately into a cell, or the gasket may be premolded onto the cell cover to form an integral sub-assembly that is inserted into the cell prior to crimping. Because of their forming requirements, elastomers and Teflon type materials are generally fabricated as separate components, whereas thermoplastic materials such as nylon and polypropylene may be premolded onto the cell covers. Thermoplastic materials are commonly preferred for reasons of their lower costs combined with their reasonably good sealing qualities. A primary requirement for any sealing material is that it be chemically stable in the presence of battery electrolytes, whether they be inorganic or organic solvent type electrolytes. The viscoelastic properties of the polymer represent one of the most important factors that control the quality of any crimp seal. The consistent manufacturing of leakproof seals requires the use of gaskets with well-controlled mechanical properties within a defined range for each specific seal geometry and application. The addition of reworked polymer (scrap recycling) to virgin polymer in thermoplastic molding operations may have an adverse effect on the resulting product, and it is a practice that should be approached with some circumspection.

The crimping operation itself is of critical importance for the fabrication of leakproof seals. Various forming operations may be employed. Among these, the direct impact crimping method is the simplest and most often used, but spin crimping may be expected to give better seals. The residual stress in impact crimped seals has a radial component that may strain the cell case laterally to a sufficient extent to generate incipient leakage paths. This does not occur with spin crimped closures. We focus on the impact crimping method below. In this operation, the upper portion of the cell case is deformed to the desired shape by forcing a suitably shaped tool onto the cell assembly held in a conforming cavity. For most cell manufacturing operations, the crimping operation is completed in a very short time, typically in a fraction of a second. The response of the metal may be discussed with reference to a typical stress–strain curve of metals (Fig. 3.12). The response of a particular metal depends upon the actual state of the metal at the time of the crimping operation. In order to obtain the desired permanent shape of the metal, identified by point A on the strain axis, the metal must be strained beyond the yield point to point B, defined by the desired residual stress, point C, and by the shape of the relaxation curve between points B and D. The designer needs to specify the permanent set, point A, and the residual stress, point C. The magnitudes of the stresses and strains at points B and D and the duration of

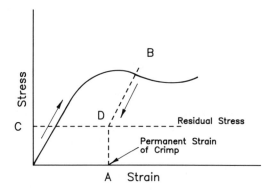

Figure 3.12. Response of metal in crimp seal formation. (Refer to text for details.)

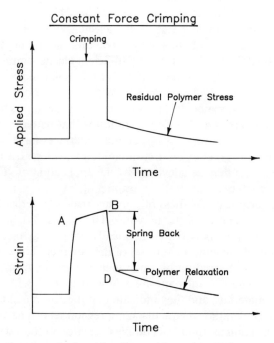

Figure 3.13. Response of polymer in crimp seal formation. (Refer to text for details.)

the crimping operation have a significant effect on the quality of the resulting seal, particularly for plastic materials such as nylon and polypropylene, but less so for elastomeric materials.

The response of the polymeric gasket may be discussed with reference to Fig. 3.13 for the case of an idealized constant-force crimping operation. Analogous considerations apply to constant-displacement and spin crimping operations. When the crimping force is applied to the seal structure, the polymer is deformed to point A, and, while the compressive force is maintained on the metal/polymer structure, the polymer is deformed plastically to point B (there is an initial elastic response that depends upon the viscoelastic properties of the gasket). When the applied force is removed, the structure relaxes to point D by virtue of the elastic component of its stress/strain characteristics. Thereafter, the polymer continues to relax, but much more slowly, owing to plastic flow in response to the residual design stress of the seal. The actual stress/strain characteristics of seal assemblies are more complicated than suggested by the above figure. They are a complicated function of the seal geometry and all the crimping operation parameters. The two most important features that need to be emphasized are the excessive, temporary deformation of the polymer and its relaxation (creep) in response to the stress maintained on the polymer once the seal has been made. It is clear that the viscoelastic properties of the gasket, the seal geometries, the seal-forming tools, and the processing conditions must be carefully selected and controlled to obtain leakproof seals. Another way to state the polymer requirement is to say that the polymer must be capable of storing sufficient elastic strain energy during the entire life of the battery if the battery is to remain leakproof. It is for this reason that elastomers and elastomer-filled plastics give better performance than do purely plastic materials. However, cost considerations militate against the use of elastomers. Another consideration is that battery leakage is more likely to occur if the mating metal/polymer interfaces have defects such as cracks, scratch marks, dents, or circumferential waviness. If such defects are present, their effects may be alleviated by the application of viscous bonding agents to the interfaces prior to the crimping operation.

The long-term leakproofness of polymer-sealed batteries depends upon the viscoelastic properties of the polymer. The effects of aging on seal deterioration, assuming the polymer is chemically stable, may be discussed most simply in terms of the compression set of the polymer. The compression set is the permanent deformation of a polymer that results from exposure to a constant stress. The general patterns exhibited in Fig. 3.14 apply to most thermoplastic materials, and they indicate that the compression set increases with time and temperature. The actual compression set of a particular material in a specific seal configuration is a complex function of the varying stress and temperature history of the seal in question. The implications are obvious for

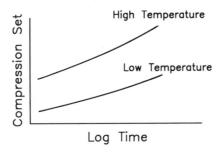

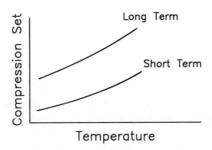

Figure 3.14. General response of polymer to aging.

the quality of polymeric compression seals and the maintenance of a leakproof seal during the life of a battery. The quality of the seals can only be expected to deteriorate with increasing battery age, the more so the longer the battery is exposed to elevated temperatures or thermal cycling. For most polymers of interest, there is also the consideration that they become brittle at low temperatures. For each application with a demanding low-temperature or high-temperature requirement, it may be necessary to design special seal structures that can operate under extreme conditions without developing leaks.

3.3.2. Hermetic Seals

Aqueous electrolyte batteries with voltages in excess of the water decomposition voltage are intrinsic gas producers. If they need to be hermetically sealed, they must be provided with a means of scavenging the gas to prevent excessive cell pressures and ruptures. Lithium batteries are different. Although electrolyte decompositions occur as a result of chemical and/or electrochemical reactions, the electrolytes are less prone to gassing than aqueous systems. Pressure-induced leakages are, therefore, less likely to occur. However, if leakages should occur in lithium batteries, serious hazards may be created, as discussed in Part 4. It is highly desirable that lithium batteries be hermetically sealed, and we need to understand how leakage paths may be formed in hermetically sealed batteries to minimize the likelihood of leaks occurring either due to faulty seal designs or due to improper handling.

A distinction may be made between three types of ambient temperature lithium batteries based on their electrolytes:

1. Solid electrolyte systems, including polymer electrolyte systems
2. Solid depolarizer systems, i.e., organic solvent electrolyte systems with solid positive electrodes
3. Liquid depolarizer systems, i.e., inorganic or mixed organic/inorganic solvent electrolyte systems that contain the active positive electrode material in solution

There are no pressurizing gases in solid electrolyte systems, except possibly at very high temperatures, and such cells may be hermetically or crimp sealed, depending on the application, with a low expectation of any leakage from the cells. Hermetic seals are preferred as a means of avoiding ingress of air and moisture that may render the cells defective. Most of the solid depolarizer lithium cells with organic solvent electrolytes do not generate gas, except for some of the very high voltage systems, and they may be hermetically or crimp sealed with little danger of bursting at normal operating temperatures. Since ingress of air or moisture would affect their performance adversely, they do require very high quality crimp seals if sealed by this method. Properly designed and well-made crimp seals appear to be satisfactory for this type of cell unless very volatile solvents are used. The liquid depolarizer lithium batteries are quite different. The strongly oxidizing depolarizers (sulfur dioxide, thionyl chloride, sulfuryl chloride, and bromine chloride) have high vapor pressures, are toxic, and are highly corrosive. For safety reasons, these batteries need to be hermetically sealed to prevent the escape of any depolarizer. Although good crimp seals can be made, they are not acceptable for liquid depolarizer lithium batteries. Hermetic glass-to-metal (GTM) seals are required.

The integrity of GTM sealed liquid depolarizer lithium batteries depends upon the ability of the GTM seals to withstand the aggressive chemical environment inside these batteries and their pressures. We present a brief review of the principal factors that need to be considered in the design of leakproof GTM seals. The possibility of forming leakage paths elsewhere in these cells will be discussed in Section 3.3.3. Leakage paths may be created in GTM seals by the processes discussed below, and it is useful to distinguish between the chemical and the mechanical factors that affect the seal integrity.

Mechanical Characteristics. A simplified seal design is shown in Fig. 3.15. Although highly oversimplified, it will suffice for our purposes. Several different GTM seal configurations have been developed and have been found to be satisfactory. A GTM seal is formed by fusing a suitable glass to a central terminal pin and cover of a cell at a high temperature in a controlled atmosphere furnace. Because of the high temperatures involved, it is desirable that the coefficients of thermal expansion of all these components be matched to avoid the development of destructive stresses in the GTM structure during the cool-down to room temperature. Since glass compositions fail quite readily in tension, such stresses need to be avoided. If, for reasons of chemical reactivity (see below), a thermal match is not feasible, components and designs are used that leave the glass in a state of compression. Although glasses may fail in compression, their resistance to compressive failure is greater than their resistance to tensile failure. The design of safe and viable GTM seals is greatly facilitated by analyses of the residual stresses in the glasses of GTM seals as a function of their geometry, material properties, and the processing conditions. Finite element computer programs are available for this purpose.[10] The effects of the geometric parameters shown in Fig. 3.15 (all dimensions are important) on the residual stresses and the effects of thermal cycling have been discussed in the literature.[11] By a careful choice of material compositions, geometries, and processing conditions, it is possible to fabricate chemically and mechanically stable GTM compression seals for liquid depolarizer lithium batteries. If the cell pressures increase to such an extent that the cell covers experience a significant deflection, tensile stresses sufficient to crack the glass

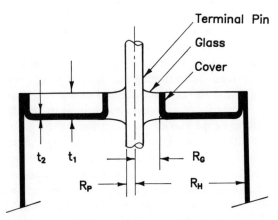

Figure 3.15. Typical glass-to-metal seal. t_2 is the thickness of the cover, t_1 is the distance from the top of the cell to the bottom of the cover, R_p is the radius of the terminal pin, R_G is the radius from the center of the pin to the inside diameter of the cover, and R_H is the distance from the center of the pin to the inside wall of the cell.

and to generate leakage paths may develop. This effect has to be considered in the analysis of the stress distributions in GTM seals as part of the design phase. The deflections can be reduced significantly by employing thicker cell covers.

Chemical Stability. Leakage paths may be generated in GTM seals if their integrity is violated by chemical reactions or if improper chemical processing conditions are employed that lead to the formation of a poor bond between the glass and the cell cover or the terminal pin. Poor bonding may also result from insufficient glass flow. Control of the seal-forming time–temperature profile, of the composition of the furnace atmosphere, and of the cool-down is of critical importance. Another consideration is that terminal pins made of refractory metals may have been formed by powder metallurgical techniques that leave the pins porous. Such pins provide potential leakage paths for both gases and electrolytes. Chemical or electrochemical attack may occur either at the glass-to-metal interfaces or at the glass surface itself with either or both of the following consequences:

1. The reduction of the glass oxides may form metallic bridges inside the glass that lead to the eventual short circuiting of the cell.
2. Chemical transformations of the glass components, starting at the glass surface, may generate local stresses due to volume changes and cause cracking and spalling of the glass.

The spalling of the glass transfers its compressive stress onto the remaining glass, which may fail as a result of the increased stress it now has to support. Both of the above processes initiate their attack from the inside of the GTM seal, and they tend to occur near the negatively polarized glass-to-metal interface. Both have a substantial random component in terms of their rates and the regions affected as the reactions proceed.

The chemical stability of a glass is determined by the thermodynamic and reaction kinetic properties of its constituents. Since one side of the glass is in contact with a metal maintained at the strongly reducing potential of the lithium electrode, one or more of the oxides in the glass may be reduced. If, in addition, electrolytic contact is maintained between the negatively polarized metal and the glass, conditions exist where glass destruction is likely to occur at a significant rate. Some selected values of the free energies of oxide reduction by lithium may illustrate how favorable are some of the reactions in lithium cells that may destroy GTM seals and generate leakage paths[12]:

Component	Free energy of reduction (kcal/mol)
Na_2O	-85
B_2O_3	-76
SiO_2	-73
Al_2O_3	-14
BeO	13
CaO	24

The large negative free energy changes associated with the reduction of the first three oxides show that GTM seals containing high concentrations of these oxides are likely to deteriorate in lithium cells. GTM seals containing aluminum, berylium, and calcium oxides in significant amounts are more likely to resist attack. The rate of attack is also affected by the composition of the electrolyte. In the more strongly oxidizing liquid depolarizers ($SOCl_2$ and $SOCl_2 + BrCl$), the rate of attack is slower than in the less strongly oxidizing system based on SO_2 as a depolarizer. There appears to be a competition between destructive reduction reactions and restoring oxidizing reactions. The present generation of low-silica GTM seals available for lithium batteries are quite effective in resisting chemical attack under all normal user conditions,[13,14] but failures may still occur and generate leakage paths if faulty processing conditions are employed or if the cells are exposed to excessive pressures.

3.3.3. Corrosion-Induced Leakage Paths

Corrosion is a well-documented phenomenon that has been treated extensively in the literature.[15] It involves the spontaneous destruction of metals by chemical and/or electrochemical reactions, and it is controlled by the thermodynamic and kinetic properties of the system involved and its components. The metals in a battery are connected either to a strong reducing agent (e.g., zinc, lithium) or to a strong oxidizing agent (e.g., manganese dioxide, thionyl chloride). In addition, they are in contact with highly corrosive electrolytic solutions. Metals in contact with negative battery electrodes are maintained at potentials that block their corrosion (but see comments near the end of this section). Metals in contact with positive battery electrodes are maintained at potentials that favor their corrosion. It is reasonable to expect, therefore, that corrosion may generate leakage paths if the battery case or cover is in contact with the positive electrode of a battery. Examples would be cylindrical alkaline manganese batteries and coin cells. The general absence of corrosion under these circumstances is attributed to the passivation of the selected metals at the positive electrode potentials. Interestingly, the earlier versions of zinc/carbon cells were constructed with zinc cans, and they all tended to develop leaks as they aged. The corrosive penetration of battery containers to form leakage paths is very unlikely with today's batteries. If it occurs, it may generally be attributed to faulty processing conditions that impair the protection normally afforded by passivation, to the use of improperly processed materials, or to the use of material from lots with the wrong composition. Some of the factors that promote corrosion of battery containers and cause leakage will be reviewed briefly. Polymeric battery housings are of no concern in this connection. They do not corrode in the conventional sense; they are more likely to fail, if they do fail, as a result of excessive mechanical stresses and/ or the gradual chemical degradation of the polymers of which they are made.

Two manufacturing steps are of particular importance for the maintenance of the passive state of positively polarized metals in batteries:

1. The cold forming of metal cans, covers, and vents
2. The welding of any parts of the containing structure*

Cold forming is important because it may create cracks and residual stresses that promote corrosion, and welds are important because they may lead to phase transformations and residual stresses in the weld fusion zones, both of which increase the susceptibility to corrosion. Battery cans are generally made by deep drawing, and residual tensile stresses in the base bend region may be sufficiently high (worst case scenario) to cause hairline cracks if the bends are very sharp or to promote corrosion to the point where leaks may develop at some later time. Residual stresses in this region may also increase the probability of cell ruptures if the cells experience overpressures. The residual stresses are a function of the state of the metal, the drawing speed, the depth of the draw, and the geometry of the case. They can be relieved by a post-draw annealing of the metals. Similar considerations apply to vent structures that may be formed by coining operations, but in this case a compromise must be reached between the degree of annealing and the hardness required to meet the specifications for a satisfactory operation of the vent. An interesting aspect of crack formation in metals under tensile stress in a corrosive medium is that crack propagation can occur at stresses below the critical failure stress. A synergistic effect exists between the mechanical and the chemical stresses. This effect needs to be allowed for in setting the design and processing specifications for the drawing/annealing operations. Experience has shown that all the above problems can be solved, and there are no reasons to expect the development of leakage paths in battery cases and covers in batteries manufactured under well-controlled conditions.

---

* Corrosion of interior welds may lead to contact failures. Although a serious defect, it is of no concern at this point since it does not cause cell leakage.

Welding operations may change the local state of a metal to such an extent that corrosion occurs when the welds are exposed to battery electrolytes. The more aggressive the welding regime is, the greater the residual stresses and the areas affected. This applies to electrical spot welds (electrical contact structures) and to inert gas welds (cell closures). In the case of electrical spot welds, the nature of the welding electrodes and their cleanliness are important. The possible transfer of welding electrode material to the weld may lead to the formation of new metallurgical phases susceptible to corrosion. A similar phenomenon occurs in spot welds on nickel-plated steel.[16] Excessive welding power has been known to puncture battery cases, forming fine pinholes that may not be noticed until delayed leakages occur. Leakages due to welding operations can be eliminated for all practical purposes by careful control of the welding operations.

The preceding effects operate in lithium batteries as well and may cause leakage. However, the chemical environment in liquid depolarizer lithium batteries and in high-voltage lithium batteries is more aggressive than in aqueous electrolyte batteries, and greater care must be exercised to avoid conditions that promote corrosion. If any water is present in lithium batteries, it may be difficult to avoid corrosive attack. An additional process occurs in lithium batteries that is not observed in aqueous electrolyte systems and may lead to the formation of leakage paths. Normally, a metal maintained at the lithium potential will be protected cathodically. In lithium batteries, if the metal in question is capable of forming a lithium alloy, a spontaneous transfer of lithium will occur from the anodic lithium to the cathodic metal.[17] This is not a corrosion process in the conventional sense, but it is an effective way of destroying the structural integrity of the metal. This process may lead to the formation of leakage paths if any lithium alloying metals are incorporated in the battery case or cover. A process similar to hydrogen embrittlement, i.e., lithium embrittlement, may occur in highly stressed (tensile) regions of battery containers and may enhance crack propagation. Lithium battery leakages have been observed, thought to be due to this process, but no conclusive evidence is available to assess its importance. We do not have any reliable guidelines yet on means of avoiding the formation of battery leakages by this process, other than to minimize residual tensile stresses in battery containers. As in the case of aqueous battery systems, careful attention to material selection and processing conditions has been effective in eliminating lithium battery leakage due to welding operations.

3.4. EXTERNALLY INDUCED LEAKAGE

Although batteries have been designed to resist considerable abuse, conditions exist where the battery integrity may be violated, causing leakage or more severe hazards. The most common types of abuse are mechanical and electrical abuse, with thermal abuse encountered less frequently. Hazards can generally be avoided by handling batteries in accordance with the manufacturers' recommendations for the different types of batteries. Experience has shown, however, that users do abuse batteries, sometimes by deliberate acts of destruction, but more often through carelessness and lack of attention to recommended handling procedures. We review briefly the processes that occur when batteries are abused and the associated hazards. Leakage is the most frequently observed event associated with abuse, but ruptures and explosions may also occur.

Mechanical Abuse. When a battery is excessively deformed mechanically by any means, crushed, punctured, or exposed to intense vibration or shock (as might occur in auto collisions), electrolyte leakage should be expected. If the cells contain any gas pressure at all, the electrolyte may be forced from the cells under sufficient pressure to spray nearby objects or personnel. The important consideration is that excessive mechanical force of any kind may break a battery and cause unavoidable electrolyte spills, creating serious hazards, especially with lithium batteries that may ignite and start fires upon rupturing. The force required to break open a cell or battery by external abuse varies greatly from one cell design to another, and it depends upon the point of attack of the force. There are no general rules available to quantify the resistance of a cell to mechanical abuse. Each specific situation must be examined in terms of the exposure conditions and the cell design involved. Special battery designs are employed in applications where excessive forces are likely to be encountered. Experience has shown that batteries can be designed to withstand gun firings without rupturing or leaking. Ordinary batteries are not that robust.

Electrical Abuse. The response of batteries to electrical abuse depends upon their type, size, and rate capability and upon the nature of the abuse. The most common electrical abuse is inadvertent short circuiting. Generally, the short circuits are of short durations and cause no problems beyond a possible impairment of the battery performance. Explosions can occur if hydrogen is present and sparks are generated (refer to Chapter 2 on lead–acid battery accidents). Prolonged, low-resistance short circuits generate considerable heat in high-rate batteries and may raise the temperature and the vapor pressure of the electrolyte to the point where cells rupture and discharge their electrolytes in vapor/liquid form. Fires may result from the shorting of high-rate, organic electrolyte lithium batteries.

The temperatures generated by short circuits may be illustrated with some estimates based on a hypothetical 3-V primary coin cell with a heat capacity of 8 cal/°C and a stoichiometric capacity of 500 mAh. We allow for the decreasing voltage as a function of the cathode composition by setting

$$V = 4(0.5 - C)^{1/2}, \qquad C = \text{Ah of capacity removed} \qquad (3.8)$$

and for the increasing impedance of the cell as a function of the depth of discharge by setting

$$R = 1.10 - \exp(-4.6C) \qquad (3.9)$$

This equation signifies that the resistance increases from 0.1 Ω at the start of the discharge to 1.0 Ω at the end of discharge. The total heat generated in the cell under the assumption of a dead short (zero external impedance) as a function of time can be found most simply by numeric calculation, as the differential equation for the time dependence of the temperature is highly nonlinear. Under adiabatic conditions, the heat generation increases the temperature of the cell as shown in Fig. 3.16. The estimated temperature rise would be less than indicated in the figure if appreciable quantities of electrolyte were to evaporate inside the cell. It may be seen that the temperature rises very rapidly during the first few minutes of the short and then levels off. If the short is maintained for an extended period of time, the temperature rises to the point where a cell rupture should be expected with the development of minor or major leaks. Short-duration short circuits, on the other hand,

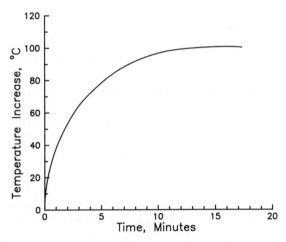

Figure 3.16. Adiabatic temperature rise of short-circuited 3-V coin cell.

would not be expected to give rise to any leakage. The short-circuit problem will be discussed in more detail in Chapter 6 on thermal runaways.

Prolonged short circuits of large high-rate batteries can be far more destructive and hazardous than those of smaller batteries since much higher power levels and greater rupturing forces are involved. An example is provided by a high-rate lithium/thionyl chloride battery designed by Dey and Hamilton.[18] The cell in question had a diameter of 3.0 in. and a thickness of 0.9 in. It delivered a short circuit current of 1500 A at a cell voltage of about 1 V. In other words, the heat generated inside the cell as a result of the short circuit amounted to about 4 kW, i.e., 0.6 kW per cubic inch. The cell vented shortly after the imposition of the short circuit, and the venting was accompanied by a continuous flow of sparks. When the same type of cell was discharged at 100 A, it vented after about eight minutes when the internal pressure reached about 250 psi.

Charging of primary cells is another common electrical abuse. Generally, it results from carelessness or deliberate attempts to act against the warnings of the manufacturers. When primary alkaline cells are charged, they may accept some charge, but most of the charging current is consumed in the generation of hydrogen and oxygen inside the cells (the gas-generating processes are described in Section 3.1). Ruptures and/or explosions are natural consequences of such abuse. The charging of a primary battery may happen in either of two ways: it may be placed in a charging device designed for some rechargeable battery and charged, or it may be placed in a high-voltage series stack of cells with the cell in question inserted with the wrong polarity (all cells are marked with their polarity to avoid this mishap). In the latter case, the wrongly placed cell will be charged when the battery is connected to a load. A variant of this abuse is the use of a high-voltage charging device to charge a lower voltage battery. As discussed in Section 3.1.2, this leads to overcharging and copious gas generation accompanied by pressure increases in unvented cells and likely cell ruptures, possibly gas explosions due to the accumulated hydrogen and oxygen.

Different processes occur in primary lithium batteries on charging, and it is useful to distinguish between the responses of solid depolarizer and liquid depolarizer lithium batteries. In the former, lithium deposition occurs on the negative electrode with the likely formation of internal short circuits. This means that any further charging current flows through the lithium shorts without causing any electrochemical reactions, only ohmic heating. Depending on the current and the impedance of the shorts, the rate of heating may or may not be sufficient to heat the cells to the point of bursting. Since the electrodeposited lithium has a large surface area and may be ignited on exposure to the air, the charging of solid depolarizer, primary lithium batteries may create potentially dangerous situations. If no internal short circuits are formed, continued charging will deplete the electrolyte of conducting species with a resulting increase in the cell impedance. If constant-current charging is employed, the temperature will rise quickly and may cause cell ruptures with a possible accompanying fire. The ion depletion process and its thermal consequences will be discussed in Chapter 6 on thermal runaways. The same phenomena occur in rechargeable lithium batteries on overcharge unless electrical or electrochemical overcharge protection is provided. Controlled voltage charging is more benign, and, depending on the rate of heat transfer from the cells, ruptures and cell leakages may be avoided.

Liquid depolarizer lithium batteries may be expected to behave differently since the electrolytes react quite readily with the electrodeposited lithium formed during charging. Their behavior will be discussed with reference to the lithium/sulfur dioxide system. For simplicity, it will be assumed that a fully charged cell is charged. The same arguments apply to the charging of fully or partially discharged cells, except that the hazards will arise at different times after the start of charging. The distinction between safe and unsafe behavior will be defined in terms of whether or not venting occurs. A cell is defined as safe if benign venting occurs and as unsafe if it reaches the temperature at which lithium melts. Molten lithium reacts rapidly with liquid depolarizers in a very exothermic reaction, and cell ruptures or explosions are likely to occur. The liquid depolarizer lithium cells are designed to vent to prevent explosions, but venting of the toxic fumes from these cells is itself a hazardous event even if no explosions occur.

The charging of lithium/sulfur dioxide cells beyond their full capacities involves the reactions:

$$Li^+ + e^- \rightarrow Li \qquad \text{at anode}$$

$$Br^- - e^- \rightarrow \tfrac{1}{2}Br_2 \qquad \text{at cathode}$$

which gives rise to the following chemical reactions and the associated heats of reaction:

$$Li + \tfrac{1}{2}Br_2 \rightarrow LiBr \qquad Q = 84 \text{ kcal/mol}$$

$$2Li + 2SO_2 \rightarrow Li_2S_2O_4 \qquad Q = 61 \text{ kcal/mol}$$

as well as several unspecified exothermic reactions between lithium and the acetonitrile solvent. The chemical reactions produce a great deal of heat in addition to the electrical heat generated in the cells. The thermal evolution of Li/SO_2 cells on charge will be discussed in more detail in Chapter 6 on thermal runaway.

Thermal Abuse. Long-term storage of batteries at elevated temperatures within their design range can be tolerated with no adverse effects other than an increased rate of self-discharge, but it may increase the likelihood of benign cell leakage. Heating of cells to temperatures above their design range is likely to cause leakage and ruptures and/or explosions if the temperatures are sufficiently high. The high temperatures accelerate the rate of gas generation in gas-producing cells and increase the vapor pressure of the battery electrolytes. In addition, polymeric seals tend to soften at elevated temperatures (thermoplastic crimp seals), and the pressure tolerances of the cells decrease. Pressure-driven leakages are likely to occur at moderately elevated temperatures via thermally generated leakage paths in crimp seals. Vented cells may also expel electrolyte under these conditions. Sealed cells will remain intact as long as the thermally induced pressure remains below the failure pressure of the seal or of the cells themselves. Exposures of batteries to flames for more than very short periods of time and disposal of batteries in fires invariably lead to very high cell pressures sufficient to breach their integrity followed by a more or less rapid expulsion of electrolyte. Depending on the rate of temperature rise, venting may reduce the rate of pressure increase sufficiently to prevent violent explosions. In aqueous electrolyte batteries of any type, ruptures and explosions are driven by the high vapor pressure of the electrolyte and by any gas generated as a result of external heating. In lithium batteries, internal exothermic reactions are triggered by the high temperatures, and they add to the externally supplied heat, leading to a more intensive energy release. Some quantitative aspects of cell ruptures will be discussed in Chapter 4.

3.5. LEAKAGE RATES

One of the premises of battery safety is that batteries are safe as long as their contents can be contained within the battery structures. The strict application of this as a design principle may not always be feasible, and in its place a less stringent rule is applied. The less stringent rule requires that batteries be designed so that if or when the battery contents escape, the potential hazards associated with their escape must be reduced to the fullest extent possible. As we have seen, it may be necessary to design leakage paths into battery structures to prevent the buildup of excessive gas pressures (Section 3.1.3). It is of interest, therefore, to examine how the leakage rates of electrolytes and gases depend upon the characteristics of various leakage paths such as pinholes and fine cracks in seals or in cell cases. This information is useful for the design of cells with controlled gas leakage rates and for an assessment of the severity of the hazards associated with battery leaks in general.

The analysis of leakage rates will be based on the standard equations of fluid flow. The flow regimes of principal interest are those of laminar flows and molecular flows (Knudsen flows). Turbulent flow regimes do not occur in slow battery leakages. For simplicity, we consider two basically different types of leakage paths: pinholes and narrow cracks. We ignore entrance and exit effects in our flow estimates. The discharge coefficients needed to allow for these effects are probably of the order of 0.5. This means that the actual flows are likely to be only about one-half of the estimated values given in the following sections. Since no information is available on the rates of flow caused by electrolytic creepage, we cannot estimate the rates of such flows, but it may be noted that they are slow.

An important limitation exists for all flows: no flow rate can exceed the acoustic velocity in the flowing medium. This signifies that if the driving pressures exceed that required to generate this maximum flow rate, they will not increase the rate of flow above the acoustic rate. If this situation exists and it is necessary to increase the flow rate to prevent excessive pressure buildups, the only available design option is to increase the cross-sectional area of the flow channel.

3.5.1. Electrolyte Leakage

There are no circumstances under which electrolyte leakage can have any beneficial effects, including leakages that result from battery venting. The question of what constitutes a tolerable leakage rate for a particular battery depends upon many factors such as the nature of the electrolyte, the degree of exposure of personnel and equipment, and the feasibility of battery maintenance to alleviate the effects of leakages. Automotive lead–acid batteries provide a good illustration. It is a rare unsealed lead–acid battery that does not leak to some degree, yet reasonable and safe maintenance procedures render such leaks tolerable with a minimal damage to equipment and no undue exposure of operating personnel.

The rate of flow of electrolyte in battery leaks under isothermal conditions is governed by the Bernoulli form of the energy conservation equation:

$$v \, dP + (1/g)u \, du + dF = 0 \qquad (3.10)$$

where v is specific volume, P is pressure, g is the engineering conversion factor, u is velocity, and F is friction. In the absence of friction, the pressure differential between the interior and exterior of a cell is converted to kinetic energy only. Given that the acoustic velocity is about 1500 m/s in aqueous electrolytes and about 1200 m/s in organic solvent electrolytes, we find that pressure differentials of about 30 psi in frictionless systems suffice to drive fluid flows at acoustic velocities. Since acoustic data are not available on the battery electrolytes of interest, we have used data on aqueous salt solutions and on organic solvents to estimate the above velocities.

The frictional term in Eq. (3.10) depends upon the geometry of the leakage path, and we consider two idealized geometries: cylindrical channels (pinholes in steel cases) and narrow slit-shaped channels (cracks in cell cases and imperfect crimp seals). Furthermore, we make several simplifying assumptions. We assume that the leakage paths are long compared with their diameters, and, as stated above, we ignore entrance and exit effects. Also, we assume that the flow channels have constant cross sections, no bends, and smooth walls. The effect of these assumptions is to overestimate the rates of flow. In fact, the leakage paths in batteries are likely to be highly nonuniform and to have rough surfaces, all of which would increase the frictional energy dissipation to values greater than estimated below.

Pinhole Leaks. These are the leaks that may occur in fine cell punctures created by pitting corrosion and/or weld burns. We idealize the pinhole geometry to that of cylindrical channels. The frictional energy terms may be derived from a momentum balance on the fluid and take the form[19]:

$$dF = fG^2 \, dL/2gR\rho^2 \qquad (3.11)$$

where f is Fanning's friction factor, G is the mass rate of flow, L is the length of the flow channel, R is the hydraulic radius (cross section/wet perimeter), and ρ is the fluid density. For laminar flow in cylindrical channels, the friction factor is

$$f = 4\mu/RG \qquad (3.12)$$

where μ is the viscosity of the fluid. Since the electrolytes are incompressible under the conditions of interest, the flow equation may be integrated to

$$(P_2 - P_1)v + (u_2^2 - u_1^2)/2g + 2GL\mu/gR^2\rho^2 = 0 \qquad (3.13)$$

Index 1 signifies the cell interior, and index 2 the cell exterior. Accordingly, $P_1 > P_2$ and $u_1 = 0$ (stagnant electrolyte inside the cells). This equation may be rearranged to

$$G^2 + 4LG\mu/R^2 = 2g\rho\Delta P \qquad (3.14)$$

where we have used $G = u\rho$ and $v\rho = 1$.

We use this equation to estimate the leakage rates driven by various pressures in pinholes of various diameters. For concreteness, we consider a sulfuric acid electrolyte containing 38% H_2SO_4 (mass fraction). This represents the approximate composition of the electrolyte in fully charged lead–acid batteries. Although other electrolytes have different properties, the differences are sufficiently small to render the estimated leakage rates below of the correct order of magnitude for most electrolytes of interest. The relevant properties of sulfuric acid at 25°C are:

Density: 1.28 g/cm^3
Viscosity: 2.25 cp

The pinholes are assumed to be present in steel cases with a thickness of 0.020 in. The estimated leakage rates are shown in Fig. 3.17. The results suggest that electrolyte leakage from pinholes with diameters of about 20 microinches and less are not likely to affect long-term battery performance adversely. If any escaping electrolyte were to come in contact with electrical components, however, corrosion might occur. Similarly, exposure of sensitive body parts may cause injuries.

Slit-Shaped Channel Leaks. This type of leakage path is representative of fine hairline cracks in battery cases and narrow channels in crimp seals

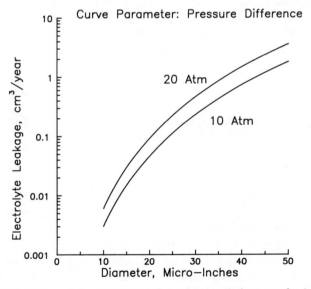

Figure 3.17. Estimated electrolyte leakage from pinholes. (Refer to text for details.)

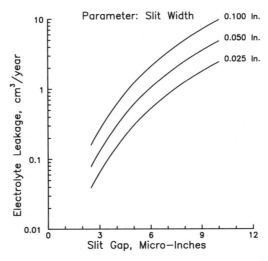

Figure 3.18. Estimated electrolyte leakage from narrow cracks. (Refer to text for details.)

with imperfect contact between the mating surfaces of the sealing grommet and the steel case or cover. The flows in such channels are very complicated, but may be idealized by the much simpler equivalent flows between two closely spaced parallel flat plates with the flow channel extending infinitely far in the lateral directions perpendicular to the flow direction. For concreteness, we assume the plates to be spaced a distance b apart and the channel to be 0.4 in. wide and 0.2 in. long in the direction of flow. The frictional term for the flow of incompressible fluids in such channels is given by (adapted from Ref. 20)

$$F = 12LGv^2\mu/gb^2 \qquad (3.15)$$

and the flow equation takes the form

$$G^2 + 24LG\mu/b^2 = 2g\Delta P/v \qquad (3.16)$$

We again use the properties of sulfuric acid to estimate the leakage rates as a function of the driving pressures and the separation of the walls of the slit-shaped channel. The estimated leakage rates are shown in Fig. 3.18. The leakage rates increase markedly with the size of the slit openings, as expected. An increase in the slit gap increases the leakage rates more than a widening of the slits. Comparisons of pinholes and slits with the same cross-sectional areas show that pinholes give higher leakage rates whenever the aspect ratio of the slits (width/gap) exceeds the length ratios of the two types of leakage paths. Leakages in crimp seals generally occur via slit-shaped channels, and, given the dependency of the leakage rates on the slit geometry, it may be seen that defects such as eccentricity and waviness may contribute significantly to an increase in the leakage rates of crimp-sealed cells. The estimated leakage rates suggest that the mating crimp seal surfaces should be separated by no more than a few microinches to avoid unacceptable electrolyte leakages.

3.5.2. Gas Leakage

Gas flows are more complicated than liquid flows for two reasons:

1. Gases are compressible and their linear velocities increase along the direction of flow as the pressure decreases.
2. If the channel dimensions are comparable with the mean free path of the gas molecules, diffuse reflection of gas molecules from the channel walls affects the flow to an increasing extent as the channel dimensions decrease (Knudsen diffusion).

The mean free path of hydrogen is about 1.5×10^{-5} cm at 25°C and 1 atm. This means that flows in channels with characteristic dimensions below about 10^{-4} cm must allow for Knudsen diffusion. If the characteristic dimensions are less than about 10^{-5} cm, at the given temperature and pressure, the flow of hydrogen is governed by pure Knudsen diffusion. It may be noted that the Knudsen diffusion equation we use below applies to channels with large length/diameter ratios and not to orifice flows (i.e., not to channels with small length/diameter ratios).

The maximum flow rate of any gas is given by the acoustic velocity in that gas at the prevailing temperature and pressure. By combining the equations for the frictionless adiabatic expansion of a gas and the equation for energy conservation, we find an expression, the critical pressure ratio, which is the pressure ratio required to achieve a Mach number of unity for the flow of the gas, i.e., its maximum flow rate[21]:

$$r_c = P_c/P_0 = [2/(k + 1)]^{k/(k-1)} \tag{3.17}$$

where k is the specific heat ratio (1.41 for hydrogen), P_0 is the upstream pressure, and P_c is the downstream pressure at Mach = 1 (critical pressure). The corresponding gas flow rate is given by

$$G_c^2 = gk(P_0/v_0) \cdot [2/(k + 1)]^{(k+1)/(k-1)} \tag{3.18}$$

where the subscript 0 signifies upstream conditions. This is the maximum possible rate of flow under frictionless adiabatic conditions. Any increase in the pressure ratio above the critical value will not increase the rate of flow. If the flow of a gas is driven by its critical pressure ratio (strongly dependent upon the molecular weight of the gas), the actual flow rates will be less than the estimated maximum values because of fluid friction. In the case of hydrogen, the maximum flow rate at 25°C is 1.65 g/cm$^2 \cdot$s. This amounts to a leakage rate of about 20 cm^3 (STP)/day for an idealized flow channel like a nozzle with a diameter of 10 μm. The flow rates given by Eq. (3.18) place an upper limit on the leakage rates that may be reached in the subsonic, pressure-driven flows discussed below.

Pinhole Leaks. We consider pinholes with the same geometry as in the case of electrolyte leakages and limit the discussion to the case of continuum flows. Knudsen flows will be discussed in a separate section. The basic differential equations that govern the laminar flow in cylindrical channels are the same for liquids and gases, but the integrated form of Eq. (3.10) for gases is different from Eq. (3.13) for liquids because of the compressibility of gases. For gases that obey the ideal gas law, we obtain the following expression for the isothermal flow of a gas:

$$G^2 \ln(P_1/P_2) + 32LG\mu/D^2 = gM\Delta P^2/2RT \qquad (3.19)$$

where M is molecular weight, R is the gas constant, and T is absolute temperature. We use this equation to obtain estimates of the leakage rates of hydrogen from cells via pinhole leaks of various sizes in metal cases assumed to be 0.020 in. thick. The viscosity of hydrogen is 0.0088 cp at 20°C. The results are shown in Fig. 3.19 for various driving pressures. The leakage rates increase with the square of the driving pressure and with the fourth power of the diameter of the pinhole. Given the fast diffusion of hydrogen in air, it is doubtful that hydrogen leakage rates of the order of 1 cm^3 (STP)/day will cause any problems in ventilated areas. If such leakage rates were to occur in sealed battery compartments, however, explosive gas mixtures might form. Thus, pinholes with diameters of 10 microinches and less would probably be acceptable for many applications from a safety point of view. As described in Section 3.1.3, leakage rates of 0.02 cm^3 (STP)/day and higher would suffice for alkaline D-size cells to prevent excessive pressure buildups due to hydrogen evolution from zinc corrosion. It may be noted that all the estimated leakage rates are less than the critical flow rates except that the flows through the 20-microinch pinhole approach the critical flow rates at driving pressures of about 20 atm.

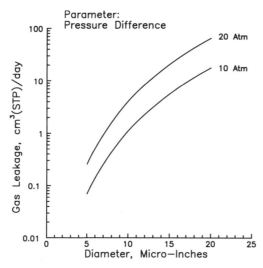

Figure 3.19. Estimated hydrogen leakage from pinholes. (Refer to text for details.)

Slit-Shaped Channel Leaks. The slit-shaped leakage paths will be idealized by the same parallel-plate configuration discussed for liquid flows. The frictional term for gases takes the form (adapted from Ref. 20):

$$dF = (Gv/2a)(P_1^2 - GL/a)^{-1/2} \, dL \qquad (3.20)$$

$$a = gMb^2/24RT\mu \qquad (3.21)$$

and the integrated energy balance becomes

$$2(RTG/M)^2[\Delta(1/P^2)/g] + (RT/M) \ln(P_2/P_1)$$

$$- (RT/2M) \ln(1 - GL/aP_1^2) = 0 \qquad (3.22)$$

for isothermal, laminar flow. Some estimated leakage rates for hydrogen in such cracks are presented in Fig. 3.20 for various sizes of cracks and driving pressures. We have used cracks that are 0.2 in. long and 0.1 in. wide for illustrative purposes. The flow rates in the figure are based on the assumption of a laminar flow regime. Since the smaller channel dimensions are comparable with the mean free path of hydrogen at low driving pressures, Knudsen diffusion will affect the flow rates in such fine channels. This means that the actual flow rates will be less than those estimated above for very fine channels.

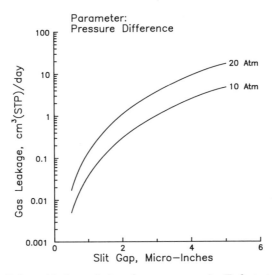

Figure 3.20. Estimated hydrogen leakage from narrow cracks. (Refer to text for details.)

A comparison of pinhole and slit leakage rates shows that pinhole leaks are more serious than slit leaks for channels with the same cross-sectional areas.

Knudsen Flows. We consider the case of hydrogen leakage in very fine pinhole channels in steel cases 0.020 in. thick with channel diameters of 10^{-5} cm and less. These are situations where the flows are governed by molecular diffusion at pressures close to 1 atm. For Knudsen flows, the flow rate is given by[22]:

$$N = (\pi D^3/24)(2RT/\pi M)^{1/2}\Delta P/RTL \qquad (3.23)$$

The estimated leakage rate of hydrogen is about 0.002 cm³ (STP)/day. At pressures above 1 atm, the mean free path decreases below the cited value, and Knudsen flows become important only for channels with dimensions of the order of 10^{-6} in. and less and at pressures greater than 10 atm. The corresponding leakage rates would be less than 10^{-3} cm³ (STP)/day. This indicates that gas leakage by molecular diffusion is a relatively unimportant problem from a safety point of view.

REFERENCES

1. P. Ripoche and M. Rolin, *Bull. Soc. Chim. Fr.* **1980**(9–10), I-386.
2. J. W. Moore and R. G. Pearson, *Kinetics and Mechanism,* John Wiley and Sons, New York (1981).

3. P. Bro and S. C. Levy, *Quality and Reliability Methods for Primary Batteries,* John Wiley and Sons, New York (1990).
4. R. O. Hammel, Sealed lead–acid batteries, in *Handbook of Batteries and Fuel Cells* (D. Linden, ed.), Chapter 15, McGraw-Hill, New York (1984).
5. J. A. Wiseman, Sealed nickel–cadmium batteries, in *Handbook of Batteries and Fuel Cells* (D. Linden, ed.), Chapter 18, McGraw-Hill, New York (1984).
6. C. E. Rogers, Permeability and chemical resistance, in *Engineering Design for Plastics* (E. Baer, ed.), Reinhold, New York (1964).
7. M. N. Hull and H. I. James, *J. Electrochem. Soc.* **124,** 332 (1977).
8. S. M. Davis and M. N. Hull, *J. Electrochem. Soc.* **125,** 1918 (1978).
9. L. M. Baugh, J. A. Cook, and J. A. Lee, *J. Appl. Electrochem.* **8,** 253 (1978).
10. Finite element analysis program ABAQUS, Hibbitt, Karlsson and Sorrensen, Inc., Providence, Rhode Island.
11. J. D. Miller and S. N. Burchett, Some Guidelines for the Mechanical Design of Coaxial Compression Pin Seals, Report SAND82-0057, Sandia National Laboratories, Albuquerque, New Mexico (1982).
12. B. C. Bunker, C. C. Leedecke, S. C. Levy, and C. C. Crafts, Glass-to-metal seal corrosion in lithium–sulfur dioxide cells, in *Power Sources 8* (J. Thompson, ed.), Academic Press, London (1981).
13. S. C. Douglas, B. C. Bunker, C. C. Crafts, and R. K. Quinn, Ampule Tests to Simulate Glass Corrosion in Ambient Temperature Lithium Batteries, Report Sand83-2301, Sandia National Laboratories, Albuquerque, New Mexico (1983).
14. R. D. Walkins, Development of CABAL Glasses for Use in Lithium Ambient Temperature Batteries, Report SAND87-0393, Sandia National Laboratories, Albuquerque, New Mexico (1987).
15. W. R. Cieslak, Corrosion in batteries and fuel cell power sources, in *Metals Handbook,* 9th ed., Vol. 13, ASM International, Metals Park, Ohio (1987).
16. W. R. Cieslak, F. M. Delnick, and C. C. Crafts, Compatibility Study of 316L Stainless Steel Bellows for XMC3690 Reserve Lithium/Thionyl Chloride Battery, Report SAND85-1852, Sandia National Laboratories, Albuquerque, New Mexico (1986).
17. A. N. Dey, *J. Electrochem. Soc.* **118,** 1547 (1971).
18. A. N. Dey and N. Hamilton, *J. Appl. Electrochem.* **12,** 33 (1982).
19. R. H. Perry and C. H. Chilton (eds.), *Chemical Engineers' Handbook,* 5th ed., Section 5, p. 18, McGraw-Hill, New York (1963).
20. R. H. Perry and C. H. Chilton (eds.), *Chemical Engineers' Handbook,* 5th ed., Section 5, p. 25, McGraw-Hill, New York (1963).
21. R. H. Perry and C. H. Chilton (eds.), *Chemical Engineers' Handbook,* 5th ed., Section 5, p. 29, McGraw-Hill, New York (1963).
22. R. D. Present, *Kinetic Theory of Gases,* McGraw-Hill, New York (1958).

4

Ruptures

Battery ruptures are events of intermediate severity, more severe than leakages, but less severe than explosions. They are characterized by the sudden bursting of a cell or battery and the rapid escape of gas and/or electrolyte to the environment, and possibly the escape of battery solids as well with some kinetic energy. The ruptures are frequently accompanied by a loud report. Some physical damage may result from ruptures, but the more serious hazards are those created by the electrolyte, which may cause personal injuries and corrosive equipment damage. The rupture of lithium batteries may cause fires.

Excessive mechanical abuse and exposure to high temperatures or fires are predictable causes of battery ruptures. Ruptures attributable to internal processes are less predictable as to the time when they occur, but since these processes commonly result from some form of abuse, their eventual occurrence is generally predictable. Ruptures are most often caused by gas pressures generated by any of the processes discussed in Chapter 3 whenever these processes are allowed to proceed to the point where the pressures exceed the bursting strength of the batteries. The internal pressures generate tensile stresses in the cell casings, and, for our purposes, we consider only uniform pressures. Situations do exist, however, where nonuniform pressures may act on the cell containers owing to localized expansion of battery solids. Although such localized expansion is important when it occurs, it is of little importance for the majority of batteries in general use. The stresses created by internal pressures depend upon the cell geometries and the cell pressures. Except for some special applications, the external pressure may be assumed to be atmospheric pressure. The response of cell structures—their strain—depends upon the mechanical properties of the stressed components and the stress distribution in the components. The geometries of principal interest are those of cylindrical and prismatic cells, their closures, and their terminal structures. In the case of lithium batteries, we also need to consider their safety vents. Except for lead–acid batteries that are made with hard rubber or polypropylene containers and many of the prismatic batteries, most of the smaller batteries are made with metal containers, either nickel-plated cold-rolled steel (CRS) or stainless steel (SS). We limit our discussion of rupture to batteries made with steel containers. Nonmetal containers may be analyzed by the same methods that are used for metal containers, but their material properties are quite different (see Chapter 3). Because of their lower strength, nonmetallic housings are thicker than steel housings and are generally provided with structural reinforcements to limit wall deflections due to internal pressures. This is particularly true for the larger sizes of rechargeable batteries.

4.1. MECHANICAL STRESS AND PRESSURE TOLERANCE OF CELLS

The mechanical strength of the metals used in battery housings depends upon their chemical compositions and microstructures as well as upon the chemical environment in which they operate. Standard handbooks may be consulted for information on the properties of the metals used in battery construction.[1,2] Since the rupturing stress of battery cases is likely to decrease as a result of long-term exposure to battery electrolytes, estimates based on the properties of virgin metals may be expected to understate the likelihood of ruptures. There is no reliable way to account for this effect in a quantitative manner at the present time, but designers allow for it by introducing empirical safety factors. It is well to keep in mind, however, that battery ruptures are extremely rare. When they occur, they are invariably due to some form of abuse, either deliberate or inadvertent. For the purposes of our discussion, we consider the mechanical properties of any particular material as given.

The response of a battery case to internal pressures prior to rupture comprises two sequential stages:

1. An initial stage of reversible elastic deformation
2. A following stage of irreversible plastic deformation

If the stresses do not exceed the yield stress, the cell case will return to its original shape when the internal pressures are reduced. Whenever the stresses exceed the yield stress, the cell case will be permanently deformed even after the stresses are relaxed. If the stresses exceed the ultimate strength of the case material, ruptures will occur. A normal design criterion for mechanical structures is that the stresses in any part of the structures, including battery cases, must not exceed the yield strength of the material in any part of the structures, after allowing for some safety factor. The estimation of the stresses in simple structures presents few problems, and analytical expressions are available for this purpose.[3] Battery cases are not simple structures. Realistic estimates of the stresses in battery cases require detailed information on their geometries, which is not available. For example, the wall thickness of cylindrical cases varies from that in the design drawings, particularly in the base region of the cells, where appreciable thinning may occur as a result of the metal forming operations. Then, there is the complicated geometry of the top covers, for which no analytical stress formulas are available. The estimation of the stresses in cell covers requires the use of numerical techniques, such as finite element methods,* but, even then, the results of such calculations should be viewed with caution.

Estimation of the onset of battery case ruptures is difficult. Large strains occur before ruptures are initiated, and allowances must be made for these changes in the case geometries, a problem that is exacerbated by the uncertain information on the starting geometries. Although finite element methods can be used to estimate the bursting pressures of geometries that approximate those of actual battery cases, a more realistic approach to the determination of the bursting pressure of battery cases is the experimental approach. Even if calculations are made, it is advisable to check them by actual measurements such as hydrostatic pressure testing. In the following, we present an elementary discussion of the elastic response of the various parts of cylindrical cells to internal cell pressures. The analysis of the response of prismatic battery cases is too complicated for the purposes of our present discussion; it requires much more detailed analyses based on finite element calculations.

Cylindrical Cell Walls. Cylindrical cell cases are thin-walled structures, and, except for bends, ends, and edges, the stresses in the cells may be considered to be isotropic. Internal cell pressures (P) create hoop stresses (H) and longitudinal stresses (L) in cylindrical cell walls (diameter D, wall thickness w) as illustrated in Fig. 4.1. Whenever the wall thickness is much smaller than the case diameter, the stresses far from the ends of cylindrical shells are given by

$$H = 0.5 \ (PD/w) \tag{4.1}$$

$$L = 0.25 \ (PD/w) \tag{4.2}$$

---

* A useful finite element analysis (FEA) program called ABAQUS is available from Hibbitt, Karlsson and Sorrensen, Inc., in Providence, Rhode Island. It requires 16 Mb of RAM and 750 Mb of hard memory. Many FEA programs are suitable for the estimation of the stresses in battery cases, but potential users should be aware that all FEA programs require a great deal of computer memory.

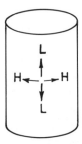

Figure 4.1. Stress components in cylindrical cell wall. *H*, Hoop stresses; *L*, longitudinal stresses.

The hoop stress is twice as large as the longitudinal stress, and longitudinal ruptures (i.e., ruptures in the longitudinal direction normal to the hoop stress), rather than transverse ruptures may be expected to occur as a result of excessive internal pressures. Longitudinal ruptures are also favored by the elongation of the metal grains in the longitudinal direction that results from the drawing of the cell cases. Since the end closures of the sealed cells provide some hoop reinforcement, the longitudinal ruptures are likely to be initiated near the middle of the cylindrical cells. Prior to rupture, the cylindrical cells will bulge outward as a result of the elastic deformation of the cell walls. The quantitative response of a cell to internal pressures may be illustrated with some estimates based on cell cases made of cold-rolled steel and stainless steel. We use approximate property values; actual values depend strongly on the specific steel used and its microstructure.

Type of steel	Yield strength (psi)	Tensile strength (psi)
CRS	40,000	60,000
316SS	30,000	87,000

For most practical purposes, the properties of the steels may be considered to be independent of temperature over the normal operating range of most batteries. The yield strength of CRS changes with temperature at a rate of about 26 psi/°F in the operating range of most batteries. There are special applications, however, where the temperature effect needs to be considered, especially high-temperature applications, where excessive cell pressures are more likely to occur. For concreteness, we consider the response of D-size cells with a diameter of 1.34 in. and various wall thicknesses. The cells will deform elastically (i.e., reversibly) as long as the hoop stress is less than the yield strength. If the stress exceeds the yield point, permanent deformations will occur. For cold-rolled steel, we find the minimum wall thickness required to resist permanent deformation as a function of the internal cell pressure (no safety factor applied):

Pressure [psi (gauge)]	Wall thickness (10^{-3} in.)
500	8.4
1000	16.8
1500	25.1
2000	33.5

Alkaline cells are generally made with nickel-plated cold-rolled steel with a thickness of about 20 mil for D-size cells, and it may be seen that they may be expected to tolerate internal pressures of about 1000 psi (gauge) without experiencing permanent deformation.

The elastic deformations at hoop stresses below the yield point may be estimated for D-size cells made with 0.020-in. thick CRS in terms of the expected increases in the cell diameters (a modulus of 6×10^6 psi was used):

Pressure [psi (gauge)]	Increase in diameter (in.)
500	0.0038
1000	0.0075
1500	0.0113

The estimated values indicate that the increases in the cell volumes may amount to 1–2% of the total cell volumes, allowing for strains beyond the elastic limit. Therefore, if the void volume is of the order of 5% of the total cell volume, the increase in the cell diameter will serve to reduce the gas pressure somewhat, and more gas will have to be generated to reach the rupturing pressure. If the pressure is caused by the expansion of a solid or liquid phase, the estimated volume changes will reduce the applied pressure much more than is the case with a gaseous pressurizing agent. Generally, however, liquid or solid expansion is not a problem since cells are designed to accommodate any such expansion. A continuing rise in pressure beyond the elastic limit will lead to an irreversible plastic deformation, and ruptures will occur if the pressures reach the point where the stress exceeds the ultimate tensile strength. The rupturing pressures estimated for a CRS D-size cell are shown in Fig. 4.2 as a function of the wall thickness (no safety factor applied). It may be seen that these cells may be expected to survive internal pressures in excess of 1000 psi (gauge) without rupturing, depending on wall thickness. Similar estimates may be prepared for the stainless steel cases that are commonly used for lithium batteries. Given the lower elastic limit of stainless steels and their higher tensile strengths, they may be expected to deform elastically more readily than cold-rolled steel but to tolerate higher internal pressures without rupturing. Indeed, some stainless steels have ultimate strengths that are more than twice those of cold-rolled steel.

Cylindrical Cell Base. Cylindrical cells have a flat base formed as an integral part of the cells in the drawing of the cans from flat sheet stock. The

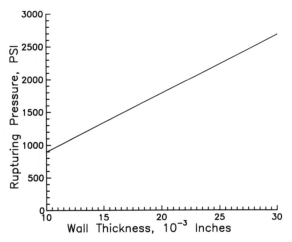

Figure 4.2. Pressure tolerance as a function of wall thickness for D-size cell of cold-rolled steel.

radius of the bend between the base and the cell wall is an important parameter that affects the stress distribution in the bend due to internal pressures. The problem may be illustrated with the idealized bend geometry shown in Fig. 4.3, where we employ a 90° bend (a practical impossibility), and we consider the case of a thin-walled shell and base. For small deflections of the base due to internal pressures, the tensile stress (T) in the base near the outer diameter may be approximated by

$$T = 0.25 \, (PD/w) \, \text{cosec} \, \alpha \tag{4.3}$$

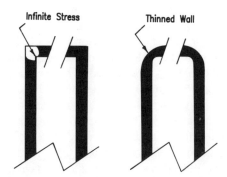

Idealized Configuration   Practical Configuration

Figure 4.3. Schematic illustration of difference between idealized and real cell base configuration (exaggerated scales).

where α is the angular deflection of the base from the 90° angle at the cell diameter. In the case of a D-size cell with a diameter of 1.34 in. and a wall thickness of 0.020 in., we obtain the results shown in Fig. 4.4. The high stresses generated by quite low cell pressures provide convincing arguments indicating that the flat base of any cylindrical cell is likely to bulge outward at even moderate cell pressures. Furthermore, the sharper the bend is, the greater the tensile stress at the inner surface of the bend and the greater the likelihood of stress-induced cracking in that area. To avoid this problem, the radius of the bend is made as large as feasible, consistent with the need for an efficient utilization of the cell volume. The bulging of the base due to internal pressures cannot be avoided even if sharp bends are eliminated, but it can be kept within reasonable bounds for all but excessive cell pressures. The bend itself is highly stressed and may rupture in response to high cell pressures, more readily so in aged cells where residual stresses may have enhanced the rate of formation and growth of corrosion cracks. The annealing of cell cases to relieve residual stresses is highly desirable.

The estimation of the pressure tolerance of a bend by calculation is a difficult problem. Not only is the stress distribution nonisotropic, but the actual geometry itself is poorly defined. The thickness of the metal in the bends is less than that of the sheet stock from which the cases are drawn, and it varies in an unknown manner with the friction experienced during the drawing operation. Furthermore, the grain structure of the metal is changed by the drawing operation, which changes the strength characteristics of the

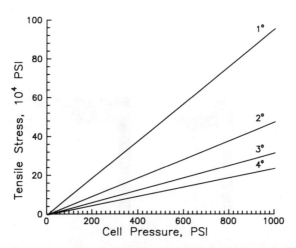

Figure 4.4. Tensile stress near edge of cell base as a function of cell pressure for different angular deflections of the base from the 90° angle at the cell diameter.

metal in the bend region. Given these uncertainties, the bursting strength of the bend structure can be determined most meaningfully by hydrostatic testing.

Cylindrical Cell Covers. The top covers of all cells are complicated geometric structures comprising either an insulating crimp seal or a welded plate with an attached insulated electrode terminal. The rupture strength of crimp seals depends upon the seal geometry and the mechanical properties of both the cover and case metals and the insulating polymer. The strength of such seals is determined most usefully by hydrostatic testing rather than by calculation. Considerable uncertainties are associated with computational estimates because of the difficulties of accounting accurately for the relative strains of the polymer and the metals and their interactions.

Welded covers with centrally located, insulated terminals (GTM seals) present an equally difficult computational problem if the covers are thin and experience a significant deflection as a result of internal pressures. Any bulging of the covers imposes tensile stresses on the GTM seals. The seals may be stressed to tensile failure at relatively low cell pressures, especially since GTM seals have low ultimate tensile strengths. This characteristic has been exploited by using GTM seals as safety vents in lithium cells, but with limited success since the rupturing pressure of such vent structures is quite variable and not easily controlled. The rupturing strength of thin welded covers with hermetic GTM seals is generally determined by hydrostatic measurements. Thick welded cell covers do not deflect to any significant extent, and the rupture strength of their GTM seals may be estimated with some degree of confidence by finite element calculations for various seal geometries, as referred to in Chapter 3, assuming no chemical deterioration of the seals.

Cover-to-case welds (Fig. 4.5) represent potential failure regions in hermetically sealed cells subject to excessive pressures or environmentally induced

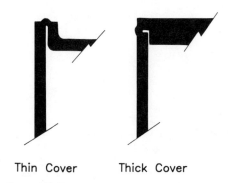

Thin Cover   Thick Cover

Figure 4.5. Two common cover-to-case structures.

corrosion. As a rule, the pressure tolerance of the welds is determined experimentally rather than by computation. The strength of virgin welds and their susceptibility to corrosion cannot be determined with any degree of accuracy by computation. Both depend critically upon the welding schedules and any subsequent annealing of the welds. It may be noted, however, that weld failures due to cell pressures are extremely rare.

4.2. SAFETY VENTS

Few batteries are entirely free of gas generation, and, for safety reasons, means may be provided for venting gases from cells or for recombining the gases inside sealed cells, as discussed in Chapter 3 for various primary and secondary batteries. The development of lithium batteries introduced more severe safety problems than had been observed with aqueous electrolyte batteries, and the need for additional safety features became apparent during their development. As will be discussed in Part IV, several safety features have been incorporated into lithium batteries to render them less hazardous and more resistant to abuse. One of these features is the use of safety vents that are actuated at predetermined pressures that depend upon the type of battery involved and its size. When the cell pressures reach the predesignated venting value, the vents rupture and relieve the overpressure. It must be understood that safety vents act as rupture disks, and they do not render lithium batteries safe. They only serve to reduce the severity of what might happen if the cell pressures were to increase to the point of rupture. The actuation of a properly functioning vent represents a controlled cell rupture that serves only to limit the pressure buildup in a cell.

Sealed lead–acid batteries generally incorporate relief valves capable of providing overpressure relief throughout the life of the batteries. The valves are actuated at pressures of about 5 atm and close again at lower pressures. The vents on lithium batteries are different. They are not valves, but rupture disks that provide no more than one-time overpressure relief. Once the valves are actuated, the cells become inoperative, but not necessarily safe. Four types of irreversible all-metal vents have been developed for incorporation in lithium batteries, and they are generally located in the base or cover of the larger cells. Small lithium cells pose much less danger than large lithium cells, and they are not provided with specially designed vents. When overpressures occur in small button or wafer type lithium cells, their covers generally pop off to relieve the excessive pressures. The four main types of lithium battery vents are illustrated schematically in Fig. 4.6. They are:

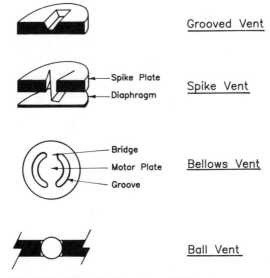

Figure 4.6. Schematic illustration of different vent types.

1. *Grooved vents.* A groove of controlled dimensions is coined into the base or side wall of a cell to create a low-strength segment designed to rupture at predetermined cell pressures.
2. *Spike vents.* The cell cover or base is made sufficiently thin to deflect under a given cell pressure to the point where the cover impinges on a spike to rupture the cover, thus creating an escape path for the pressurizing medium.
3. *Bellows vents.* The base of a cell is shaped into a single-stage bellows structure with two diametrically opposed bridges in the groove that defines the inner motor part of the bellow. When the pressure reaches a predesignated value, the motor part flips into its outer stable position, causing the bridges to be strained to rupture and creating escape routes for the pressurizing medium.
4. *Ball vents.* A ball is pressed into a suitably sized hole in the cell cover to give an interference fit adequate to form a leakproof seal and to retain the ball in place under normal operating conditions, but to yield under excessive pressures. Steel balls are used in lithium cells and nylon balls or plugs in alkaline cells, if provided with vents.

A potential problem with overpressure vents is that their apertures may be too small to allow the pressurizing medium to escape fast enough to prevent continuing pressure increases. Even if the apertures are large enough, they may become clogged with solid matter and become inoperative. An additional consideration in the design of venting structures is that an external high-conductance path must also be provided. It is not sufficient to vent only the individual cells in a multicell battery. Either a sufficient volume must be provided inside the battery compartment to contain any toxic materials at acceptably low pressures, or the battery itself must be vented to prevent an excessive pressure buildup inside the battery case.

REFERENCES

1. *Metals Handbook,* 9th ed., *Mechanical Testing,* Vol. 8, J. R. Newby, coordinator, ASM International, Metals Park, Ohio (1987).
2. H. E. Boyer (ed.), *Atlas of Stress–Strain Curves,* ASM International, Metals Park, Ohio (1987).
3. R. J. Roark, *Formulas for Stress and Strain,* 4th ed., McGraw-Hill, New York (1965).

5

Explosions

Batteries contain energy-rich materials comprising fuels (anodes) and oxidizers (cathodes) in close proximity in confined spaces (cells). If for any reason the fuel and the oxidant in a cell are brought into intimate physical contact or the electrodes are shorted externally, an exothermic reaction will occur that produces an amount of thermal energy equivalent to the energy stored in the cell. In addition, parasitic reactions may occur that produce heat inside a cell. If the rate of heat transfer from the cell is less than the rate of heat generation, the cell temperature will rise and accelerate the rate of the chemical reactions, and, if the situation is permitted to continue, it may reach the point where ruptures or explosions occur.

Historically, explosions have been associated primarily with lead–acid batteries, although explosions may also occur as a result of abuse or mishandling of other types of batteries. The growing use of lithium batteries has increased concern about battery explosions, a concern justified by the much greater energy stored in lithium batteries compared with that stored in aqueous electrolyte batteries and by the actual occurrence of lithium battery explosions. It may be noted, however, that the majority of the lithium battery explosions have occurred with special purpose batteries and with experimental prototypes. Few explosions have occurred with lithium batteries designed for consumer applications. The concerns about the safety of lithium batteries are focused primarily on the larger size batteries that are not in general consumer use. The small lithium batteries used by consumers are quite abuse resistant and provide much greater heat transfer areas relative to the cell volumes than do large batteries, which renders the small batteries less prone to thermal runaways and subsequent ruptures or explosions. Explosive incidents with small lithium batteries are extremely rare, but ruptures may occur as a result of abuse and cause some local damage. Serious abuse, such as disposal in fires or charging of primary lithium batteries, may cause violent ruptures and/or explosions regardless of the size of the batteries.

Ruptures and explosions are distinctly different events that are difficult to distinguish from one another in the absence of information on the specific processes involved in the observed events. The difficulty of distinguishing between the two is compounded by the fact that thermal runaways that cannot be contained inside a cell are generally followed by violent ruptures and in some cases by external explosions, depending on the nature and condition of the material ejected from the cell. We consider *ruptures* to involve no more than the gross and sudden violation of the mechanical integrity of a cell caused by an excessive exterior or interior mechanical force capable of straining the cell container beyond its ultimate strength. The force may be created by physical, chemical, or electrochemical processes. *Explosions* may be classified as *deflagrations* or *detonations,* and they are driven by the rapid release of an excessive amount of thermal energy either inside or outside a cell. Internal explosions need not be, but generally are, preceded by the rapid heating of a cell followed by the rupture or fragmentation of the cell. Any fragments formed in cell explosions may possess considerable kinetic energy and may contribute to the damage and injury potential of explosions. *Deflagration* involves the propagation of an exothermic combustion wave through a flammable gas or vapor mixture at subsonic velocities either inside or outside a cell. *Detonation* involves the propagation of a supersonic shock wave in a flammable gas or vapor mixture. With some exceptions, the detonation of solids in batteries is a very rare event. Once a sufficient force is generated to rupture a battery and flammable gases, liquids, or vapors are ejected from the battery, the combustion zone may propagate beyond the confines of the battery, causing either an external deflagration or explosion or a simple flame attached to the battery. The nature of the external process depends upon the nature of the flammable material, its mixing pattern with the air, and the velocity of the flame front in the mixture. The rupture itself may trigger a secondary shock wave, whether or not an internal shock wave has preceded the rupture. The term *conflagration* is sometimes used to describe rapidly growing fires external to a battery. In the case of lithium batteries, the ejection of lithium from a cell may trigger an intense external fire. If any metallic lithium remains inside a ruptured cell, it is likely to ignite and form a flaming jet extending beyond the confines of the cell.

Batteries at risk of explosions are principally the larger size, rechargeable, aqueous electrolyte batteries that may be mismanaged or abused in a manner that leads to the generation and accumulation of excessive amounts of hydrogen in confined spaces. All the larger size primary and secondary lithium batteries are also at risk. Small primary lithium batteries are unlikely to explode spontaneously unless grossly abused. Rechargeable lithium batteries with metallic lithium negative electrodes have a higher risk potential because of the increased chemical reactivity of the finely divided lithium that is formed on recharge. This finely divided lithium may react quite readily with reactive electrolyte solvents. The safety hazards of lithium batteries are discussed in considerable detail in Part IV.

5.1. THE CAUSE OF BATTERY EXPLOSIONS

Battery explosions require the presence of a chemical reaction system capable of generating sufficient thermal energy within a very short period of time to raise the temperature of the system to the point where spontaneous ignition occurs, or they require the presence of an explosive reaction mixture capable of being ignited by some external or internal means and the presence of such an internal or external trigger. The only significant source of chemical energy that may support explosions in aqueous electrolyte battery systems is the oxidation of any hydrogen that may be formed by electrochemical or corrosion processes. If oxygen and hydrogen are both present inside such batteries, internal explosions may occur if a trigger is available. External explosions may occur if hydrogen escapes from a battery without being diluted below its explosive concentration limit and if a trigger is present. Lithium batteries provide a richer variety of energy-producing chemical reactions, as will be discussed below. Not much is known about the details of the explosion mechanisms in lithium batteries, but a great deal is known about hydrogen explosions. They provide a good illustration of some of the mechanistic details of explosions in general and will be discussed in Section 5.2 below.

The simplest and best known examples of battery explosions are those associated with automotive lead–acid batteries. Although they are rare, they are well known because the automotive lead–acid batteries are widely used and the manufacturers keep emphasizing by various means the need for careful handling to avoid the accumulation of hydrogen formed during the charging of the batteries and the need to prevent the formation of sparks in the vicinity of lead–acid batteries. Sealed lead–acid batteries and small cylindrical lead–acid batteries are much less likely to create explosive situations. The same type of problem may occur with other rechargeable aqueous electrolyte batteries of the larger varieties, but rarely with the smaller size batteries. Primary aqueous electrolyte batteries with zinc electrodes may generate hydrogen by zinc corrosion, and if this hydrogen is permitted to accumulate in battery compartments or elsewhere after escaping from crimp-sealed cells, explosions may be triggered by sparks or other hot objects and cause some local destruction and injuries.

The conditions that lead to the generation of hydrogen in primary and secondary aqueous electrolyte cells were discussed in Chapter 3. If any of the hydrogen-generating processes are maintained for a sufficient length of time and in the absence of a means for inactivating the hydrogen by chemical or electrochemical recombination or by slow venting in a well-ventilated space, potentially explosive situations may be created. Whenever hydrogen coexists in a mixture with oxygen or air within specified concentration limits (see Section 5.2), there is a possibility that an explosion may occur. All that is needed is a trigger to set it off. The only significant triggers in the case of aqueous electrolyte batteries are sparks and hot wires or other hot objects. Such triggers may be present both inside and outside a battery, more often outside as a result of careless handling of wires and metallic objects near a battery.

Lithium batteries provide several strongly exothermic reactions that may supply sufficient energy to drive explosions both inside batteries and outside ruptured batteries. One similarity with aqueous electrolyte systems is that hydrogen may be generated inside or outside lithium batteries whenever lithium comes in contact with water. Based on the exothermic reactions that may occur in lithium batteries, we distinguish between two principal types of lithium battery systems:

1. Solid depolarizer, organic electrolyte systems
2. Liquid depolarizer, inorganic electrolyte systems

Most of the lithium batteries available in the general consumer market are of the former type. They are generally small batteries with a limited damage and injury potential. Batteries of the second type are not generally available in the consumer market. They are employed principally in special applications. Lithium batteries with sulfur dioxide depolarizers are of an intermediate type since they contain a liquid depolarizer and an organic solvent, generally acetonitrile. As a consequence, they share the characteristics of both types.

The following exothermic reactions may occur inside solid depolarizer, organic solvent electrolyte lithium batteries:

1. Solvent reduction by lithium
2. Solvent oxidation by depolarizer

Both are spontaneous reactions whose rates increase markedly at elevated temperatures. They occur whether the batteries are in use or on open circuit.

If the battery temperatures increase beyond their normal operating range by any means, for example, by short circuiting, external heating, or some other means, and under conditions of an inadequate heat transfer from the battery as the temperature increases, the rates of either or both of these reactions may increase to the point where the battery enters a thermal runaway. If sufficient reactive materials are present in the battery to sustain the reactions and to raise the temperature to the melting point of lithium (181°C), the thermal runaway is likely to induce an explosion. Not only is molten lithium far more reactive than solid lithium, but it may contact the cathode and short-circuit the cell and initiate a direct chemical reaction between anode and cathode, itself a potentially explosive reaction. Thermal runaway is not the same as an explosion. It is simply a process that leads to an uncontrolled increase of the battery temperature. Only if the temperature reaches the point where the rates of one or more of the exothermic reactions become so fast that they trigger the onset of deflagration or detonation kinetics do we have an explosion. Thus, thermal runaway may induce explosions but is not itself an explosion. Since elevated temperatures increase the vapor pressure of the electrolytes, this effect may be exploited to trigger the actuation of safety vents to limit the severity of thermal runaways and possibly prevent explosions. The efficacy of such venting depends upon the relative rates of venting and heat generation in the materials that remain inside the cells. If venting occurs or if the cells rupture, internal explosions may or may not be prevented, and any escaping flammable solvent and/or molten lithium may create conditions favorable for an external fire or explosion. In rechargeable lithium batteries based on metallic lithium negative electrodes, the presence of the finely divided lithium generated by charging increases the rate of reaction between lithium and solvent, and these batteries have a lower abuse tolerance than do primary lithium batteries.

The liquid depolarizer, inorganic electrolyte batteries do not contain any flammable solvents, and the principal reactions to be considered are:

1. Depolarizer reduction by direct reaction with lithium
2. Reactions between lithium and discharge products

The discharge product of greatest importance in the case of lithium thionyl chloride batteries is sulfur. In the case of the intermediate type, lithium sulfur dioxide batteries, the discharge product is lithium dithionite. The direct reaction between lithium and depolarizer is of critical importance. The two are in intimate physical contact in the cells, separated only by a reaction-inhibiting salt layer on the lithium. At temperatures above the normal operating range of the batteries, the inhibitory effect is diminished and the rate of the spontaneous reaction increases. At higher temperatures, the heat evolution from this reaction may exceed the rate of heat transfer from the cell, and the cell is likely to enter thermal runaway if sufficient amounts of unreacted material are present in the cell to sustain the heating. Such cells are likely to explode unless vented during an early stage of the thermal runaway. The most common triggers for thermal runaways are external heating by any means, sustained short circuiting, and cell reversals. If the internal cell temperatures reach the melting point of lithium, soluble depolarizer lithium cells are likely to explode or rupture violently. In lithium/thionyl chloride cells, the molten lithium may react explosively with any sulfur formed during the discharge, and in lithium/ sulfur dioxide cells the lithium may react explosively with the lithium dithionite formed during discharge. If the reversal of soluble depolarizer cells leads to a substantial deposition of lithium in the Teflon-bonded, porous carbon cathodes normally used in these cells, the possibility exists that a solid-phase reaction of considerable violence may be triggered by elevated temperatures and possibly by shock. The lithium/sulfur dioxide batteries have some unique characteristics due to the presence of acetonitrile in the electrolyte, as will be discussed in more detail in Part IV. Lithium reacts spontaneously, but slowly, with acetonitrile at normal operating temperatures to form methane. This leads to a gradual increase in the cell pressure that renders the cells more susceptible to rupture or explosions because of the enhancing effect of pressure on the kinetics of the explosion reactions, as will be discussed below.

5.2. THE EXPLOSIVE PROCESS

Considerable information is available on explosions, their mechanisms, and effects,[1,2] but little of that information is specific to the flammable and potentially explosive materials present in batteries. An exception is hydrogen, about which a great deal is known. We base much of our discussion on the behavior of hydrogen. Explosions of other battery materials bear some resemblance to hydrogen explosions. Qualitatively, the phenomena and physical processes are the same, but the chemical reactions are different and so are their quantitative characteristics.

The essential feature of an explosion is the rapid release of thermal energy from a reacting mixture. Two basic ingredients, a fuel and an oxidizer, must be intimately mixed for an explosion to occur, and an energy source must be present to trigger their reaction. Solid- and liquid-phase reactions that are not oxidation–reduction reactions but that may proceed with explosive violence are known, but they are of little relevance to batteries. Pure fuel by itself will only burn at the interface of fuel and oxidant if properly ignited. Ordinary flames illustrate that process. The concept of a proper ignition is an important one. For example, the use of a small flame to ignite a flowing combustible gas may result in the extinction of the trigger rather than ignition of the gas.

There are two common ways of triggering explosions in hydrogen/air, flammable gas/air, and flammable vapor/air mixtures that may be associated with batteries (note that other oxidants may be substituted for air, e.g., pure oxygen, chlorine, and other oxidizing agents):

1. Spark ignition
2. Thermal ignition

At temperatures below about 400°C, mixtures of hydrogen and oxygen (or air) are stable, but if they are exposed to an electric spark of sufficient intensity, an explosion will be triggered if the gas mixture has a composition within its explosive range (see below). Spark energies of the order of 0.01 watt-seconds may suffice to ignite stagnant hydrogen/oxygen mixtures if the power level of the sparks is of the order of 10^6 W/cm^3. This translates to a 10-volt spark delivering one milliamp in one tenth of a millisecond in a volume of one tenth of a cubic millimeter, a modest triggering requirement. It simply reflects that hydrogen/oxygen and hydrogen/air mixtures can be made to explode with great ease. Thermal ignition is similar except that an appreciable volume of the gas mixture needs to be heated to temperatures of 400°C or higher. This may be accomplished by contacting the gas mixture with a hot object with a temperature above 400°C, for example, a hot wire, under conditions such that normal convective or conductive heat transfer will bring the gas temperature to the ignition temperature. It may be noted that catalytic surfaces or catalytic spots may trigger hydrogen/air explosions as a result of the self-heating of the catalytic surface by the catalytic oxidation of the hydrogen. The heating of the gas mixture as a whole by any means to temperatures above 400°C will also trigger a spontaneous explosion. The minimum temperature required to trigger a spontaneous explosion, generally in air, is referred to as the autoignition temperature (AIT). The AIT varies considerably for different gases and vapors. For most organic solvent vapors, it lies in the range of about 300–500°C for their mixtures with air at a pressure of one atmosphere. At low temperatures, many organic solvents have low vapor pressures, and air mixtures of their vapors may have partial pressures that fall below their lower explosive range. This defines a temperature below which such vapors do not ignite.

Most flammable vapors and gases have composition ranges within which they form explosive mixtures and outside which they do not explode. The composition range for any one material is defined by its lower explosion limit

(LEL) and its upper explosion limit (UEL), and these limits generally refer to air mixtures of the flammable materials. The value for hydrogen in air at 0°C is 9.5 vol % for the LEL and 71 vol % for the UEL. As the temperature increases, the explosive range increases: the LEL decreases at the rate of 0.75 vol % per 100°C and the UEL increases at the rate of 2.5 vol % per 100°C. Low temperatures are no guarantee that gas explosions will not occur, but see comments above in regard to organic solvents with low vapor pressures. Since moisture inhibits the propagation of the chain reactions that drive explosions, the explosive composition ranges of most flammable gases and vapors decrease somewhat in humid air. Other effects may also be inferred from the kinetics of the explosion reactions: the greater the ignition energy (thermal ignition) and the greater the pressures of the mixture, the wider the explosion ranges. If the explosions are triggered in oxygen rather than in air, such as might happen inside batteries where hydrogen and oxygen are both produced on charge in the absence of a recombination mechanism (other than explosive recombination, that is), the LEL changes very little since the oxidant is present in excess in both air and pure oxygen, whereas the UEL increases since the mixture is oxidant starved in air, but its concentration is increased by the use of pure oxygen.

The efficacy of an ignition source depends upon its location within a flammable mixture. Whatever the velocity of the reaction wave generated at the ignition source, the greater the mass of flammable material traversed by that reaction wave in a given time, the greater will be the violence of the resulting explosion. Thus, centrally located igniters tend to develop more violent explosions than do peripherally located igniters.

Once ignited, a reaction wave is formed in a flammable mixture that propagates through the mixture at either subsonic velocities (deflagration) or at supersonic velocities (detonation). Depending on circumstances, deflagrations may develop into detonations. The rate of propagation of a reaction wave in any mixture is determined by the chemical kinetics of the reaction system, which, in turn, depend upon the composition of the system, its temperature and pressure, and the shape and size of the confining volume and are also affected by the presence of solid materials that may quench some of the intermediate reaction chains. Heat transfer is also an important consideration, whether it be by radiation, convection, or conduction, and the state of flow of the system is important. The propagation of reaction waves is quite different in turbulent systems and in quiescent systems. For internal battery explosions, flammable mixtures may be considered to be quiescent, but flammable mixtures that escape from ruptured batteries are likely to be in a state of turbulent flow.

In deflagrations, the rate of pressure equalization across the flame front occurs at the velocity of sound in the system, and, as a consequence, the pressure drop across the flame front is relatively small, typically about 8:1 for a great variety of flammable gases and vapors. This means that the pressure wave that accompanies the flame front is of moderate intensity. In the case of detonations, the rate of pressure equalization is less than the velocity of the reaction wave. Consequently, the pressure drop across the reaction wave is correspondingly greater than in the case of subsonic waves. Pressure ratios of about 40:1 are typical for many gases and vapors. This means that the associated pressure waves, the shock waves, can be quite intense and destructive.

The violence of an explosion is determined by the magnitude of the pressure wave created by the explosion and its characteristic rate of pressure increase (dP/dt), the brisance of the explosion. This follows from the observation that a cell exposed to a slowly increasing pressure will absorb some of the energy associated with the pressure wave by virtue of being strained prior to rupture. In the case of a rapidly applied pressure, a cell is likely to rupture before it can accommodate the force of the pressure wave by a prerupture strain. In agreement with this observation, different rates of pressure increase generate different fragmentation patterns in battery explosions. Very rapid pressure increases create extensive fragmentation of cells, and slow rates of pressure increases generate few, if any, fragments, the cells being more likely to rupture without fragmentation. It is interesting to note that lithium battery explosions fall primarily into the second category. Internal explosions in lithium batteries are probably deflagrations rather than detonations.

The maximum rate of pressure increase associated with explosions in simple confined volumes (V) is given by the empirical relationship[3]

$$(dP/dt)_{\max}\, V^{1/3} = K \text{ (a constant)} \tag{5.1}$$

The constant K is a characteristic of each specific gas/vapor composition for a given ignition source. It differs for differently shaped volumes. The K value for hydrogen in air is about 550 Bar $\cdot$ m/s when ignited by a 10-watt-second spark source (note: 1 Bar = 1 Megabar). The K values for most organic solvent vapors fall in the range of 40–70 Bar $\cdot$ m/s. The parametric plot in Fig. 5.1 illustrates the effect of the cell volumes on the violence of explosions for various K values. It may be noted that we have extrapolated experimental data from large and simple volumes to the small void volumes typical of batteries. Thus, the information provided by the plot is no more than the best available estimate in the absence of actual (dP/dt) data on batteries. An additional consideration is that the shapes of the void volumes in batteries

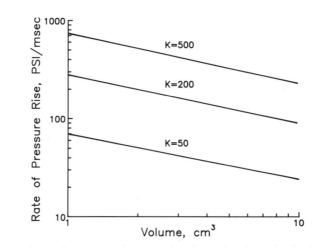

Figure 5.1. Estimated rate of pressure rise of explosions in confined volumes.

are not simple. The complicated void volumes in batteries are likely to give smaller K values than simply shaped volumes. The curves shown in Fig. 5.1 represent what are probably upper limits of the violence of battery explosions. The well-known violence of hydrogen explosions is clearly exhibited by their high K value. Flammable vapors of organic solvents give far less violent explosions. It may be seen from the figure that explosions in small confined volumes tend to be characterized by much more rapid pressure rises than explosions in larger volumes. If the initial pressure of an exploding gas/vapor mixture increases, both the violence (dP/dt) of the explosion and the maximum pressure of the explosion increase. If the ignition energy of a given ignition source increases, only the violence (dP/dt) of the explosion increases, not its maximum pressure. The nature of the oxidizing agent in exploding mixtures has a dramatic effect on the violence of an explosion. The K value of hydrogen increases from 550 Bar·m/s in air to about 2900 Bar·m/s in oxygen.[3] A similar effect may be expected for organic solvent vapor explosions, except that much smaller K values are involved. The increases in the K values are a direct consequence of the effect of the concentration of the oxidizing agent on the kinetics of the chemical reactions associated with the explosions.

The physical damage of explosions is caused primarily by the destructive force created by the pressure wave of explosions when the wave collides with various objects. If the shock waves are generated inside prepressurized batteries, the stresses created by the pressure of the shock wave plus the background pressure may exceed the ultimate strength of the battery containers. Since battery explosions tend to occur after some prior heating due to short circuiting, high external temperatures, or exothermic parasitic reactions inside the batteries, prepressurization is the rule rather than an exception. In some cases, the batteries may be prepressurized by internal gas generation (see Chapter 3) or as a result of the presence of a compressed gas such as sulfur dioxide. When deflagrations or detonations are triggered inside prepressurized batteries, the pressures created by the flame fronts or shock waves may exceed the rupturing strength of the batteries, as may be seen from the data in Fig. 5.2, where we show the effect of the initial cell pressures on the pressures likely to be associated with deflagrations and detonations. As expected, detonations are far more likely to cause ruptures than deflagrations. We noted earlier that some lead–acid batteries have safety valves that are actuated at pressures of about 5 atm. The data in Fig. 5.2 indicate that if detonations occur in such batteries (ignition of electrochemically generated hydrogen and oxygen), these valves provide no protection. However, the estimated pressures indicate that it may be possible to contain deflagrations inside batteries, at least inside small cylindrical batteries. Since deflagrations may develop into detonations and since prepressurization is likely to be present, designs based on that expectation cannot be assumed to be safe. A case of some interest is that of lithium/sulfur dioxide batteries. They are pressurized by liquefied sulfur dioxide, the depolarizer, and have internal pressures in the fully charged state of about 2 atm at room temperature, about 10 atm at 70°C, and about 18 atm at 100°C. If these batteries are preheated by any means to temperatures approaching 100°C, even mild deflagrations would create reaction waves with

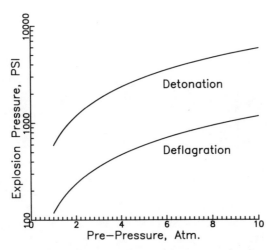

Figure 5.2. Estimated explosion pressures in prepressurized systems.

characteristic pressures of the order of 2000 psi, quite enough to rupture any cell, even if we allow for the greater dynamic rupture strength of steels exposed to a very rapidly increasing stress relative to their static rupture strength.

Battery ruptures may create secondary shock waves by the simple physical process of the rapid compression of the air surrounding a bursting battery. In mild ruptures, no more than an intense sound wave may be created, but in violent ruptures, strong shock waves may be created. The overpressures associated with shock waves in air at 20°C and a pressure of 1 atm are given in Fig. 5.3 as a function of the Mach number of the shock wave. (The figure is adapted from information presented in Ref. 4.) The Mach number is the ratio of the shock wave velocity to the speed of sound. The overpressures of even mild shock waves with Mach numbers close to unity are of the order of 15 psi. The significance of overpressures of this magnitude may be established by comparing this pressure with the pressures required to produce physical damage of various degrees of severity[5]:

Overpressure (psi)	Level of damage
0.5–1.0	Shattering of windows
1.0–2.0	Buckling of metal panels; failure of wood siding panels at joints
2.0–3.0	Shattering of concrete block panels
7.0–8.0	Shearing and flexural failure of unreinforced brick walls

Clearly, considerable physical damage as well as personal trauma may be caused by battery explosions.

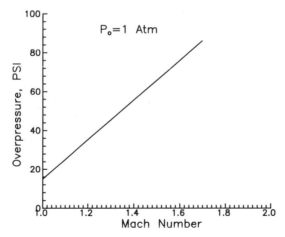

Figure 5.3 Overpressures of shock waves in air at 1 atm and 20°C.

REFERENCES

1. B. Lewis and G. von Elbe, *Combustion, Flames and Explosions of Gases,* 3rd ed., Academic Press, New York (1987).
2. G. F. Kinney and K. J. Graham, *Explosive Shocks in Air,* 2nd ed., Springer-Verlag, New York (1985).
3. W. Bartknecht, *Explosions,* Springer-Verlag, New York (1980).
4. R. G. Stoner and W. Bleakney, *The Attenuation of Spherical Shock Waves in air, J. Appl. Phys.* **19,** 670 (1948). [Cited by G. Zabetakis, Flammability Characteristics of Combustible Gases and Vapors, Bulletin 627, U.S. Bureau of Mines, U.S. Department of the Interior, Washington, D.C. (1965).]
5. G. Zabetakis, Flammability Characteristics of Combustible Gases and Vapors, Bulletin 627, U.S. Bureau of Mines, U.S. Department of the Interior, Washington, D.C. (1965).

6

Thermal Runaway

Heat may be generated in batteries by chemical and electrochemical reactions and by physical processes. If the rate of heat generation exceeds the rate of heat removal, the temperature of the battery will increase. As the temperature increases, so do the rates of any exothermic reactions and so does the rate of heat transfer. Chemical reaction rates tend to increase exponentially with temperature, and heat transfer rates increase linearly with temperature at moderate temperatures. At higher temperatures where radiation contributes to heat transfer, the rate of heat transfer has a component that increases with the fourth power of the temperature. Two thermal time patterns may be considered (Fig. 6.1A and B). Actual histories fall somewhere between the two extreme patterns shown in the figure. During the initial phase of pattern A, more heat is generated than can be removed by heat transfer, and the battery temperature increases. Then at time t_1, the rate of increase of heat generation reaches a maximum value, and the rate of increase then decreases until at time t_2 the heat generation attains its maximum value, beyond which it decreases. It may decrease because of the exhaustion of the reactive materials or because some external driver of heat generation has ceased to function. At time t_3, the rate of heat generation becomes equal to the rate of heat transfer, and the battery temperature decreases beyond that time. If the maximum temperature reached by the battery is less than that required to induce venting, rupture, or explosions, the battery has experienced what we call a benign, temporary thermal runaway. The temperature history of the battery is controlled by the areas enclosed between the two heat rate curves. Net heating occurs when heat generation exceeds heat transfer, and net cooling occurs when heat transfer exceeds heat generation, but there may be time lags between the two phases. The two areas represent the energies involved in heating and cooling, respectively. Pattern B represents a somewhat different thermal his-

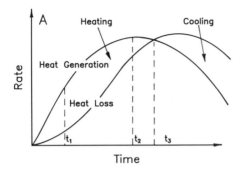

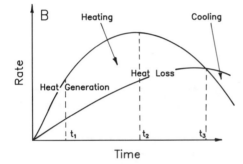

Figure 6.1. General thermal patterns in cells with heat generation and dissipation. (Refer to text for details.)

tory. In this case, the rate of heat generation exceeds the rate of heat transfer for a substantial length of time before the temperature begins to decrease. Otherwise, the two patterns display the same qualitative features. In pattern B, the heating area enclosed between the two curves represents an amount of energy in excess of that required to raise the temperature to the point where venting, rupture, or explosions may occur. When this happens, we speak of a thermal runaway proper. It is defined by two quantities:

1. The temperature required to trigger an adverse event
2. The energy required to reach that temperature

Thermal runaways need not occur fast. The temperature of a battery may increase slowly until it eventually reaches event-triggering values. This is more likely to happen with well-insulated, multicell batteries than with poorly insulated batteries. If an event occurs as a result of cell heating, heating may or may not continue. If the initial event is a gentle venting or a mild rupture and heating continues, explosions may occur. Lithium batteries are more likely to experience such a sequence of events than aqueous electrolyte batteries.

The same physicochemical processes are involved in benign and proper thermal runaways. The differences between the two are differences of degree only. Thermal runaways do not occur as a result of normal battery operations. They are invariably due to some form of abuse or mishandling. Even then, they are rare events. The heat-generating processes that contribute to thermal runaways and that may occur singly or in various combinations are of the following kinds:

1. High rates of discharge
2. Short circuits
3. Charging and reversals
4. Chemical reactions
5. Ion depletion

Since batteries are designed to tolerate some abuse in all the above categories with relative impunity, we are concerned primarily with excessive abuse and the likelihood that it may cause thermal runaways proper. In the following, we discuss each of the several processes from a general point of view.

6.1. HIGH DISCHARGE RATES

Whenever a battery is discharged, some of its energy is consumed in maintaining the flow of current through the battery. This energy appears as waste heat, and, at high rates of discharge, the heat may be sufficient to raise the battery temperature to unacceptably high levels and create high pressures, to trigger exothermic reactions, or to cause electrolyte evaporation and escape from vented cells. The energy loss is manifested by an operating voltage that is less than the open-circuit voltage of the battery at the applicable state of discharge and temperature. For the same reason, the charging voltage is greater than the open-circuit voltage during charging operations. Another source of heat is always present that is associated with the chemical reactions involved in battery charging and discharging, the entropic heat.

The instantaneous heat generation is given by

$$q = I\eta - JT\Delta S \tag{6.1}$$

where q is the rate of heat generation, I is the current, T is the absolute temperature, $\eta = V_0 - V$, is the overvoltage, ΔS is the entropy change, and J is a conversion factor. For most batteries, the entropic contribution to heat generation is small relative to the ohmic contribution, and it may be neglected for our purposes. The overvoltage is a function of the current, the state of discharge, and the temperature of the battery. For a given set of conditions, it may be approximated by the Tafel equation:

$$\eta = a + b \ln I \qquad (6.2)$$

where a and b are constants. The energy is dissipated nonuniformly inside a battery and generates a complicated temperature profile. The higher the current, the more nonuniform the temperature profile. Part of the heat is absorbed as sensible heat, part is consumed in phase transformations (e.g., evaporation), and part is absorbed in shifting chemical equilibria (e.g., sulfuryl chloride dissociation). We refer the reader to the works listed in the bibliography at the end of this chapter, comprising various analyses that have been presented of the current distribution, heat generation, and temperature profiles in batteries, for a full appreciation of the quantitative aspects of battery behavior. For our purposes, a much simplified analysis will suffice since we are primarily interested in the qualitative aspects of thermal runaways.

In order to illustrate the effects of high discharge rates, we make several simplifying assumptions:

- Constant current discharge
- Overvoltage independent of temperature and state of discharge
- No thermal gradients inside the battery
- Constant heat capacity (C)
- Adiabatic system
- Negligible entropic heat effect
- No parasitic chemical reactions

If these assumptions were not made, the analysis would become unduly cumbersome for our purposes. Readers interested in more detailed analyses should refer to the literature cited in the bibliography at the end of this chapter. Under the stated simplifying assumptions, the heat balance takes the simple form

$$C \, dT/dt = I\eta \qquad (6.3)$$

It may be noted that among the different discharge modes of constant load resistance, constant current, and constant power, the former is the least severe and the latter the most severe from the point of view of heat generation in a battery. We have chosen to consider the case of intermediate severity, the constant current case. It is the simplest case to handle computationally. The other cases may be examined in a similar manner, but the computations become quite involved and add little to our understanding of the processes we are interested in. For concreteness, we consider a 10-Ah D-size battery with an effective heat capacity of 150 cal/°C and an overvoltage given by $(a,b) = (0.44, 0.35)$ in Eq. (6.2). This reflects a reasonably realistic overvoltage for a battery with a moderate rate capability. A worst case scenario is one in which the full rated battery capacity is available at all discharge currents, a situation that is realized only at low rates of discharge. Based on all the stated assumptions, we obtain the results shown in Fig. 6.2 for the temperatures reached at the end of life of the battery. It may be seen that the temperature may reach about 110°C when such batteries are discharged at high sustained rates at 25°C. This suggests that thermal runaways are unlikely to occur for room temperature discharges of aqueous electrolyte batteries. In fact, D-size, aqueous electrolyte batteries do not generally deliver a useful service at rates much above 2 A, and no excessive temperatures would be generated for discharges within their rated service temperature range. High-rate lithium batteries of the same size might present a problem, especially the soluble depolarizer lithium batteries. With rate capabilities extending into the 10–20-A range, the discharge of such batteries at elevated temperatures may generate sufficient heat to raise the cell pressures to the point of venting, rupturing, or explosions or to raise the temperature to the point where exothermic reactions may be triggered and cause a thermal runaway. Another consideration is that the current distribution becomes highly nonuniform at high discharge rates, and highly localized heating may occur and generate hot spots that may trigger exothermic reactions in lithium batteries. Ohmic heating may trigger chemically sustained thermal runaways.

6.2. SHORT CIRCUITS

Short circuits represent an extreme case of high-rate discharges. The discharge rates are limited primarily by the internal impedance of the batteries.

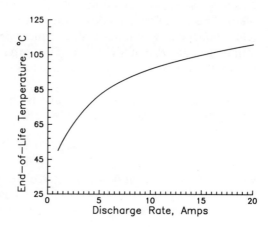

Figure 6.2. Estimated end-of-life temperatures of D-size cell on high rates of discharge under adiabatic conditions. (Refer to text for details.)

The most severe situation occurs when batteries are shorted by external impedances that approach zero, a most unlikely case. Cases of greater practical importance are those in which the external impedance may be of the order of 0.001–0.01 Ω. In such cases, some heat is generated outside the batteries, but most of it is generated inside the batteries. In the case of internal short circuits, all the heat is generated inside the batteries. Massive internal short circuits do not generally occur unless a battery is severely damaged by some external force, in which case hazards other than those attributable to thermal runaways are likely to be present. Most commonly, internal short circuits are created by charging or reversal processes and are of moderate impedance and short durations owing to local burnouts of the shorts. Internal short circuits are very rare in well-designed and properly operated batteries, although some rechargeable solid depolarizer lithium batteries may be more prone to internal shorting than rechargeable aqueous electrolyte batteries. The effects of non-uniform current distributions and heat generation become more severe in short-circuited cells than in cells on lower but still high rates of discharge. Internal short circuits may generate intense local hot spots capable of triggering exothermic chemical reactions, particularly in lithium batteries.

We discuss short circuiting based on the same simplified model that was employed for the high-rate discharges, and we assume the worst case scenario of a zero-impedance external short circuit. We also assume that the shorts are maintained until the full battery capacity has been exhausted. Given these assumptions, the rate of heat generation becomes

$$q = V^2/R \qquad (6.4)$$

where V is the full battery voltage (assumed constant), and R is the internal impedance (assumed constant). We consider both the adiabatic and the non-adiabatic cases and explore the effect of various heat transfer rates on the thermal history of the batteries in the latter case. It is again emphasized that reliable quantitative estimates require more detailed models and process descriptions than are employed here. The purpose of our simplified analyses is just to illustrate some of the qualitative aspects of thermal runaways.

Adiabatic Process. When no heat is transferred from the battery, the temperature of the battery is given by

$$C \, dT/dt = V^2/R \qquad (6.5)$$

where dT/dt is constant under the stated assumptions. The temperatures at the end of life (EOL) are shown in Fig. 6.3 as a function of battery voltage for various effective heat capacities. Since lithium batteries have lower effective

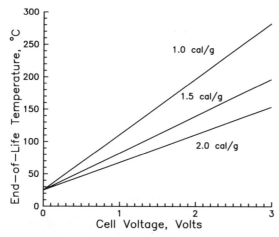

Figure 6.3. Estimated end-of-life temperatures of D-size cell on short circuit under adiabatic conditions for various effective heat capacities. (Refer to text for details.)

heat capacities than aqueous electrolyte batteries, they are more likely to experience higher temperatures than aqueous electrolyte batteries. Their higher voltages render them even more prone to excessive heating. The estimated EOL temperatures are considerably greater than for high-rate discharges, as would be expected, and they suggest that short circuits are quite likely to generate thermal runaways in high-voltage batteries under adiabatic conditions. It must be recognized, however, that the estimated temperatures represent extreme cases. In actuality, the rates of heat generation would be lower than those calculated from Eq. (6.5) for several reasons: the impedance of a short circuit is rarely less than 0.01 Ω, and the shorts are generally of short duration. Furthermore, no discharges occur under fully adiabatic conditions, as there is always some heat loss from hot batteries. At high rates of discharge, the available battery capacity is much reduced, and the discharges are terminated long before the batteries reach the temperatures shown in Fig. 6.3. Under truly adiabatic conditions, however, the discharge rates are irrelevant; the total energy of a battery is dissipated as internal heat and would give the EOL temperatures shown in the figure.

Instead of the EOL temperatures, we may estimate the time to reach a specified temperature, in particular, the temperature at which a secondary event may be triggered, such as venting, rupturing, or explosions. Based on the heat generation given by Eq. (6.5) for the same cell considered earlier and an effective heat capacity of 150 cal/°C, we obtain the curves shown in Fig. 6.4 for the time to reach 100°C starting from room temperature (25°C) for cells with various internal impedances and voltages. The strong dependence

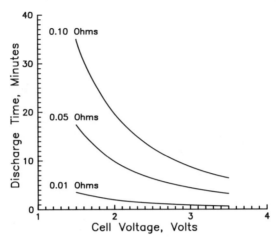

Figure 6.4. Estimated time to reach 100°C on adiabatic short circuit of D-size cell for various internal impedances and voltages. (Refer to text for details.)

of the heating rate on the cell voltage and impedance is clearly evident. The capacities consumed in reaching 100°C from room temperature are shown in Fig. 6.5 in terms of the depth of discharge at that temperature. Under the given assumptions, these depths of discharge depend only on the cell voltage. Similar estimates may be prepared for any type of battery and for various

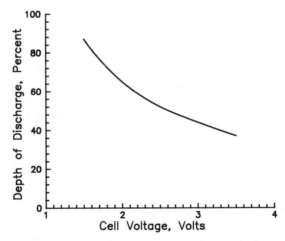

Figure 6.5. Capacity required to reach 100°C from 25°C on adiabatic short circuit of D-size cell. (Refer to text for details.)

combinations of internal and external impedances, including their dependence on the state of discharge of a battery. The inclusion of these dependencies renders the computations more elaborate and more accurate but does not alter the qualitative features displayed by the simplified models used above.

Nonadiabatic Process. Batteries are generally installed in devices that permit some heat to be transferred from the batteries to the containing structures. This reduces the temperatures below the values obtained under adiabatic conditions. For very high rate applications, containing structures may be used that are designed to increase the rate of heat transfer from the batteries and to prevent thermal runaways. We examine the effect of heat transfer on the thermal history of short-circuited batteries, again based on the simplified model used above and the assumptions stated there. Since the rate of heat transfer is proportional to the outer surface area of a battery and the rate of heat generation is proportional to the volume of a battery, the size and shape of a battery play an important role in the thermal history of a short-circuited battery. For our purposes, we introduce an overall heat transfer coefficient that incorporates the effects of the battery size and shape, conduction, convection, and radiation. At temperatures much above or much below the normal operating range of batteries, radiation must be allowed for explicitly since radiative heat transfer is proportional to the fourth power of temperature. We consider a cylindrical D-size cell with the properties given in the preceding sections. The results of our analysis should be viewed for their qualitative significance only; quantitative estimates require the more complicated models referred to earlier.

The rate of heat transfer is given by

$$q = k\Delta T \tag{6.6}$$

where k is a battery-specific and container-specific constant. When we neglect entropic and other chemical contributions to the rate of heat generation, the heat balance equation takes the simple form

$$IV = J[k\Delta T + C(d\Delta T/dt)] \tag{6.7}$$

This equation integrates to

$$\Delta T = [1 - \exp(-kt/C)]IV/Jk \tag{6.8}$$

where t is time in consistent units and $\Delta T_{t=0} = 0$. For numerical estimates, we employ overall heat transfer coefficients in the range of 0.05–0.15 cal/ $(s \cdot {}^\circ C)$ and assume that the battery is in thermal contact with a heat sink maintained at a constant temperature. The significance of the given heat transfer coefficient values will be discussed later. The time to EOL for the constant current discharges is given by

$$t_{EOL} = (R/V)(\text{cell capacity}) \qquad (6.9)$$

and with this substitution we find

$$\Delta T_{EOL} = (1/J)\,(V/C)\,(Rk/VC)^{-1}\,\{1 - \exp[-(\text{cap})(Rk/VC)]\} \qquad (6.10)$$

where (cap) signifies battery capacity. The estimated increase in the cell temperatures reached at end of life is plotted in Fig. 6.6 as a function of V/C with Rk/VC as a parameter for a 10-Ah D-size cell. The effect of heat transfer is illustrated in Fig. 6.7, where we plot the temperature rise as a function of time for the 10-Ah cell, assuming a heat capacity of 150 cal/°C and a heat transfer coefficient of 0.10 cal/(s · °C). It may be seen that heat transfer is reasonably effective in decreasing the temperature of the cells to values below the adiabatic temperatures given by Eq. (6.3), provided the rate of heat transfer is sufficiently high. The lowering of the end-of-life temperatures relative to the adiabatic case was found to be:

Cell voltage	Temperature lowering (°C)
1.5	27
2.0	29
2.5	29
3.0	30
3.5	31

The heat transfer analysis may be employed to estimate the design parameters required to render a cell less prone to thermal runaway if the temperatures at which venting, rupture, or explosions occur are known. As an illustration, consider the short-circuited cells just discussed and assume that the cells will be safe if the temperature increases need to be limited to maximum values of 100°C or 150°C, respectively. The calculations give the results shown in Fig. 6.8 in terms of the parameters V/C and Rk/VC. Applying these parameters to the case of a 3.5-V cell with a heat capacity of 150 cal/°C gives the results shown in Fig. 6.9. Each curve in the figure delineates the boundary between safe and hazardous cell designs. As expected, an increase in either the internal impedance or the heat transfer coefficient moves a design toward greater safety. Similar plots may be prepared for any particular battery, but we emphasize again that more detailed models are required for actual design work. The above estimates present worst case scenarios for relatively moderate rates of discharge. For very low impedance short circuits where internal temperature gradients become important, it is necessary to allow for internal heat

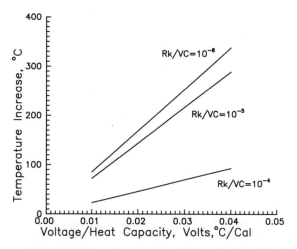

Figure 6.6. Parametric representation of temperature rise on short circuit with heat loss. (Refer to text for parameter definitions.)

transfer as well. We have chosen to cast our discussion within a much simplified framework to render the presentation as transparent as possible and to accentuate the basic qualitative features of thermal runaways.

Since heat transfer is an important factor in the control (if possible) of thermal runaways, we explore what may be some realistically attainable heat transfer coefficients. Here, we consider heat transfer by conduction. The con-

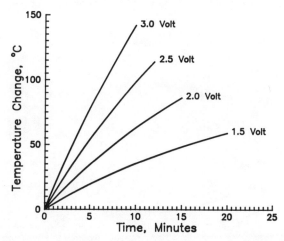

Figure 6.7. Effect of heat transfer on temperature rise of short-circuited cells with various voltages. (Refer to text for details.)

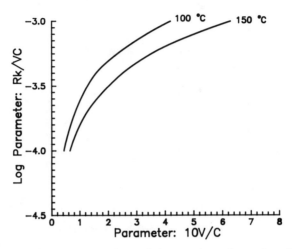

Figure 6.8. Heat transfer design correlation to limit temperature rise on short circuit. (Refer to text for details.)

tributions of convection and radiation will be discussed in Section 6.3 below. For concreteness, we examine the heat transfer coefficients applicable to D-size cells in intimate contact with a constant temperature reservoir under the following conditions:

1. The battery has a 0.002-in. thick insulating sleeve with a thermal conductivity of about 0.1 Btu/(h · ft² · °F)/ft that is in perfect thermal contact with the constant temperature reservoir. The steel case of the

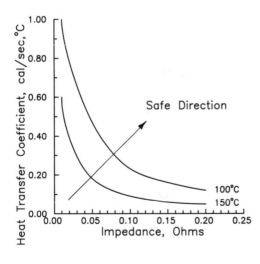

Figure 6.9. Heat transfer coefficients required to limit temperature rise as a function of cell impedance. (Refer to text for details.)

battery is 0.020 in. thick and has a thermal conductivity of about 26 Btu/(h · ft² · °F)/ft. The principal thermal resistance resides in the plastic sleeve, and the overall heat transfer coefficient is about 5 cal/ (s · °C).

2. The battery is contained in a plastic housing with a wall thickness of 0.10 in. and the same thermal conductivity as the plastic sleeve above. The outside of the plastic housing is in perfect thermal contact with the constant temperature reservoir. A 0.005-in. thick air layer is present between the battery and the plastic housing. The thermal conductivity of air is about 0.014 Btu/(h · ft² · °F)/ft. In this case, the two materials have comparable thermal impedances, and the overall heat transfer coefficient is about 0.08 cal/(s · °C).

The overall coefficients used in our heat transfer estimates ranged from 0.05 to 0.15 cal/(s · °C), and they reflect the characteristics of the second configuration above. That configuration is representative of many practical battery applications. However, we did assume the presence of a constant temperature external reservoir, which is not representative of battery applications. In order to obtain practically valid estimates of heat transfer from batteries, the impedance of the coupling between the battery housing and its thermal environment needs to be included in the analysis. We do this for the case of thermal runaways driven by ion depletion in the next section.

The high value of the heat transfer coefficient applicable to the first configuration above shows the importance of the thermal coupling between batteries and their immediate environment. It suggests that it may be possible to lessen the likelihood of thermal runaways by designing battery configurations to maximize the rate of heat transfer from a battery. Indeed, in very high rate applications, the effective cooling of a battery becomes a design imperative, with considerations of internal heat transfer being included as well.

6.3. CHARGING AND OVERCHARGING

Excessive heat generation due to charging and overcharging as a possible cause of thermal runaway is more of a problem with primary batteries than with secondary batteries. Aqueous electrolyte secondary batteries have been designed to accept some overcharge with relative impunity. Overcharging of such batteries leads to the evolution of hydrogen and oxygen, which may be chemically recombined inside the batteries or vented to the atmosphere. In the case of recombination, the rate of heat generation is controlled by the overcharging current, and it is thereby maintained at levels that do not cause thermal runaway. Rechargeable lithium batteries pose more of a problem since they possess no inherent chemical recombination mechanism.

Overcharging of secondary lithium batteries and charging of primary lithium batteries present the same basic problem: lithium ion depletion of the electrolyte at the negative electrode and various irreversible oxidation reactions at the positive electrode. In the case of soluble depolarizer lithium batteries, the electrodeposited lithium may react with the depolarizer to regenerate lithium ions or to form insoluble species. In the case of organic solvent electrolyte systems, lithium ion depletion occurs. If the charging is voltage limited, the increasing impedance of the electrolyte due to ion depletion will reduce the charging current to the point where ohmic heating may be kept within tolerable bounds and the likelihood of a thermal runaway is reduced. The case of constant current charging is more serious since resistive heating increases as the electrolyte impedance increases. We examine the possibility of thermal runaways due to the charging of primary lithium batteries in some detail. The same type of analysis is applicable to the overcharging of secondary lithium batteries. The constant current mode of charging will be selected for discussion for reasons of simplicity.

The method of analysis will be illustrated with a concrete example involving the charging of D-size Li/SO_2 cells at 25°C starting from a 100% state of charge (SOC). The extension of the analysis to the charging of cells from a lower SOC involves no new principles and will not be discussed. The following reactions need to be included in the thermal analysis:

Charging reactions (beyond the state of full charge):

$$Li^+ + e^- \rightarrow Li$$

$$Br^- \rightarrow \tfrac{1}{2}Br_2 + e^-$$

Chemical reactions:

$Li + \tfrac{1}{2}Br_2 \rightarrow LiBr$ $\qquad Q = 83.7$ kcal/mol

$2Li + 2SO_2 \rightarrow Li_2S_2O_4$ $\qquad Q = 61.0$ kcal/mol

$Li + AN \rightarrow$ Various organic products, various heats (AN = acetonitrile)

and we require quantitative expressions for:

1. Electrochemical heat generation
    (a) Reaction entropy
    (b) Overvoltage energy
    (c) Resistive heating

2. Heat evolution due to the chemical reactions
3. Heat transfer from the cell
   (a) Convection
   (b) Radiation
   (c) Conduction

For simplicity, we assume that there are no thermal gradients within the cells (a reasonable assumption for all but the highest charging rates), and we consider the cells to be sitting upright in a confined space that allows for convective cooling inside a compartment whose walls are maintained at a constant temperature of 25°C. We also assume that the battery housing is fabricated from an insulating material, which means that conductive cooling can be neglected. The incorporation of conductive cooling in the analysis is straightforward and may easily be included when required.

Based on the properties of typical D-size Li/SO_2 cells, we find the following expressions for the various heat generation and heat loss terms as a function of the charging current (I):

1 (a) Reaction entropy

$$q_1 = 3.69 \cdot I \cdot T \cdot 10^{-4} \text{ cal/s} \qquad (6.11)$$

1 (b) Overvoltage heat

$$q_2 = I \cdot (0.185 + 0.078 \ln I) \text{ cal/s} \qquad (6.12)$$

1 (c) Resistive heat

$$q_3 = rdI^2/4.186aC \text{ cal/s} \qquad (6.13)$$

where r is the specific resistivity, d is the thickness of the electrolyte layer, a is the superficial electrode area, and C is the electrolyte concentration.

2 (a) Lithium/bromine reaction heat

$$q_4 = 8.37 \times 10^4 \, k_4 A[(C_0 - C)/2]^s \exp(-E_4/RT) \text{ cal/s} \qquad (6.14)$$

where k_4 is the chemical reaction rate constant, A is the surface area of electrodeposited lithium, C_0 is the initial electrolyte concentration, E_4 is the activation energy, and s is the order of the chemical reaction rate.

2 (b) Lithium/sulfur dioxide reaction heat

$$q_5 = 6.1 \times 10^4 \, k_5 A \, \exp(-E_5/RT) \, \text{cal/s} \qquad (6.15)$$

where k_5 is the chemical reaction rate constant, and E_5 is the activation energy.

2 (c) Lithium/acetonitrile reaction heat

$$q_6 = k_6 2AQ \, \exp(-E_6/RT) \, \text{cal/s} \qquad (6.16)$$

where k_6 is the chemical reaction rate constant (lumped reactions), Q is the combined heats of the organic reactions, and E_6 is the activation energy (lumped parameter).

3 (a) Natural convection

$$q_7 = 0.38 \times 10^{-4} \pi D L^{3/4} \Delta T^{5/4} \, \text{cal/s} \qquad (6.17)$$

where D is the cell diameter, and L is the cell height.

3 (b) Radiative heat transfer

$$q_8 = 6.84 \times 10^{-13} \pi D (L + D/2)(T^2 + T_0^2)(T + T_0)\Delta T \, \text{cal/s} \qquad (6.18)$$

where T is the cell wall temperature, and T_0 is the temperature of the cell compartment (emissivity and absorptivity factors of 0.5 were used).

4. Reactive surface area of electrodeposited lithium. Since the active surface area of the lithium is unknown, we make the simplifying assumption that it is a constant per unit mass of electrodeposited lithium: K cm^2/mol Li. This leads to the equation:

$$dA/dt = K(I/F - A\{k_4[(C_0 - C)/2]^s e^{-E_4/RT} + k_5 e^{-E_5/RT} + k_6 e^{-E_6/RT}\}) \qquad (6.19)$$

where F is the electrochemical equivalence factor, and t is time. This equation reflects the rate of change of the surface area due to growth by electrodeposition and loss by chemical reaction.

A total energy balance leads to an expression for the rate of change of the cell temperature as a function of time and charging current given that the heat capacity of the cell is $W = W(t, T)$:

$$dT/dt = \Sigma q_i/W \qquad (6.20)$$

As may be seen from the expressions for heat generation and heat transfer, this is a highly nonlinear equation that cannot be solved analytically. Another difficulty is that the required phenomenological coefficients are not available. Given this situation, we consider three limiting cases that can be solved (numerically). While the limiting cases represent extreme situations, they still retain some practical value in defining boundaries of cell behavior. The three limiting cases are:

1. Steady-state conditions
2. Maximum chemical reaction rates
3. Vanishing chemical reaction rates

Actual battery behavior may be expected to fall within the ranges defined by the extreme limiting cases and their associated hazard potentials. The regions of hazardous operation may be characterized in terms of two criteria:

1. If the cell temperatures remain below the venting temperature, the hazard potentials are low, but the hazards may still be serious.
2. If the temperatures reach the melting point of lithium (181°C), the hazard potentials are high and the hazards are serious.

The energy balance equation was solved numerically for the three limiting cases, with some additional simplifying assumptions, and the results are presented below.

The steady-state temperature (Fig. 6.10) increases rapidly with the charging current and reaches an estimated value of 95°C at a charging current of about 0.7 A. This is close to the venting temperature of D-size Li/SO$_2$ cells. At charging currents of about 2 A, the estimated steady-state temperature reaches the melting point of lithium, creating a very hazardous situation. Although the hazard potential is less at lower than at higher charging currents, it must be emphasized that the charging of any primary battery is hazardous, as is overcharging of rechargeable lithium batteries. In the case of Li/SO$_2$ and other cells, additional factors need to be considered, such as the presence of unstable reaction products not included in the above analysis.

The estimated cell temperatures as a function of time and charging current under the assumption of maximum chemical reaction rates are shown in Fig. 6.11. It may be seen that the cells may be expected to vent in just a few minutes when charged with currents of about 2 A and above. If for some reason the vents do not operate and charging continues, the electrically driven

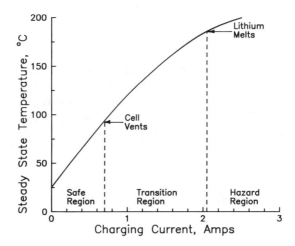

Figure 6.10. Steady-state temperature of cell on charging. (Refer to text for details.)

thermal runaway continues, and the cell temperature will reach the melting point of lithium. At a charging current of 5 A, explosions may be expected within about 12 minutes.

In the case of vanishing chemical reaction rates, there is no regeneration of electrolytic charge carriers, and as the concentration of the electrolyte decreases, the cells enter an electrically driven thermal runaway under constant current conditions. Assuming that a sufficient driving voltage is available

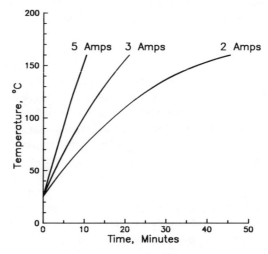

Figure 6.11. Temperature of cell under conditions of maximum chemical reaction rate for various charging currents. (Refer to text for details.)

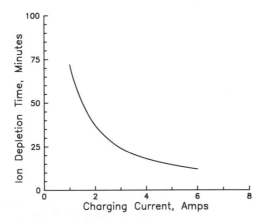

Figure 6.12. Estimated time to deplete ionic charge carriers as a function of charging current. (Refer to text for details.)

from the current source, the cells enter an explosive mode as the electrolyte becomes depleted in charge carriers. The estimated time for this to happen is shown in Fig. 6.12. It may be seen that high charging currents would be expected to trigger explosions in just a few minutes. In actual cases, complete ion depletion is unlikely to occur since charging devices do have voltage limiting characteristics. This leads to a lowering of the charging currents as the electrolyte becomes depleted in ion carriers.

The cases of greater practical interest are intermediate between the limiting cases discussed above, and all the given heat sources contribute to cell heating to some degree. This general problem cannot be solved exactly since it is not feasible to obtain the required data, either by computation or by experimentation. Also, simplifying assumptions had to be introduced in the analysis to render it tractable. This means that the given results are of value primarily as a qualitative guide. It is generally necessary to supplement computational analyses of battery hazards with experimental evaluations under the appropriate conditions for various applications of interest.

BIBLIOGRAPHY

A considerable body of literature is available on the current distribution, heat generation, and temperature profiles in electrochemical cells, and we present a selection of references that may be useful to readers interested in pursuing these subjects in more detail. The text by Newman is devoted to the fundamentals of electrochemical processes and is basic reading for anyone interested in the quantitative analysis of the behavior of cells and batteries.

The other references deal with specific systems. The various chapters in the standard battery handbook (Ref. 1 in Chapter 1) may be consulted for both detailed and general information on most of the available battery systems.

John S. Newman, *Electrochemical Systems,* Prentice-Hall, New York (1973).

D. Bernardi, E. Pawlikowski, and J. H. Newman, A general energy balance for battery systems, *J. Electrochem. Soc.* **132,** 5 (1985).

K. W. Choi and N. P. Yao, Heat transfer in lead–Acid batteries designed for electric vehicle propulsion applications, *J. Electrochem. Soc.* **126,** 1321 (1979).

J. Lee, K. W. Choi, N. P. Yao, and C. C. Christianson, Three-dimensional thermal modeling of electric vehicle batteries, *J. Electrochem. Soc.* **133,** 1286 (1986).

H. Gu, T. V. Nguyen, and R. E. White, A mathematical model of a lead–acid cell: Discharge, rest and charge, *J. Electrochem. Soc.* **134,** 2953 (1987).

R. M. La Follette and D. N. Bennion, Design fundamentals of high power density, pulsed discharge lead acid batteries. I. Experimental, *J. Electrochem. Soc.* **137,** 3693 (1990).

R. M. LaFollette and D. N. Bennion, Design fundamentals of high power density, pulsed discharge lead acid batteries. II. Modeling, *J. Electrochem. Soc.* **137,** 3701 (1990).

Z. Mao, R. E. White, and B. Jay, Current distribution in a HORIZON™ lead–acid battery during discharge, *J. Electrochem. Soc.* **138,** 1615 (1991).

D. Fan and R. E. White, A mathematical model of a sealed nickel–cadmium battery, *J. Electrochem. Soc.* **138,** 17 (1991).

D. Fan and R. E. White, Mathematical modeling of a nickel–cadmium battery, *J. Electrochem. Soc.* **138,** 2952 (1991).

K. W. Choi and N. P. Yao, A mathematical model for porous nickel electrodes in zinc/nickel oxide cells, in *Proceedings of the Symposium on Battery Design and Optimization,* Proc. Vol. 79-1, p. 62, The Electrochemical Society, Pennington, New Jersey (1979).

Z. Mao and R. E. White, Mathematical modeling of a primary zinc/air battery, *J. Electrochem. Soc.* **139,** 1105 (1992).

A. N. Dey and P. Bro, Primary Li/SOCl₂ cells. IV. Cathode reaction profiles, *J. Electrochem. Soc.* **125,** 1574 (1978).

K. Y. Kim, H. V. Venkatasetty, and D. L. Chua, Studies on thermochemical and electrochemical reactions and heat distribution in Li/SOCl₂ battery system, in *Proceedings of the 29th Power Sources Symp., Electrochemical Society, Pennington, New Jersey (1981). Conference,* Atlantic City, New Jersey, June 1980.

S. Szpak, C. J. Gabriel, and J. R. Driscoll, Catastrophic thermal runaway in lithium batteries, *Electrochim. Acta* **32,** 239 (1987).

T. I. Evans and R. E. White, A thermal analysis of a spirally wound battery using a simple mathematical model, *J. Electrochem. Soc.* **136,** 2145 (1989).

T. I. Evans, T. V. Nguyen, and R. E. White, A mathematical model of a lithium/thionyl chloride primary cell, *J. Electrochem. Soc.* **136,** 328 (1989).

Y. I. Cho and D.-W. Chee, Thermal analysis of primary cylindrical lithium cells, *J. Electrochem. Soc.* **138,** 927 (1991).

R. Cohen, A. Melman, N. Livne, and E. Peled, Heat generation in lithium–thionyl chloride cells, *J. Electrochem. Soc.* **139,** 2386 (1992).

E. E. Kalu and R. E. White, Thermal analysis of spirally wound Li/BCX and Li/SOCl₂ cells, *J. Electrochem. Soc.* **140,** 23 (1993).

D. M. Bernardi, H. Gu, and A. Y. Schoene, Two-dimensional mathematical model of a lead–acid cell, *J. Electrochem. Soc.* **140,** 2250 (1993).

III

AQUEOUS ELECTROLYTE BATTERIES

An appreciation of the nature of battery hazards requires a knowledge of the processes that occur when batteries are abused or mishandled, which have been briefly reviewed in Part II, and it requires a knowledge of the materials present in batteries and the health hazards that may be created if they should escape from a battery. In this part, we review the basic chemistry of aqueous electrolyte batteries and some properties of the principal constituents of the aqueous electrolyte batteries in common use from the point of view of their health hazards. We do not discuss specialty batteries such as reserve batteries and fuel cells nor the more esoteric systems that are not in common use. Comprehensive information on batteries in general may be found in the *Handbook of Batteries and Fuel Cells* (see Ref. 1 in Chapter 1). Battery manufacturers may be contacted for much useful information on their products, both performance data and recommended handling procedures. Sources of information on the toxicity and exposure limits of some battery materials are given in Refs. 2–6 in Chapter 7. We have included toxicity values in Chapters 7 and 8 expressed as the time-weighted average (TWA) and the threshold limit values (TLV) as a guide to the relative toxicity of some battery materials. It may be noted that the values refer to exposures from airborne materials, a form of exposure unlikely to be encountered by battery users. Statutory exposure limits are not available for many of the materials contained in batteries. This is generally an indication of a low toxicity. The Federal (U.S.) maximum acceptable contaminant levels (MCL) of battery materials in drinking water are included for some substances; very few values are available. A considerable body of data is available on material toxicity based on animal experiments (see Ref. 6 at the end of Chapter 7 for brief summaries). In the absence of TWA, TLV, and MCL information, this body of data is a useful source of information on the relative toxicities of many materials. In assessing the significance of these and other toxicity values, however, it is important to recognize that there are no absolute measures of toxicity (refer to our comments "Some Toxicity Considerations" in Section 2.1) and that results from animal experiments are not directly translatable to humans. Since the improvement of environmental quality is a continuing effort, the given exposure limits may change, and limits may be imposed on substances for which there are no present limits. Readers are encouraged to check current federal and state regulations for up-to-date information on toxicity and permissible exposure levels.

7

Primary Batteries

Many different sizes and shapes of batteries are available in each of the different categories of battery systems. We are not concerned with these differences. Rather, we focus attention on the substances contained in the batteries and their properties. Most of the materials used in batteries are innocuous, but some are toxic and/or corrosive and may cause harm if released to the environment. It is desirable that users know what the batteries contain and that they have some information on the hazards to which they may be exposed if the contents of a battery should escape, why the contents may escape, and how accidents may be prevented (see Parts II and V). Based on such knowledge, they may take steps to avoid accidents, if possible, and to protect themselves from injuries if accidents should occur. It is reasonable to assume that the severity of any hazard is likely to increase as the amounts of hazardous materials present in a battery and ejected from a battery increase, and it is desirable, therefore, to know how much of such materials may be present in a battery. We give a rule of thumb to estimate the approximate amounts of active materials that may be present in batteries based on their capacity.

The chemical composition of a battery changes with its state of discharge. Generally, the overall hazard potential of fresh batteries is greater than that of discharged batteries. However, if any toxic materials are present, their toxicities may not change much with changes in their chemical state as a result of charge or discharge as their toxicities are generally determined by their *in vivo* biochemical transformations. Electrolyte is consumed during the discharge of an aqueous electrolyte primary battery, and, as a result, the amount of electrolyte present in discharged batteries is less than in fresh batteries. Alkaline cells may even become almost dry as a result of discharge. This means that hazards associated with the electrolyte may be less with discharged than with fresh cells. However, since the state of discharge of a battery involved in an accident is generally unknown, it is prudent to assume that the hazard level is that of a fresh battery. An exception is aged and/or discharged batteries that may be more highly pressurized than fresh batteries owing to accumulated gas formed by corrosion reactions. Such batteries may be more hazardous than fresh batteries.

7.1. ZINC/CARBON BATTERIES (LECLANCHÉ BATTERIES)

Zinc/carbon batteries are available in a great variety of sizes of both cylindrical and prismatic shapes in either of two interior configurations. The conventional bobbin cells are housed in cylindrical zinc containers, with zinc being the active anode material, and they contain a centrally located bobbin type cathode with manganese dioxide as the cathode-active material. The cells are sealed with asphaltic materials. The other configuration, the inside-out structure, employs a cell case made of impermeable carbon that also serves as current collector for an outer cathode. The zinc anode is located in the center of the cell. The inside-out cells have crimp-like seals. The cells are provided with multilayered outer wraps composed of various insulating materials and may have an outer steel jacket. The outer jacket carries the cell identification and the required labeling information, and it indicates the cell polarity. All the zinc/carbon cells have positively polarized cell tops and negatively polarized cell bottoms. The batteries are used extensively in very many consumer applications. On a total volume, worldwide basis, the zinc/carbon batteries are used more widely than any other type of battery, although they are gradually being displaced by the improved version, the so-called zinc/chloride batteries, and by alkaline manganese batteries.

The electrochemical system may be represented by

$$Zn/NH_4Cl, ZnCl_2, H_2O/MnO_2, C \qquad 1.5\ V$$

The nominal voltage of 1.5 V decreases markedly during discharge and during sustained high-rate drain. Thus, the cells do not have a very good voltage regulation, and this limits their usefulness in many consumer applications. The anodes consist of pure zinc metal with small amounts of cadmium and lead. The cadmium reduces the corrosion rate of zinc in ammoniacal electrolytes, and lead renders zinc more easily formable. The electrolyte is a concentrated, slightly acidic solution of ammonium chloride with added zinc chloride to improve its conductivity. It is generally present as a starchy paste to give it sufficient rigidity to serve as a separator as well as electrolyte. The higher rate versions have paper type separators. Unspecified corrosion inhibitors are added to the electrolyte in low concentrations to reduce the corrosion of zinc. The earlier versions of zinc/carbon cells contained small amounts of mercuric salts as corrosion inhibitors, but this use of mercury has been much reduced because of mercury's toxicity and adverse environmental effects. The cathodes are made from a mixture of manganese dioxide, carbon black, and electrolyte, with carbon black constituting about 10% by weight, or more, of the mixture. The greater amounts of carbon are used for cells designed for pulse duty and higher rate applications. Various grades of manganese dioxide have been used by different manufacturers, ranging from the classical use of naturally occurring manganese oxide with about 70% of MnO_2 to the modern use of refined, high-grade manganese oxides containing about 95% MnO_2. Various types of mineral impurities and some heavy metals in low concentrations are present in the manganese oxides, the amounts of impurities depending on the degree of upgrading of the natural ores that are used. In bobbin cells, the positive current collector is a central rod of carbon that has been impregnated with a waxy substance to render it impermeable to air (oxygen enhances the corrosion of zinc), and the zinc case is the negative current collector. In the inside-out cells, the carbon case serves as the positive current collector, and the vane-shaped central zinc anode is connected to a brass current collector. The cells are configured to have the same polarity as the bobbin cells, i.e., a positive top and a negative base. All current collectors are connected to exterior steel elements that serve as cells terminals. Zinc/carbon cells may also be constructed in flat-pack configurations for special high-rate applications. Their components are basically the same as those of the cylindrical cells.

The discharge reaction of zinc/carbon cells is given by

$$Zn + 2MnO_2 + 2NH_4Cl \rightarrow 2MnOOH + Zn(NH_3)_2Cl_2$$

If the cells are discharged to very low voltages, some lower oxides and hydroxides of manganese may also be formed. In the absence of detailed information from the manufacturers, the amounts of electroactive materials present in a cell may be estimated from the equivalent weights of the materials and the nominal cell capacities. The equivalent weights are:

- 3.2 g/Ah for manganese dioxide (MnO_2)
- 1.2 g/Ah for zinc metal (Zn)

Since the rated nominal capacities take into account the inefficiencies of cell discharge, the amounts actually present in a cell would be somewhat greater than estimated from the nominal cell capacities. The amount of electrolyte present in a cell varies quite a bit with the cell design, but it would rarely exceed one-fourth of the cell volume; it may be less.

The only significant side reaction in zinc/carbon cells is the corrosion of zinc, which leads to the formation of hydrogen and various zinc salts. Since the cells are vented, by virtue of their construction, most of the hydrogen is likely to escape to the atmosphere without generating excessive cell pressures. If the gas leakage paths are blocked, the pressures may increase to the point where a noticeable bulging may be observed.[1]

The health hazards of zinc/carbon cells are low and are associated primarily with the ammonium chloride electrolyte and to a lesser extent with the manganese compounds and with the small amounts of lead and cadmium that may be present in the zinc anodes. The ammoniacal paste is the component to which users are most likely to be exposed. It is an acidic mixture and a noticeable tissue irritant. The metallic components and their compounds are of concern primarily from the point of view of their possible ingestion and as a source of contamination of ecosystems following uncontrolled disposal of the batteries.

7.2. ZINC/CHLORIDE BATTERIES (HEAVY-DUTY ZINC/ CARBON BATTERIES)

The zinc/chloride batteries are a near relative of the zinc/carbon batteries. They differ from the latter only in that they employ an electrolyte consisting of a concentrated solution of zinc chloride. The electrolyte is more acidic than that used in zinc/carbon cells, and it is more highly conducting. The cathodes are made from high-grade manganese dioxide rather than from manganese ores. Some chemically upgraded manganese oxide ores may be used in the cathode mixture. The cathodes contain carbon as a conductive additive. The separators are made from various grades of papers. In most other respects, the cell designs are comparable to those of the zinc/ carbon cells.

The electrochemical system may be represented by

$$Zn/ZnCl_2, H_2O/MnO_2, C \quad 1.5 \text{ V}$$

As their alternate name signifies, the zinc/chloride cells are designed for higher rates of discharge than the conventional zinc/carbon cells, and they find application in consumer devices where high drain rates are required.

The discharge of zinc/chloride cells gives somewhat different reaction products than does that of zinc/carbon cells:

$$Zn + 2MnO_2 + 2H_2O \rightarrow Zn(OH)_2 + 2MnOOH$$

Deep discharges lead to the formation of lower oxides of manganese.

The hazards associated with zinc/chloride cells are essentially the same as for zinc/carbon cells. The only difference is one of degree in that the zinc chloride electrolyte is more aggressive than the ammoniacal electrolyte.

7.3. ALKALINE MANGANESE BATTERIES

Alkaline manganese batteries are available in cylindrical, prismatic, and coin configurations. The prismatic batteries are generally constructed from cylindrical unit cells. The cylindrical cells far outnumber the other shapes. Most of the alkaline manganese cells are housed in crimp-sealed steel containers, and the cylindrical cells are provided with insulated metal jackets or polymeric shrinkwraps. The polarity of the batteries is clearly marked on the jackets, the top being the positive terminal and the base (i.e., the case) the negative terminal. There is no jacket on the coin cells; the cell designations and their polarity are embossed in the base of the coin cells. Their polarity is the opposite to that of the cylindrical cells; the top is negative and the base positive. Few alkaline manganese coin cells are available. The majority of coin cells used today are based on the alkaline silver oxide system or on solid depolarizer lithium systems. Alkaline manganese batteries are used in a great variety of consumer applications. Together with the zinc/carbon batteries, they dominate as the power source of choice for the majority of portable consumer devices.

The electrochemical system may be represented by

$$Zn/KOH, H_2O, ZnO/MnO_2, C \qquad 1.5 \text{ V}$$

The nominal voltage is 1.5 V, and it decreases slowly during discharge. The batteries have a much better voltage regulation than zinc/carbon batteries and an excellent current-carrying capability; that is, the voltage decreases less during discharge and as a result of heavy loads than is the case with zinc/carbon cells, hence their popularity in consumer applications. The anodes are made from high-purity zinc metal and may contain small amounts of unspecified corrosion inhibitors that have replaced the mercury formerly used for this purpose. The electrolyte is a concentrated solution of potassium hydroxide in water, about 40% KOH by weight, and it contains dissolved zinc oxide in the form of a hydrated zinc hydroxide or as a zincate. A gelling agent such as carboxymethylcellulose may be added to the electrolyte to increase its viscosity. The separators are thin porous sheets of inert cellulosic or other polymeric materials. The cathodes are made from high-purity manganese oxide blended with small amounts of carbon. The cathodes may contain trace amounts of heavy metal impurities.

The overall discharge reaction is

$$Zn + 2MnO_2 + 2H_2O \rightarrow Zn(OH)_2 + 2MnOOH$$

The zinc is converted to a hydrated zinc hydroxide and zincate, and the manganese dioxide to manganese oxyhydroxide and manganese hydroxide, the amount of the latter increasing the more deeply discharged the cells. The amounts of electroactive materials present in a cell may be estimated from their equivalents weights:

- 3.2 g/Ah for manganese dioxide (MnO_2)
- 1.2 g/Ah for zinc metal (Zn)

The amounts actually present are slightly greater than estimated from the nominal capacities of the batteries since the rated capacities are less than the capacities that correspond to the stoichiometric amounts of the active materials in the cells. Note that an allowance must be made for multicell batteries, as the above figures relate to the amounts present in each individual 1.5-V cell. The amount of electrolyte present in a cell may be estimated from the cell volume; it is generally less than one-fourth of the total cell volume.

The side reaction of principal concern is the reaction of zinc with the alkaline electrolyte, which produces zinc hydroxide and hydrogen. At normal cell temperatures, the hydrogen evolves slowly, and much of it escapes innocuously to the atmosphere via the crimp seals. Since the gas may generate high cell pressures, the puncturing of alkaline cells may generate forceful sprays of electrolyte with damaging effects.

The health hazards of alkaline manganese batteries are associated primarily with the alkaline electrolyte, which causes severe chemical burns. In distinction to acid burns, there is no immediate burning sensation associated with alkali burns to alert victims to damage in progress.

7.4. SILVER OXIDE BATTERIES

Silver oxide batteries are used in many consumer applications. They are available in a variety of sizes and dominate in applications that require small button and coin cells. They are not available in the larger cylindrical sizes, primarily because they cannot compete with other systems on a cost basis. The principal ingredient, silver oxide, is far more expensive than the cathode materials used in other cells with comparable performance. The cost factor is much less important in miniature cells, where performance tends to be a deciding factor in the selection of batteries for various applications. In their design, construction, and composition, the cells are equivalent to alkaline manganese batteries, except that silver oxide has replaced manganese oxide and separators are used that have a greater oxidation resistance than the cellulosics used in the alkaline manganese batteries. The cells are fabricated with crimp-sealed steel housings, and the cell identifications and the cell polarities are embossed in the base of the cells. The cell tops are the negative terminals and the cases the positive terminals on these as on most other coin and button cells.

The electrochemical cell may be represented by

$$\text{Zn/KOH, } H_2O, \text{ ZnO/Ag}_2\text{O, C} \qquad 1.6 \text{ V}$$

The exact voltage of the cells may differ from the nominal voltage in a manner that depends upon the exact composition of the silver oxide. Some cells may contain various amounts of divalent silver in the oxide. The more divalent silver there is in a cell, the higher its voltage. Values as high as 1.76 V may be obtained. Divalent silver tends to be unstable and may be reduced spontaneously by reaction with the electrolyte, with the evolution of oxygen. For most practical purposes, a cell voltage of 1.5–1.6 V is a reasonable expectation. The silver/zinc cells discharge with relatively constant load voltages and have a high rate capability, both of which are highly desirable characteristics for many consumer applications.

The anodes are made from high-purity zinc, and the electrolytes are concentrated solutions of potassium hydroxide, about 40% by weight. Sodium hydroxide may replace the potassium hydroxide in cells designed for long-term applications. The system is more stable in NaOH solutions than in KOH solutions, but some of the high rate capability is lost. The electrolytes contain small amounts of unspecified zinc corrosion inhibitors that have replaced the mercury formerly used in the zinc anodes to reduce their spontaneous reaction with the electrolyte. A gelling agent may also be used to immobilize the electrolyte. The separators are generally made from composites of polyolefins and cellulosics and may contain nylons and/or poly(vinyl alcohol) to increase their wettability. The cathodes are made from silver oxide with small amounts of graphite or carbon as a conductive additive. Some cells contain duplex cathodes made from silver oxide with an admixture of manganese dioxide to reduce cost and to provide an end-of-life indicator. The lower voltage manganese dioxide provides a defined voltage drop near the end of life.

The overall discharge reaction is given by

$$\text{Zn + Ag}_2\text{O + } H_2O \rightarrow \text{Zn(OH)}_2 + 2\text{Ag}$$

The formation of metallic silver increases the electronic conductivity of the cathode during discharge and contributes to the high rate capability of the cells, giving them also a very good voltage regulation. The amounts of silver oxide and zinc present in a cell may be estimated from the nominal capacity of a cell and the equivalent weights of the active materials:

- 4.3 g/Ah for silver oxide (Ag_2O)
- 1.2 g/Ah for zinc metal (Zn)

The actual amounts are likely to be somewhat greater than estimated from the nominal cell capacities since the latter allow for the inefficiencies of the cell discharges. The amount of electrolyte in a cell may be expected to be no more than about one-fourth of the cell volume, and most of it will be held in the porous anode and in the separators. Very little electrolyte is present as unconstrained liquid.

The principal side reaction of concern is the corrosion of zinc in the alkaline electrolyte, which generates hydrogen and zinc hydroxide and zincates. Most of the hydrogen escapes to the atmosphere via the crimp seal, but, in very tightly sealed cells, the rate of escape may be so slow that appreciable cell pressures are generated. A secondary side reaction is the decomposition of silver oxide. The high voltage of silver oxide cells reflects the intrinsic instability of silver oxide, which may generate measurable oxygen partial pressures in the cells. The strong oxidizing power of silver oxide favors the decomposition of oxidizable separators such as cellulosic separators. However, cellulosics may serve as scavengers of soluble silver ions and prevent their diffusion to the anode, where they would be reduced to silver, which would increase the corrosion rate of zinc and possibly form short circuits between the anode and the cathode.

The health hazards of silver/zinc cells are low and are associated primarily with the alkaline electrolyte and the chemical burns it may generate in exposed body tissues. Tissue damage from alkali burns tends to extend into the lower epithelial layers rather than being limited to superficial regions. The eyes, in particular, are vulnerable to serious damage from exposure to alkaline electrolytes. Silver and silver oxide are relatively innocuous materials. Since most of the silver/zinc cells in common use are quite small, the extent of any physical damage that might result from cell ruptures would be limited.

7.5. MERCURIC OXIDE BATTERIES

Until recently, mercuric oxide cells were widely used in consumer electronics, in electronic instrumentation, and in medical devices because of their high volumetric energy density and very stable, constant voltage. They were available in many sizes and shapes—button cells, coin cells, cylindrical cells, and prismatic batteries with various interior configurations. The recognition of the health hazards posed by the release of mercury and its compounds into the environment coupled with the uncontrolled disposal of mercury batteries has led to reductions in the use of mercuric oxide and mercury in batteries. At the present time, mercury batteries occupy a very small segment of the battery market; they are used only in a small number of specialty applications. Their construction is basically the same as that of the silver oxide cells: porous anodes made from high-purity zinc with some mercury to reduce zinc corrosion, a concentrated potassium hydroxide solution saturated with zinc oxide as electrolyte, with or without a carboxymethylcellulose or other thickener, a cellulosic or polyolefin separator, and a densely compacted mercuric oxide cathode with some added carbon or silver as conductive additive, all contained in a crimp-sealed steel housing.

The cell may be represented by

$$Zn(Hg)/KOH, H_2O, ZnO/HgO, C \text{ (or Ag)} \qquad 1.35 \text{ V}$$

The cell voltage is remarkably stable and the cells have been used as voltage reference sources. During discharge at moderate rates, the voltages remain relatively constant, and the cells have a good voltage regulation. Depending on their interior structures, the cells may be designed for either high-rate or low-rate applications.

The overall discharge reaction is

$$Zn + HgO + H_2O \rightarrow Zn(OH)_2 + Hg$$

The formation of liquid mercury during discharge may lead to shorting, but this problem can be avoided by the addition of silver to the cathodes to bind the mercury or by the use of silver oxide as an additive to the cathodes. The use of silver oxide destroys the good voltage regulation of the cells since silver oxide discharges at a higher voltage than mercuric oxide, but it generates the silver needed to immobilize the liquid mercury formed during the subsequent discharge of mercuric oxide. The amounts of electroactive materials present in a cell may be estimated from the nominal cell capacity using their equivalent weights:

- 4.0 g/Ah for mercuric oxide (HgO)
- 1.2 g/Ah for metallic zinc (Zn)

The amount of electrolyte may be estimated to be about one-fourth of the cell volume, possibly less. Most of the electrolyte is likely to be immobilized in a gelled form, absorbed by the separator, and held in the porous matrix of the anode.

The mercuric oxide cells are very stable, and the only side reaction of importance is the corrosion of the zinc anode, which produces hydrogen gas. As with other crimp-sealed cells, most of the hydrogen escapes slowly to the atmosphere via the crimp seals with no adverse effects, but the cells are likely to be pressurized by hydrogen.

The health hazards of the cells are high and are associated with the alkaline electrolyte, as described for the other alkaline cells, and with the toxic effects of mercury and the mercuric compounds present in the cells. In many alkaline cells, there is a tendency for any escaping electrolyte to form small amounts of salt deposits at the crimp seal owing to the neutralization of escaping electrolyte by atmospheric carbon dioxide. The alkaline deposit is relatively harmless unless it is ingested or brought into contact with the eyes. It does not significantly affect the cell performance and may be removed with a moist cloth followed by drying of the cell with a dry cloth or absorbent paper.

7.6. CADMIUM/MERCURIC OXIDE BATTERIES

The instability of cells with zinc electrodes is due primarily to the corrosion of zinc and the associated generation of hydrogen. Although the hydrogen poses some hazard, that hazard is generally of limited concern in the case of small, well-made cells. A greater concern is the effect of zinc corrosion on the long-term performance of the cells, especially for low-rate applications. This problem has led to the design of alkaline mercuric oxide cells in which the zinc anodes have been replaced with cadmium anodes that have a greater long-term stability than zinc, but also a lower voltage. However, for the same reason that mercuric oxide cells with zinc anodes are being phased out of noncritical applications, namely, because of their toxicity, so the mercuric oxide cells with cadmium anodes are being phased out of noncritical applications. In terms of cell sizes and shapes, the cells are analogous to the mercuric oxide cells with zinc anodes and need no elaboration at this point. The only exception is that Cd/HgO cells may be hermetically sealed; the gas evolution in these cells is extremely low.

The cells may be represented by

$$Cd/KOH, H_2O/HgO, C \text{ (or Ag)} \qquad 0.91 \text{ V}$$

The low cell voltage is a measure of the intrinsic stability of the system. As in the case of all mercuric oxide cells, the discharge voltages remain relatively flat until the end of life is approached. Most of the cells have a low rate capability, by design, since they are used primarily for long-term applications where high drain rates are not required.

The overall discharge reaction is

$$Cd + HgO + H_2O \rightarrow Cd(OH)_2 + Hg$$

As in the case of other mercuric oxide cells, liquid mercury is formed during discharge, and it may be immobilized by the use of silver in the cathode. The lower limits of the amounts of active materials present in a cell may be estimated from the nominal cell capacity and their equivalent weights:

- 2.1 g/Ah for cadmium metal (Cd)
- 4.0 g/Ah for mercuric oxide (HgO)

The volume of electrolyte in a cell may be estimated to be about one-fourth of the cell volume. The electrolyte concentration would be about 40% by weight in fresh cells. Since water is consumed during discharge, the cells tend to dry out during discharge, and, at the end of discharge, there may be very little free electrolyte left in a cell. Thus, the likelihood of exposure to spilled electrolyte from ruptured or leaky cells decreases as the cells approach end of life. Contact with any of the interior components, however, still poses a risk.

The cadmium/mercuric oxide cells are very stable and are not subject to any side reactions of any consequence.

There are health hazards associated with all the electroactive components of Cd/HgO cells: their anodes, electrolytes, and cathodes. Among these, the hazards associated with the electrolyte are probably of the greatest practical concern for the general user. Exposure to the alkaline electrolyte may cause severe tissue damage. However, ingestion of mercury, cadmium, or any of their compounds is a distinct hazard, and contact with these materials should be avoided.

7.7. ZINC/AIR BATTERIES

The uniqueness of zinc/air cells resides in their cathodes. Instead of using a prepackaged oxidation agent, as in most other batteries, the cells exploit oxygen from the air as cathode-active material. Hence, the cells must be open to the atmosphere during active discharge. The elimination of the self-contained cathode-active material provides additional space for the anode-active material, and, as a result, zinc/air cells have a considerably greater capacity and energy density than other batteries of the same size.

The batteries come in a variety of sizes ranging from large industrial multicell batteries with capacities in excess of 1000 Ah to small button cells with capacities of the order of 80 mAh. The button cells are available in the general consumer market for a variety of applications, and we limit our discussion to that type of zinc/air batteries. The cells are similar to conventional button cells in appearance in that the active structures are contained in crimp-sealed, steel containers. The cell cover is the negative terminal and the cell case the positive terminal, as with other button cells. Their interior structures differ from those of conventional cells primarily in the cathode assemblies. Most of the interior volume is occupied by the porous zinc electrode, which is filled with the concentrated potassium hydroxide electrolyte saturated with zinc oxide as in conventional cells with zinc electrodes. The separator and the catalytic air electrode occupy a small fraction of the cell volume. The cathode itself is a complex structure comprising current collector, catalyst, hydrophobic section, and an inner air space. The function of the hydrophobic section is to prevent flooding of the cathodic catalyst/current collector assembly. The outer region of the latter must remain accessible to air and its inner region accessible to electrolyte for the cells to function properly. The cells are provided with a small aperture to allow air to enter them, but the apertures should remain sealed prior to discharge to prevent zinc corrosion by ingress of oxygen. The discharge rate of the cells is limited by the rate of oxygen availability, which is fixed by the size of the air aperture. Most zinc/air cells are designed for low-rate applications, but high-rate versions are available.

The electrochemical cell may be represented by

$$\text{Zn/KOH, H}_2\text{O, ZnO/catalytic collector/O}_2 \quad 1.45 \text{ V}$$

Although the open-circuit voltage is relatively high, the operating voltages are close to 1.1–1.2 V, depending on the cell design and the loads. The anodes and electrolytes are essentially the same as in other alkaline cells with zinc electrodes: pure zinc, concentrated potassium hydroxide, about 30% KOH by weight saturated with zinc oxide, unspecified corrosion inhibitors, and cellulosic or polyolefin separators. The cathode structure comprises various polymeric materials, including halocarbon polymers, and small amounts of heavy metal catalysts and/or manganese oxides and carbon.

The overall reaction involves the oxidation of zinc

$$\text{Zn} + \tfrac{1}{2}\text{O}_2 + \text{H}_2\text{O} \rightarrow \text{Zn(OH)}_2$$

The reaction product is a mixture of zinc hydroxide and potassium zincates. The discharges proceed at relatively constant voltages for any given load, but the voltages change appreciably with the discharge rates for all but the high-rate cells. The amount of zinc present in a cell may be estimated from the rated capacity of the cells and the equivalent weight of zinc:

$$1.2 \text{ g/Ah for zinc (Zn)}$$

As in the case of all other cells, the actual amount will be somewhat greater because of the inherent inefficiencies of the discharge process and the allowance that must be made for zinc corrosion. The amount of electrolyte in a cell may be estimated to be about one-fourth of the cell volume, but it changes during discharge owing to the consumption of water and the possible ingress or egress of moisture.

The most important side reactions in zinc/air cells are the reaction of zinc with the alkaline electrolyte and zinc corrosion due to the presence of oxygen in the cells. In addition, two other processes occur that are important for the electrolyte balance and composition: (i) atmospheric carbon dioxide may enter the cells and react with potassium hydroxide to form insoluble potassium carbonate and (ii) ingress of moisture from humid atmospheres increases the electrolyte volume, and egress of water from the electrolyte in dry atmospheres tends to dry out the cells. These processes affect primarily the performance of the cells and may cause cell sweating.

Zinc/air cells have few health hazards. These are associated primarily with the alkaline electrolyte and the injuries that may result from direct contact with the electrolyte and from handling components that contain electrolyte. Since the cells are open to the atmosphere, no pressure buildup occurs unless the air apertures are obstructed. The likelihood of cell ruptures is vanishingly small. The catalytic materials are present in small amounts and do not give rise to any present significant hazards unless ingested.

7.8. PRIMARY BATTERY MATERIALS AND TOXICITY

We present only a cursory review of toxicity relevant to the most important battery materials. Readers in need of more information should consult the references cited at the end of the chapter.[2-6] Some additional information on the toxicity associated with lithium batteries is included in Part IV. Caution should be exercised in interpreting toxicity data based on animal studies alone in the absence of clinical information on the effects of poisonous substances on humans. Readers are advised to ascertain whether or not the information they obtain relates to humans or to animals.

Battery containers and other inert components such as separators, seals, insulators, and contacting structures are not included in our discussion. They are of little importance as contributors to battery toxicity. Exceptions are nickel and chromium, which are allergens. They may give rise to dermatitis in sensitive individuals following contact with battery containers made of nickel-plated cold-rolled steel or stainless steel (contains nickel and chromium). It may be noted that only a small percentage of the population is sensitive to these allergens.

7.8.1. Cadmium

Chemical Species: Cd, CdO, Cd(OH)$_2$

Batteries: Cadmium/mercuric oxide batteries (Mercad batteries)

Physical Forms: The cadmium-containing materials are present as a dense metal or as porous powder compacts. The ratio of hydroxide to metal increases with the depth of discharge of the batteries. The hydroxide is a loosely compacted material with a low cohesivity. All the cadmium compounds are insoluble in the electrolyte.

Toxicity: Cadmium and its compounds are highly toxic materials. They affect primarily the kidneys and the pulmonary system and have been classified as potential carcinogens.

$$\text{TWA: } 0.01 \text{ mg/m}^3 \text{ (CdO dust)}$$

$$\text{TLV: } 0.01 \text{ mg/m}^3$$

$$\text{MCL: } 0.01 \text{ mg/liter}$$

Simple Symptoms: Ingestion leads to vomiting, diarrhea, and stomach pain. Inhalation of cadmium dusts, primarily in industrial environments, causes metal fume fever. The symptoms include fever, chills, headache, and chest pain and appear several hours after exposure. General battery users are not likely to be exposed to this type of poisoning.

7.8.2. Manganese

Chemical Species: MnO$_2$, MnOOH, Mn(OH)$_2$

Batteries: Manganese is present in the great majority of primary batteries available in the consumer market, principally zinc/carbon, zinc/chloride, and alkaline manganese batteries. Zinc/air cells and other primary alkaline cells may contain some manganese oxide in their depolarizers.

Physical Forms: The manganese compounds in batteries are insoluble, reasonably cohesive, porous bodies and are likely to contain carbon as a conductive additive. The more deeply discharged the cells, the more manganese hydroxide they contain relative to manganese dioxide.

Toxicity: Manganese is an essential nutritional element in low concentrations. In higher concentrations, manganese and its compounds are toxic. Their toxicity affects the pulmonary system and the central nervous system. Because of the importance of manganese as a battery material, we cite two references dealing specifically with manganese poisoning.[7,8] General battery users are not likely to be exposed to toxic concentrations of manganese species.

$$TLV = TWA: 5 \ mg/m^3 \ (dust)$$

Simple Symptoms: Ingestion of large amounts causes gastrointestinal disturbances. Inhalation of dusts or fumes, the occurrence of which is confined essentially to industrial locations where manganese-containing substances are processed, causes metal fume fever. Symptoms include fever, chills, headache, and chest pain. Chronic exposure may lead to mental disturbances (manganese madness), manifested by motor control problems and aberrant behavior patterns.

7.8.3. Mercury

Chemical Species: Hg, HgO, Zn(Hg)

Batteries: The use of mercury in batteries has decreased significantly in recent years. It is used principally in special purpose zinc/mercuric oxide batteries and in cadmium/mercuric oxide batteries. Some mercury may still be used in low concentrations in batteries with zinc electrodes.

Physical Forms: Mercuric oxide is present in batteries as an insoluble, densely compacted material with a low porosity that may contain some carbon or silver. On discharge, the oxide is converted to liquid metal. If silver is present, the liquid mercury is amalgamated with the silver to a solid. In liquid form, mercury may be dispersed as droplets in the electrolyte absorber and in the zinc hydroxide portion of discharged zinc electrodes.

Toxicity: Mercury and its compounds are highly toxic. They affect most of the vital organs adversely, including the kidneys, the central nervous system, the pulmonary system, and the gastrointestinal tract.

TWA: 0.05 mg/m^3

TLV: 0.05 mg/m^3

MCL: 0.002 mg/liter

Simple Symptoms: The classical "mad hatter" syndrome is a well-known manifestation of central nervous system poisoning by mercury vapor inhalation. The initial symptoms of mercury vapor inhalation are those of metal fume fever: fever, chills, headache, and shortness of breath. Ingestion of mercury and mercury compounds leads to stomach aches, weakness, and general lassitude prior to the later development of more severe manifestations of poisoning.

7.8.4. Silver

Chemical Species: Ag, Ag$_2$O, AgO

Batteries: Silver/zinc cells, mostly button and coin type cells. Metallic silver may be present in cells with mercuric oxide electrodes.

Physical Forms: The metal is present as a relatively cohesive porous solid with interspersed silver oxide. The amount of silver relative to oxide increases with the depth of discharge of the cells. Some silver may be embedded in the separator/absorber structure. The alkaline electrolyte contains a low concentration of soluble silver species.

Toxicity: Silver and silver oxide have no appreciable toxic effects. There are no reasons for thinking that battery users may be exposed to any adverse effects attributable to silver.

TWA: 0.01 mg/m^3 (Ag)

TLV: 0.01 mg/m^3 (Ag$_2$O)

Simple Symptoms: Chronic industrial exposures may lead to argyria: the formation of local, and possibly general, gray-blue skin patches.

7.8.5. Zinc

Chemical Species: Zn, ZnO, Zn(OH)$_2$, ZnCl$_2$

Batteries: Zinc is present in most of the button, coin, and cylindrical cells available in the consumer market and some specialty batteries: zinc/carbon, zinc/chloride, alkaline manganese, silver/zinc, zinc/air, and zinc/mercuric oxide batteries. Zinc chloride is present in heavy-duty zinc/carbon cells.

Physical Form: The metal is present as a reasonably cohesive, porous, solid body of zinc powder filled with zinc hydroxide. The zinc hydroxide is an amorphous solid with a low cohesivity, and it contains an appreciable amount of electrolyte. The more deeply discharged the cells, the more zinc hydroxide and the less metallic zinc they contain. Some zinc hydroxide is dissolved in the alkaline electrolyte.

Toxicity: Zinc is an essential nutritional element in low concentrations. It has a very low toxicity. Ingestion of 12 g of elemental zinc during a two-day period gave no evidence of kidney, liver, or other problems. Zinc chloride forms an acidic electrolyte and may cause chemical burns equivalent to those caused by sulfuric acid (refer to Section 8.9.2 for a discussion of acid burns).

TWA: See Section 7.9.3

TLV: See Section 7.9.3

Simple Symptoms: Inhalation of zinc dusts in industrial environments may produce metal fume fever, manifested by fever, chills, sweating, and general weakness starting 4–8 h after exposure and lasting 1–2 days. Battery users are not likely to be exposed to any form of zinc poisoning.

7.9. AQUEOUS ELECTROLYTE TOXICITY

The electrolytes employed in the majority of batteries available today are either strongly acidic solutions or strongly basic solutions (i.e., alkalies), and most of the reported battery injuries are caused by exposures to either of these two types of electrolytes. Exposures may be due to battery mishandling or abuse or, less frequently, spontaneous battery leakage. The causes of ruptures, explosions, and leakage that may lead to exposures have been discussed in Part II. External injuries are more common than internal injuries due to ingestion and inhalation. Inhalation may occur only if batteries rupture with sufficient violence to form breathable, airborne electrolyte sprays or mists or if individuals expose themselves to the fine acid mists that may form during the charging of vented lead–acid batteries. Ingestion is very rare but may occur as a result of improper cleanliness following the handling of leaky batteries, ruptured batteries, or contaminated battery parts. Ingestion of small cells by infants is a potential source of internal injury.

Since electrolyte exposures may cause serious injuries and since simple procedures are available to lessen the severity of both external and internal exposures, we include brief descriptions of the immediate remedial actions that can be taken to lessen that severity.

7.9.1. Ammoniacal Electrolytes

Chemical Species: The ammoniacal electrolytes are made by dissolving ammonium chloride (NH_4Cl) in water to which some zinc chloride ($ZnCl_2$) may be added. Both salts hydrolyze in water to form acidic solutions. The ammoniacal electrolytes should not be confused with alkaline ammonium hydroxide (NH_4OH) solutions formed by dissolving ammonia (NH_3) in water.

Batteries: All the zinc/carbon batteries (Leclanché batteries) contain ammoniacal electrolytes, except for the heavy-duty zinc/chloride batteries, a high-rate variant of zinc/carbon batteries that contain zinc chloride solutions in water.

Physical Form: The ammoniacal electrolytes are generally present as thick pastes with low fluidity.

Toxicity: The toxicity of ammoniacal solutions derives from two characteristics: the acidic nature of the electrolyte pastes and their nonvanishing vapor pressure of ammonia. The acidic paste generates a low-intensity acid burn on contact with the skin. Inhalation of ammonia causes a reaction between the NH_3 and the water in the mucuous membrane of the respiratory tract and the water in the eyes to form an alkaline solution that exposes the inhaler to potential alkaline tissue damage. The toxicity of the ammoniacal electrolytes is relatively low. Alkaline injuries are discussed in Section 7.9.2 below.

TWA (NH_3): 50 ppm

TLV (NH_3): 25 ppm

Simple Symptoms: Since ammonia is a strong irritant upon inhalation, even low-level exposures tend to evoke evasive action before serious injuries can occur. Ammonia is absorbed very rapidly and in large amounts in water, and if extended exposures occur, then, despite spontaneous, profuse watering of the eyes by tear formation, alkaline eye injuries may occur. The rupture of zinc/carbon cells does not generally lead to sufficiently high ammonia concentrations to be injurious unless evasive action cannot be taken.

7.9.2. Alkaline Electrolytes

Chemical Species: The most commonly used alkaline electrolyte is a concentrated solution of potassium hydroxide (KOH) in water, generally a 30–40% solution. Other, less frequently used alkalies are sodium hydroxide (NaOH) and lithium hydroxide (LiOH), which may be used in combination with potassium hydroxide. They are all strong alkalies and have similar physiological effects.

Batteries: Alkaline electrolytes are used in all the common primary cells, except for the zinc/carbon cells, and in rechargeable batteries with aqueous electrolytes, except for the lead–acid batteries.

Physical Form: The alkaline electrolytes are employed as concentrated aqueous solutions in liquid form, in gelled form, or as thick pastes. Much of the electrolyte is confined within the porous structures of the electrodes and the electrolyte absorbers. In cells with zinc electrodes, the electrolytes are saturated with dissolved zinc oxide.

Toxicity: Alkalies are very destructive of biological tissues of any kind. Chemical attack begins within seconds of exposure and involves both lysis and saponification of the tissues. The exposed tissues become soft and gel-like. The nature of the chemical transformations renders the tissues more permeable to alkali penetration with a consequent rapid penetration to the deeper-lying tissues and their destruction. The rate of penetration depends upon the nature of the cation of the alkalies and increases in the following order:

$$K^+ < Na^+ < Li^+ < NH_4^+$$

with ammonium hydroxide having the highest rate of penetration.

$$TLV: 2 \text{ mg (KOH)}/m^3$$

Eye exposures may occur in various ways—for example, as a result of electrolyte sprays from ruptured cells or by rubbing the eyes with hands or handkerchiefs that have been in contact with an alkaline electrolyte. It is a particularly serious event and may cause severe impairment or permanent loss of eyesight. It is characteristic of alkali exposures that there may be no immediate irritation to indicate that exposure has occurred, unlike exposures to ammonia. Exposure to alkalies may have delayed effects such as opacification and ulceration, and possibly catastrophic corneal damage, even after exposures to dilute alkalies lasting no more than a few minutes. Hence, immediate remedial action is required. It is imperative that exposed eyes be thoroughly and continuously irrigated with copious amounts of clean water for an extended period of time (20–30 min is recommended), as soon as exposure has occurred, and that medical attention be obtained without delay. No attempt should be made to employ any neutralizing agents.

Ingestion of alkali is unlikely to occur, except possibly as as a result of gross mismanagement of batteries. Low rates of ingestion may accompany the swallowing of leaky cells. If any swallowed cells rupture while in the digestive system, a rare event, massive ingestion would occur. Internal injuries caused by alkalies can be very serious, depending on the amounts of alkali involved, particularly since there may not be any immediate symptoms to indicate that exposure has occurred and because of the delayed effects of alkali penetration.

Ingestion may occur as a result of improper cleanliness after handling of ruptured batteries or battery parts that contain alkali. The primary effect of alkalies that enter the oral cavity and the digestive tract is on the esophagus, where they cause the rapid development of necrosis. Secondary effects occur in the stomach and the intestinal tract. The stomach acids cannot neutralize anything but very minute quantities of alkalies and cannot be relied upon to counteract any ingested alkali. Whenever ingestion has occurred, the exposed individual should be given water or milk to dilute the alkali, and immediate medical attention is required. No attempts should be made to neutralize ingested alkalies with acidic media or to induce vomiting.

Inhalation of alkaline fumes or mists may occur as a result of cell ruptures, but such events are rare. If they occur, they may cause serious delayed injuries such as pulmonary edema. Immediate medical attention should be sought if such exposures have occurred.

Skin exposures are much more frequent than any other exposure to alkalies and may result from handling of ruptured batteries, from electrolyte residues on leaky cells, or from normal maintenance of ventable, rechargeable batteries with alkaline electrolytes. Unless removed as soon as exposure has occurred, the alkali will gradually destroy the exposed and underlying tissues.

Again, thorough and extended irrigation with water is required to reduce the severity of any injuries.

Simple Symptoms: Exposures to alkalies are insiduous because they do not generally evoke any immediate sensations with well-defined symptoms. An exception is exposure to ammonia, which signals its presence by a distinctive odor and respiratory irritation. The absence of any immediate symptoms of alkali exposure may permit the alkali to penetrate exposed tissues and cause serious, irreversible injuries as a consequence of failure to institute immediate remedial action. One easily recognized symptom is present if the fingers are exposed to alkalies: the alkalies render the fingers slippery soon after exposure, a sure sign of alkali attack. The slipperiness tends to persist for some time, until the skin has regenerated. In the case of low-intensity exposures of the hands and fingers, the slipperiness may disappear after a few hours or within a day.

7.9.3. Zinc Chloride Electrolytes

Chemical Species: $ZnCl_2$ dissolved in water. Zinc chloride is hydrolyzed by water to form acidic solutions.

Batteries: The electrolyte is used in the heavy-duty variety of zinc/carbon cells, that is, the zinc/chloride cells that are commonly available in the consumer market. Ordinary zinc/carbon cells may contain some zinc chloride in combination with ammonium chloride.

Physical Form: The electrolyte is generally present as a freely mobile liquid more or less fully contained within the porous structures of the electrodes and absorbers.

Toxicity: The physiological effects of zinc chloride electrolytes may be attributed separately to the possible toxic effects of the zinc species and to the chemical burns caused by the electrolyte as an acid. The toxicity of zinc species was discussed in Section 7.8.5. The effect of acidic media will be discussed in Section 8.9.2.

$$\text{TWA: } 1.0 \text{ mg/m}^3 \text{ } (ZnCl_2)$$

$$\text{TLV: } 1.0 \text{ mg/m}^3 \text{ } (ZnCl_2)$$

Simple Symptoms: The principal symptoms of exposure are those of acid burns, discussed in Section 8.9.2 below.

REFERENCES

1. S. M. Ashraf, *Res. Ind.* **33,** 55 (1988); **37,** 158 (1992).
2. Casarett and Doull's Toxicology, *The Basic Science of Poisons,* 3rd ed., Macmillan, New York (1986).
3. M. J. Ellenhorn and D. G. Barceloux, *Medical Toxicology. Diagnosis and Treatment of Human Poisoning,* Elsevier, New York (1988).
4. F. W. Oehme (ed.), *Toxicity of Heavy Metals in the Environment,* Marcel Dekker, New York (1979).
5. N. I. Sax and R. J. Lewis, Jr., *Dangerous Properties of Industrial Materials,* Van Nostrand Reinhold, New York (1989).
6. J. B. Sullivan, Jr. and G. R. Krieger (eds.), *Hazardous Materials Toxicology. Clinical Principles of Environmental Health,* Williams and Wilkins, Baltimore (1992).
7. G. C. Cotzias, *Physiol. Rev.* **38,** 503 (1958).
8. S. Abd El Naby and M. Hassanein, *J. Neurol. Neurosurg. Psychiat.* **28,** 282 (1965).

8

Rechargeable Batteries

A great number of rechargeable battery systems have been developed over the years, but few have attained commercial significance. Lead–acid, nickel/cadmium, nickel/iron, and silver/zinc batteries are exceptions. They are available in many different sizes and play important roles in various applications. The lead–acid and the nickel/cadmium batteries are the most important ones in terms of market share and range of applications. The nickel/iron batteries are limited to specialty applications, generally requiring high-capacity. The silver/zinc battery is of some importance, but it is also limited to special applications. It is unsuitable for general consumer applications in all but small battery sizes because of the high cost of silver. The material cost of silver is relatively unimportant in small cells.

The emergence of portable power tools and consumer electronic devices with high power and energy requirements during the last few decades has increased the need for improved rechargeable batteries. Although primary batteries can satisfy both the power and the energy requirements, their use in these applications is prohibitive on a cost basis. A primary battery may be equivalent to a rechargeable battery delivering 5–10 deep cycles, but for rechargeable batteries with greater cycle lives and the requisite energy per duty cycle and the required low weight, the economics are very much in their favor. It may be noted that rechargeable batteries generally have cycle lives that exceed several hundred or more cycles, but their energy densities are low. Lead–acid and nickel/cadmium batteries are used for most of these devices today, but the need persists for lightweight, rechargeable batteries with a greater energy density per duty cycle. This need, as well as the need for improved batteries for electric vehicle propulsion, has led to a continuing search for and development of improved rechargeable batteries. The systems under consideration include various types of rechargeable aqueous electrolyte systems, nonaqueous lithium systems, and various high-temperature systems. In this chapter, we discuss rechargeable aqueous electrolyte systems that are in use at the present time and some that have reached the stage of development where they may find use in various consumer applications in future years.

The following is a short list of rechargeable, aqueous electrolyte battery systems that are in use at the present time or that have been developed for special applications:

PbO_2/Pb	$NiOOH/Fe$
$NiOOH/Cd$	$NiOOH/H_2$
$NiOOH/MH$	Ag_2O/H_2
$NiOOH/Zn$	Ag_2O/Fe
Ag_2O/Zn	O_2 (Air)/Fe
MnO_2/Zn	Cl_2/Zn
O_2 (Air)/Zn	Br_2/Zn

Detailed technical descriptions of these and other systems are available in the *Handbook of Batteries and Fuel Cells*.[1] The seven systems in the left-hand column will be discussed in this chapter from the point of view of their toxicity hazards. The physical processes in rechargeable batteries and their associated hazards were discussed in some detail in Part II.

8.1. LEAD–ACID BATTERIES, VENTED AND SEALED

Lead–acid batteries are used for traction, starting, lighting, standby power, and a variety of industrial applications as both medium- and large-size batteries. Small lead–acid batteries are used in several portable consumer products. The most commonly available battery is the prismatic automotive starter battery, and, as discussed in Chapter 2, it is the battery involved in the greatest number of reported battery accidents. The prismatic batteries range in size from batteries with capacities of about 5 Ah for small motorcycle starter batteries to those with capacities of about 400 Ah for heavy-duty tractor starter batteries. Even larger batteries are used for submarine propulsion. The automotive starter batteries for passenger vehicles have capacities in the range of 40–120 Ah, with most of them having capacities of about 80 Ah. The batteries are generally used in multicell configurations with nominal voltages of 6 V or 12 V. The small cylindrical cells have capacities of about 3 Ah or more. Because of the low energy density of the lead–acid system, it is not suitable for miniature cells. Miniature cells have an intrinsic low packaging efficiency, and miniature lead–acid batteries cannot compete with miniature nickel/cadmium batteries.

The interior construction of the batteries varies a great deal and depends upon their intended applications. The small cylindrical cells have spirally wound electrodes, and the prismatic batteries have flat–plate or tubular electrode structures. The electrode designs also differ as a function of the expected depth of discharge of the electrodes on cycling. Automotive starter batteries are normally discharged to a small fraction of their available capacity, and they are maintained at essentially full charge during their lifetime. Traction batteries, on the other hand, are discharged very deeply and utilize a large fraction of their available capacities on most of their cycles. They have a much more robust construction in order to survive the repeated deep discharges with minimal mechanical and electrical deterioration. The battery containers are also different for the different types of batteries. The prismatic batteries and their components are fabricated from hard rubber or polypropylene with appropriate polymeric sealants, and the cylindrical cells are fabricated with steel containers. In all of them, the battery cases are electrically isolated from the active electrodes. There are two main classes of lead–acid batteries: vented batteries and sealed batteries. The vented batteries allow pressurizing gases to escape from the cells, and they have removable fill caps for replenishing water lost as a result of charging and spontaneous decomposition. The sealed batteries (maintenance-free batteries) have design features to regenerate any water lost by decomposition. In addition, they have various safety features such as overpressure vents with flame arrester and antiflashback structures. The small cylindrical cells have only overpressure relief vents. Another distinguishing feature is the state of the electrolyte. In many of the smaller cells and the sealed cells, the electrolyte is immobilized by gelling or by reducing its volume to the point where it is fully absorbed by the separators and the porous electrodes. Gas recombination is much facilitated by the starved electrolyte condition. Other cells have a fully mobile electrolyte in flooded cells. The starved cells are generally capacity-limited by the amount of electrolyte in the cells (the electrolyte is an active reaction partner; see the cell reaction below). The constructional details and the characteristics of lead–acid batteries are discussed in Refs. 1 and 2.

The electrochemical cell in the charged state may be represented by

$$PbO_2/H_2SO_4, H_2O/Pb \qquad 2.0 \text{ V}$$

The cells have a high rate capability at normal temperatures and discharge at reasonably well-regulated load voltages of 1.6–1.9 V on continuous loads. The negative electrodes and their current collectors are made of metallic lead, and the positive electrodes are made of lead dioxide with lead current collectors. The electrodes contain various lead salts whose composition depends upon the state of charge of the electrodes. The metallic lead generally contains some alloying constituents. In the small cylindrical cells with spirally wound electrodes, the current collectors are made of high-purity lead with no alloying components. The compositions of the separators and interior cell insulators vary for the different cells. The small sealed cells generally have matted fiberglass separators bonded with organic binders (e.g., acrylic binders) and polypropylene liners to isolate the electrodes from the steel cases. The larger batteries have separators made from phenolic and cellulosic materials, from various polymers [porous rubbers, polyolefins, poly(vinyl chlorides)], or from glass fibers. Glass fiber separators are preferred for heavy-duty batteries such as traction batteries. The electrolyte in all lead–acid batteries is a concentrated solution of sulfuric acid in water. It may contain low concentrations of phosphoric acid and various proprietary additives. Small amounts of various inorganic materials such as finely divided silica may be used to immobilize the electrolyte in cells with gelled electrolytes ("gelcells").

The overall discharge/charge reaction may be represented by

$$PbO_2 + Pb + 2H_2SO_4 \underset{charge}{\overset{discharge}{\rightleftarrows}} 2PbSO_4 + 2H_2O$$

Several intermediate species are involved in the discharge and charge reactions, but they are not important for our present purposes. Sulfuric acid is consumed and water is generated during discharge, which leads to a significant change in the acid concentration, and hence, a change in its density that can be used to monitor the state of discharge (or charge) of a battery. Some batteries have built-in monitoring units that display the condition of the electrolyte. This is a considerable advantage from a safety point of view since it obviates the need to remove electrolyte samples from a battery to check its condition by external means. The sulfuric acid concentration is about 40% by weight in a charged battery and about 15% in a fully discharged battery. The amounts of active materials in a battery cannot be estimated with much confidence from its rated capacity since the rated capacity may be considerably less than the capacity corresponding to the equivalent amounts of materials in the battery. The following is an approximate rule of thumb for automotive starter batteries. For a charged battery, the total battery weight is distributed among its principal components as follows:

Active electrode materials	1/3 of total weight
Electrolyte	1/4 of total weight
Current collectors	1/5 of total weight
Separators and case	1/6 of total weight

The composition of the lead used in the batteries is of concern from a safety point of view. It differs among batteries designed for different applications. In most of the larger batteries, various alloying constituents are added to improve the formability of the lead and to improve its chemical properties. The most commonly used additives are antimony (with possible arsenic impurities) or calcium with small amounts of tin. There has been a trend away from antimony during the last few decades, and a trend toward the use of the more benign calcium. Maintenance-free batteries are made with lead–calcium alloys with no or only small amounts of antimony. Other metals may also be present in small amounts. The cylindrical cells with spirally wound electrodes employ high-purity lead in their electrodes and current collectors.

The principal chemical instability of lead–acid batteries is associated with the decomposition of water. Water decomposes electrolytically at voltages of about 1.23 V and above, and it is exposed to about 2 V in lead–acid batteries. At this high voltage, water decomposition would be expected to proceed at a rapid rate. However, water decomposition is kinetically inhibited on lead and lead dioxide electrodes, hence the apparent stability of lead–acid batteries. On charging and overcharging, the inhibitory effect is countered by the high charging voltages, and significant amounts of hydrogen and oxygen are formed (see Section 3.1.2). In vented cells, the gases escape via the apertures in the fill caps, and in sealed cells the gases are recombined to regenerate water. A relatively innocuous side reaction is the slow decomposition of organic separator materials due to the strongly oxidizing effect of lead dioxide. It generates carbon dioxide and is of little concern except for a slow pressure buildup that may be controlled by overpressure vents in sealed cells and by simple venting in vented cells.

A side reaction of greater concern is the formation of stibine (SbH_3) in lead–acid batteries made with lead alloys that contain antimony. It is a toxic, odorless gas, and it may contain arsine (AsH_3) owing to the arsenic impurity generally present in antimony. Arsine is also odorless. Neither gas announces its presence until some time after exposure. The gases are formed by the reaction of antimony and arsenic and their compounds with the atomic hydrogen that is generated as an intermediate in hydrogen evolution. These gases are not present in lead–acid batteries made with lead that does not contain antimony.

Experience shows (Chapter 2) that the principal physical hazards of lead–acid batteries are the hazards caused by the generation of hydrogen, its accumulation inside or outside the cells, and its ignition followed by explosion. Exposure to sulfuric acid is a serious health hazard, whether a result of an explosion or due to careless handling or faulty operating procedures. The acid causes severe chemical burns and tissue damage. It also attacks and decomposes clothing. Another health hazard comes from the lead itself and its compounds; they are all toxic materials. The risks of massive lead exposures are relatively low for the general battery user, except that manual contact with the lead terminals or other lead-containing parts of the batteries may lead to contamination and subsequent ingestion, unless proper cleanliness is practiced. The toxic health hazards of lead–acid batteries are associated with the principal battery constituents: lead and its compounds, sulfuric acid, and arsine and stibine.

8.2. NICKEL/CADMIUM BATTERIES, VENTED AND SEALED

Together with lead–acid batteries, nickel/cadmium batteries dominate the field of rechargeable battery applications today. Nickel/cadmium batteries are used in a wide variety of both stationary and portable applications where a long cycle life, high power, and a reliable performance are needed. They are the preferred power source for portable consumer products that require rechargeable batteries, such as power tools and communication equipment. Since their energy density is greater than that of lead–acid batteries, they lend themselves better to the construction of small rechargeable batteries, and they are available with capacities from 0.02 Ah to about 10 Ah for many consumer applications in both button and cylindrical shapes. For industrial and other applications, they are generally used as prismatic multicell batteries with capacities from about 5 Ah to more than 1000 Ah and with voltages as high as about 250 V.

All the button and cylindrical cells are housed in crimp-sealed containers, made of nickel-plated steel or, in special cases, of stainless steel. The button cells have no overpressure relief mechanism other than that afforded by the crimp seals, but the larger cylindrical cells may incorporate a pressure relief valve. The polarity of the cells is marked on the case itself or on their labels. Button cells have positive cases and negative top covers, whereas cylindrical cells have negative cases and positive top covers.

The prismatic batteries are housed in cases made of nylon, polypropylene, or nickel-plated steel; stainless steel may be used for special applications. All the larger batteries employ steel cases for structural stability. The plastic cases are generally reinforced with various rib structures to make them more rigid. In all the prismatic batteries, the cases are at a floating potential, and isolated terminals are located at the top of the batteries. Although nickel/cadmium batteries are thermodynamically quite stable, charging and overcharging generate hydrogen and oxygen, just as in the case of lead–acid batteries (as discussed in Sections 1.3.3 and 3.1.2). The adverse consequences of excessive cell pressures are avoided by having either ventable cells or sealed cells with provisions for gas recombination. The small sealed cells for consumer applications rely on gas recombination at the electrodes to prevent excessive pressures by maintaining the cells in a starved electrolyte condition, or they rely on the escape of the gases through the crimp seals. The majority of the prismatic batteries employ vented designs where excess pressures are relieved by the escape of the pressurizing gases through the fill caps. As a consequence, water must be added to the batteries periodically to prevent the cells from drying out. Detailed descriptions of nickel/cadmium batteries, their construction, and their characteristics are available in Refs. 1 and 3.

In the charged state, the Ni/Cd cell may be represented by

$$Fe(Ni)/NiOOH/KOH, H_2O/Cd/Fe(Ni) \qquad 1.2 \ V$$

Although the nominal voltage is 1.2 V, the actual load voltages and open-circuit voltages depend upon the composition of the electrolyte, which changes during charge and discharge, upon the temperature, and upon the loads. Under most conditions, the batteries operate with relatively flat load voltages, and they have a good voltage regulation, attributable to their high rate capability. In the charged state, the positive electrodes are composed of mostly NiOOH and the negative electrodes of cadmium metal. The manufacturing of Ni/Cd batteries is quite involved,[1] and they are commonly prepared in the fully discharged state and sold in that state after conditioning at the end of the manufacturing process. The long-term stability of the nickel/cadmium system is a consequence of its low voltage. Its open-circuit voltage is close to or slightly less than the decomposition voltage of water. An additional factor that contributes to its stability is the practical insolubility of the active materials in the electrolyte. The positive electrode in the fully charged state is a strong oxidizing agent, however, and may contribute to a slow deterioration of the cells. The electrolyte in charged cells is an aqueous solution of potassium hydroxide (KOH) with a concentration of about 30% KOH by weight.

The electrode structures differ for various applications. The traditional designs employ pocket type electrodes in which the active materials are contained in flat metal pockets. This type of design is still in use for stationary applications and gives a highly reliable performance. The need for more rugged designs and higher rate capabilities led to the development of sintered, flat-plate electrodes where the active materials are contained within the pore structure of thin, sintered nickel plaques. This type of design is widely used in mobile applications. The sintered electrode structures are also employed in the cylindrical cells in spirally wound configurations. In the button cells, the active materials in powder form are compacted into the case and cover, respectively. Polymer-bonded and fiber electrodes represent relatively new technologies that are used in some Ni/Cd batteries. The current collectors in Ni/Cd batteries are generally made of nickel-plated steel or nickel. The separators are made of cellulosic materials, felted or woven nylon, or polypropylene, depending on the applications of the batteries.

The overall discharge/charge reaction is:

$$2NiOOH + Cd + 2H_2O \underset{\text{charge}}{\overset{\text{discharge}}{\rightleftarrows}} 2Ni(OH)_2 + Cd(OH)_2$$

A lower bound on the approximate amounts of active materials in a balanced cell may be estimated from its rated capacity using the equivalent weights of the materials:

- 3.4 g/Ah for nickel oxyhydroxide (NiOOH)
- 2.1 g/Ah for cadmium metal (Cd)
- 0.7 ml/Ah for 30% electrolyte (KOH in H_2O)

The indicated amount of electrolyte is the minimal amount required to satisfy the reaction stoichiometry. More is used in flooded cells. The electrolyte may contain some lithium hydroxide (LiOH). Since water is consumed during discharge, the cells tend to dry out, and the likelihood of electrolyte leaks or spills and user exposure to electrolyte is less with discharged cells than with charged cells. Chemical species other than those shown in the cell reaction above are also present in Ni/Cd batteries in low concentrations. Carbon may be added to pocket type positive electrodes to increase their conductivity, and barium and cobalt salts may be added as conditioners. The negative electrodes contain unspecified metal salts in low concentrations to modify the crystal habits of the electroformed materials. The additives are of minor concern from a safety point of view, given their low concentrations and the predominant role of the active materials themselves as a cause for concern.

In contrast to most rechargeable batteries, the Ni/Cd batteries exhibit few intrinsic instabilities. A potentially troublesome side reaction for long-term applications is the slow chemical deterioration of the separator materials, especially cellulosic and nylon separators. Its impact on battery safety is minor, unless short circuits are generated, in which case thermal runaways may be triggered. Another possible side reaction is the enhanced rate of self-discharge due to any residual nitrate that may be present as an impurity from the electroforming process used in cell manufacturing. This has no effect on the safety of the batteries. The more serious side reactions that do impact battery safety are the gas-generating reactions that occur on charge, overcharge, and cell reversal, i.e., the generation of hydrogen and oxygen.

The hazards of Ni/Cd batteries are of a physical and a chemical nature. The physical hazards are due primarily to gas generation and thermal runaways and were discussed in Part II. The most important toxicity hazards are associated with the aggressive alkaline electrolyte and with the cadmium. The general user is more likely to suffer injuries from exposures to the electrolyte than to be poisoned by cadmium ingestion. The toxicity of cadmium and its compounds is of concern primarily from an environmental point of view whenever Ni/Cd batteries are discarded in a casual manner rather than being properly recycled. Manufacturing workers are also exposed to the possibility of cadmium poisoning, but we do not discuss the health hazards associated with the manufacturing of batteries.

8.3. NICKEL/METAL HYDRIDE BATTERIES

Nickel/metal hydride (Ni/MH) batteries have been developed as an environmentally more benign alternative to nickel/cadmium batteries. They also have the advantage of a greater energy density, 25–75% greater than that of Ni/Cd batteries, depending on their designs. The Ni/MH batteries are a recent addition to the battery family and are expected to increase in importance during coming years, especially for portable consumer applications and some special applications. In addition to a high energy density, they have long cycle lives, typically 500–1000 cycles at deep discharges. The smaller cells for consumer applications are available in most of the standard cylindrical sizes with spirally wound electrodes and nickel-plated steel cases. The cells are crimp-sealed and provided with resealable overpressure vents. The cells have the conventional polarity: case negative and top positive. The larger batteries are housed in prismatic steel cases with insulated nickel terminals crimp-sealed into the top covers with polymer grommets. The case is at a floating potential. The prismatic batteries have capacities as high as 500 Ah for special applications and are available in various series-connected high-voltage configurations. The prismatic batteries are also sealed and provided with overpressure vents as a safety feature.

The charged electrochemical system may be represented by the cell

$$\text{NiOOH/KOH, } H_2O\text{/MH} \qquad 1.2 \text{ V}$$

The nominal voltage, 1.2 V, is the average voltage of cells discharged at their rated currents at room temperature. Discharges begin at higher voltages; the voltages then level off and decrease more slowly before falling steeply as the cells approach the end of their useful capacity. The voltage characteristics depend upon the chemical composition of the metal hydride and upon its hydrogen content.

Except for the negative electrode, the interior construction of the cells is similar to that of Ni/Cd cells. The positive electrodes are essentially the same as in Ni/Cd batteries, and the electrolyte is also the same, about 30% KOH, and may contain some lithium hydroxide (LiOH). The separators are fabricated from various felted or woven polymer materials, as in Ni/Cd batteries, and are likely to contain small amounts of wetting agents to facilitate electrolyte absorption, especially if polypropylene separators are used. The negative electrodes are fabricated from sintered, porous bodies of various metals, and, to increase the access of hydrogen to the metals, they are rendered hydrophobic by wetproofing agents such as halocarbon binders and surface coatings.

The overall chemical reactions during discharge are:

Positive electrode:

$$NiOOH + H_2O + e^- \rightarrow Ni(OH)_2 + OH^-$$

Negative electrode:

$$MH + OH^- \rightarrow M + H_2O + e^-$$

Water is consumed at the positive electrode and regenerated at the negative electrode, but there is no net consumption of water during discharge as in the case of Ni/Cd and most other batteries. The reverse reactions occur on charge. Various intermediate species are formed during the reactions, but they are not important for our present purposes. Except for the positive electrode, no general rules can be given for the amounts of active materials in the cells. The designs of the cells and the compositions of the negative electrodes vary considerably among cells made by different manufacturers. The amount of nickel in the positive electrode (an excess over the stoichiometric requirement) may be estimated to be approximately:

- 3 g/Ah of nickel in active material in pressed electrodes
- 5 g/Ah of nickel in active material in sintered electrodes

In addition, nickel and steel are present as collectors and structural elements. The additional amounts vary for different electrode designs but are of the same order of magnitude as the amounts given above for the active materials. The negative electrodes are generally fabricated from either of two basically different metal compositions:

- AB$_5$ structures: R–Ni–Co–Al–Mn (R = La or misch metal, a rare earth composition)
- AB$_2$ structures: Ti–Zr–V–Ni

The various metal components serve important metallurgical and chemical functions such as microstructural phase control, corrosion inhibition, hydrogen capacity enhancement, and hydrogen pressure limitation. A virtue of Ni/MH cells compared with Ni/H_2 cells is that a small volume of a metal hydride is capable of confining a very large volume of hydrogen at relatively low pressures.

The Ni/MH system is moderately stable at normal operating temperatures. The active material in the charged positive electrode decomposes slowly in alkaline media with the generation of oxygen, as in the case of Ni/Cd batteries, and the negative electrode undergoes slow corrosion reactions. The titanium-based electrodes are more stable than the rare earth-based electrodes. The latter metals are pyrophoric in air, but the titanium-based electrodes are not. An index of the stability of the various Ni/MH systems is the self-discharge rate of the fully charged cell. In the more stable cells, it amounts to about one percent per day at room temperature. The cells are necessarily pressurized by virtue of the metal/hydrogen equilibrium, and their pressures increase on charge and decrease on discharge.

The physical hazards of Ni/MH batteries are associated mainly with the pressurized gases in the cells: hydrogen in equilibrium with metal hydride and oxygen and hydrogen generated on charge, overcharge, and reversal. Since the batteries are likely to be deeply discharged in normal usage, reversal may occur in cells of high-voltage Ni/MH batteries. An interesting feature of the Ni/MH system is that the negative electrode is a scavenger of hydrogen, and it provides a built-in means of controlling the hydrogen pressure of the cells, unless an excess of hydrogen is present relative to the hydrogen capacity of the negative electrode. If the rate of hydrogen generation on charge exceeds the rate of hydrogen absorption by the metal hydride, excessive pressures may be generated temporarily in the cells. Since hydrogen absorption is an exothermic reaction, the pressures may be further increased due to self-heating of the cells on charge. The generation of oxygen on charge and overcharge does not pose a serious problem in well-made cells since the oxygen is likely to react with the metal hydride to regenerate water. The physical hazards due to hydrogen are the same as for lead–acid batteries: the formation of explosive gas mixtures due to overpressure venting of hydrogen into poorly ventilated battery enclosures and their subsequent ignition with a following explosion. If for any reason a charged cell should rupture, the metal hydrides may ignite and create a fire hazard. Both the hydrogen and the metals are flammable. The toxicity hazards are associated primarily with the electrolyte. Exposures to the potassium hydroxide electrolyte may cause severe chemical burns. The likelihood of such exposures is quite low with Ni/MH batteries unless they rupture, are abused in some manner, or have defective seals. The metals in the negative electrode are relatively innocuous.

8.4. NICKEL/ZINC BATTERIES

Nickel/zinc batteries represent an attempt to make a rechargeable battery with the excellent qualities of the nickel electrode and the high voltage of the zinc electrode in order to obtain a high cycle life and a high energy density, the poor cycle life of zinc electrodes notwithstanding. Considerable efforts have gone into the development of this system over the years, but it has achieved no significant commercial success so far. Work is still in progress to improve the system, especially for electric vehicle propulsion. The practical energy density of prototype batteries is almost twice that of Ni/Cd batteries, a very attractive property, but their cycle life is short compared to that of most other rechargeable batteries. If their cycle life and reliability can be improved, they may become important in both consumer and specialty applications. Since the system holds some promise, we include a brief description of its properties based on prototype batteries.

The construction of the batteries is similar to that of Ni/Cd batteries and needs little elaboration. Both vented and sealed designs exist, with provisions for venting gases via fill caps or for recombining the gases formed on charge and overcharge at the electrodes. An important safety feature is the incorporation of small catalytic converters connected to the positive electrodes to scavenge hydrogen generated by zinc corrosion. In order to prevent or at least decrease the amount of hydrogen formed at the negative electrode on charge and overcharge, the cells may be designed with an excess negative electrode capacity relative to the positive electrode. Any oxygen generated on the positive electrode on charge and overcharge may be scavenged in starved electrolyte cells by the excess amount of zinc deliberately incorporated in the negative electrode.

The batteries employ flat-plate electrodes in prismatic steel cases. The cases are electrically floating, isolated from the electrode terminals, which are located on the top of the batteries. The electrodes are similar in construction to the ones used in Ni/Cd batteries and may be of the pocket, sintered, or plastic molded type with nickel or nickel-plated steel current collectors.

In the charged state, the electrochemical cell is

$$NiOOH/KOH, H_2O/Zn \qquad 1.73 \text{ V}$$

The working voltage on most continuous loads is about 1.5–1.6 V and remains relatively flat during discharge. The batteries have a high rate capability and deliver well-regulated load voltages within their rated design ranges. The electrolyte is a solution of potassium hydroxide (KOH) in water with a concentration of about 30% KOH by weight. The electrolyte may also contain lithium hydroxide (LiOH) in low concentrations.

The discharge/charge reactions are identical to those of nickel and zinc electrodes in other alkaline cells:

$$2NiOOH + Zn + 2H_2O \underset{\text{charge}}{\overset{\text{discharge}}{\rightleftharpoons}} 2Ni(OH)_2 + Zn(OH)_2$$

Water is consumed during discharge, and starved electrolyte cells tend to dry out during discharge. Because of the high cell voltage, water may be expected to decompose spontaneously in nickel/zinc cells. In contrast to lead–acid batteries, there is no intrinsic inhibition of gas evolution in nickel/zinc cells, and unspecified gas evolution inhibitors are generally added to the electrodes and/or to the electrolyte.

The solubility of zinc hydroxide in the electrolyte creates a unique stability problem not present in primary cells. The dissolution of zinc hydroxide leads to a redistribution of the zinc on recharge that causes structural and geometric changes of the negative electrodes that affect their performance adversely. Furthermore, the solubility of the zinc hydroxide predisposes the system to the formation of zinc dendrites with an increased likelihood of the formation of short circuits between the electrodes and a shortened cycle life. Various multicomponent separator structures have been used with some success to impede dendrite formation. The most effective separators comprise multilayered membrane/absorber structures of materials such as cellulosics, nylons, poly(vinyl alcohols), and others.

The physical hazards of nickel/zinc batteries are the same as those of other rechargeable aqueous electrolyte systems, namely, the hazards caused by any hydrogen and oxygen that may be formed in the cells on charge and overcharge or as a result of zinc corrosion. These processes were discussed in Part 2. The chemical toxicity hazards are due primarily to the alkaline electrolyte. Exposures to potassium hydroxide solutions, whether due to careless handling or leaky or ruptured cells, may cause serious injuries. The other active materials in the cells—nickel, zinc, and their chemical compounds— have very low hazard ratings. The structural materials do not have any significant toxicity ratings. The toxicity of the unspecified, proprietary corrosion inhibitors is not known. They are likely to be present in low concentrations, however, which would reduce the hazard levels, unless they are ingested, in which case the likely concomitant ingestion of electrolyte would be a more serious concern.

8.5. SILVER/ZINC BATTERIES

Rechargeable silver/zinc batteries are well established and play important roles in many applications where high power and energy density are at a premium. Their energy density is almost three times that of nickel/cadmium batteries. The batteries are substantially modified versions of primary silver/ zinc batteries adapted to function in the rechargeable mode. Since the chemical compositions of the two types of silver/zinc systems are essentially the same, their chemical health hazards are the same, but not their physical hazards. The rechargeable batteries have a significantly greater physical hazard potential. Because of the high cost of silver, the batteries are used to a limited extent in portable consumer products, except for some of the smaller cells in which the cost of silver is a relatively small fraction of the total cell cost. The principal uses are in military applications. The batteries range in size from about 0.2 Ah for the small button cells to about 5000 Ah for the large prismatic, multicell batteries for submersible applications. The small cells are available in button shapes and in cylindrical shapes with spirally wound electrodes and are housed in crimp-sealed steel cases. They rely on the crimp seal for the relief of overpressures. Most of the small- and medium-sized prismatic batteries are housed in cases made of various plastic materials such as nylon, polysulfone, poly(propylene oxide), or other polymers, all of a vented design and furnished with leakproof fill caps. Most of the larger batteries are housed in steel cases or substantially reinforced plastic cases. All the prismatic batteries employ flat-plate electrodes.

.The electrochemical system in the charged state may be represented by the cell

$$AgO/KOH, H_2O/Zn \qquad 1.85 \text{ V}$$

The cells have a high rate capability, but their voltage characteristics differ from those of most of the other rechargeable battery systems. The discharge curves exhibit a short initial high-voltage plateau at about 1.7 V, depending on the load, and a main discharge plateau at about 1.5 V. Apart from the initial high-voltage segment, the cells have a good voltage regulation and discharge at fairly constant load voltages during the major portion of their discharge. The interior construction is different from that of the primary silver/ zinc cells and resembles more closely the construction of Ni/Cd cells. The positive electrodes consist of an oxide-filled, sintered silver grid, and the negative electrodes are made of highly porous zinc bodies with a considerable excess of zinc. The electrolyte is a concentrated solution of potassium hydroxide containing about 40% KOH by weight.

The discharge of the cells occurs in two distinct steps according to the overall reactions:

$$2AgO + Zn + H_2O \rightarrow Ag_2O + Zn(OH)_2 \qquad 1.85 \text{ V}$$

$$Ag_2O + Zn + H_2O \rightarrow 2Ag + Zn(OH)_2 \qquad 1.55 \text{ V}$$

The reverse reactions occur on charge. Generally, the conversion of monovalent silver (Ag_2O) to divalent silver (AgO) on charge is not complete, hence the relatively short initial high-voltage plateau on discharge. This high-voltage plateau becomes shorter, the longer the charged cells are stored prior to discharge. The high rate capability of the cells may be attributed, in part, to the formation of metallic silver in the positive electrodes during discharge. The silver increases the electronic conductivity of the positive electrodes. The reactions show that water is consumed during discharge. This means that the cells tend to dry out during discharge, especially cells that are electrolyte-starved to begin with, and the likelihood of electrolyte spillage is much less with discharged cells than with charged cells.

The most important chemical species present in the cells are silver and zinc, their oxides and hydroxides, and potassium hydroxide. The positive electrodes may contain organic binders in low concentrations, and the negative electrodes may contain some lead and cadmium, and possibly small amounts of proprietary additives, all of which serve as corrosion inhibitors. The electrolyte is likely to contain unspecified zinc corrosion inhibitors in low concentrations. The amounts of active materials in a cell are considerably greater than the amounts that correspond to the rated cell capacity. The positive electrodes may contain as much as twice the amount of silver and the negative electrodes as much as three times the amount of zinc required to deliver the rated cell capacity. Some typical values are:

- 4.0 g/Ah of silver (Ag)
- 3.5 g/Ah of zinc (Zn)

The cells are generally designed to operate in a starved electrolyte condition to facilitate gas recombination at the electrodes. Depending on the porosity of the electrodes and the amount of separator/absorber used in the cells, the minimal amount of 40% electrolyte is likely to be of the order of 0.8 ml/Ah, and more is generally used. The separators are made of cellulosics, organic polymers, and possibly fibrous inorganic materials.

The silver/zinc batteries are the least stable of the commonly used rechargeable aqueous electrolyte batteries, and this affects their hazard potential. Chemical instability is the price paid for the high energy density. Another consequence of their high chemical reactivity is a relatively short cycle life. This is a general pattern among rechargeable batteries, including rechargeable lithium batteries: the greater the energy density, the less stable the batteries and the shorter the cycle life. Because of their chemical instability, the prismatic silver/zinc batteries are generally sold in the dry, electrolyte-free condition, either fully charged or uncharged, and the electrolyte is supplied in a separate kit. This necessitates the filling of the batteries prior to use, a procedure that may expose users to electrolyte spills. All the crimp-sealed button and cylindrical cells are sold in the sealed, fully assembled condition.

Chemical instabilities are associated with both the positive and the negative electrodes. The strongly oxidizing silver oxide reacts spontaneously with water to generate oxygen, and, because it is somewhat soluble in the electrolyte, it may also react with the separators, causing their oxidative degradation and the formation of metallic silver in the separator/absorber structures. The negative electrodes corrode in alkaline solutions with the evolution of hydrogen. Since the presence of oxygen and hydrogen in silver/zinc cells contributes significantly to their hazard potential, the cells are designed to recombine the gases to the fullest extent possible on the electrodes or to minimize the generation of hydrogen on the negative electrode during charge and overcharge by providing an excess negative capacity. Another potentially serious problem with rechargeable silver/zinc cells is caused by the solubility of zinc hydroxide in the electrolyte. It leads to shape changes of the negative electrode due to the relocation of the active material on cycling, and, more importantly, it increases the likelihood of zinc dendrite formation and short circuits. Because of the high rate capability of the cells, internal short circuits may lead to thermal runaways, unless the shorting dendrites are burned out by high local currents. The possibility of improving the long-term performance and of lowering the hazard potential due to short circuits is contingent on the possibility of blocking the transfer of soluble silver and zinc species between the electrodes. Progress has been made in this direction by the development of complex, multilayer separator structures that comprise oxidation-resistant, semipermeable membranes, cellulosic films and absorbers, and ion-exchange membranes. Inorganic separators made from materials such as asbestos and potassium titanate have also been used with some success. However, note that asbestos is a hazardous material.

A compromise, with a reduction in energy density, but a gain in stability, is obtained by replacing the zinc electrodes with cadmium electrodes. The resulting silver/cadmium cells have about two-thirds the energy density of the silver/zinc cells, but longer cycle lives. The use of cadmium increases the toxicity rating of the cells. The silver/cadmium cells are used in special applications but are not generally available in the consumer market.

The physical hazards of rechargeable silver/zinc batteries are related to the formation of hydrogen by the corrosion of zinc and to the generation of hydrogen and oxygen on charge, overcharge, and reversal. These are generic problems characteristic of all rechargeable, aqueous electrolyte batteries, and we refer to our discussion of these problems and the associated hazards in Part II. Because of the high voltage of the silver/zinc system and the associated chemical instability of the active materials, these problems are more severe in this system than in other rechargeable, aqueous electrolyte systems. However, most of the rechargeable silver/zinc batteries are used in special applications and operated by trained service personnel, which minimizes the likelihood of accidents. Few accidents that involve silver/zinc batteries have been reported. The chemical hazards are due primarily to the alkaline electrolyte. Exposure to concentrated potassium hydroxide solutions causes serious chemical burns with accompanying tissue damage. The principal concern in exposures is that of physical contact. The likelihood of ingestion of any of the toxic materials is very low as long as proper handling and cleanliness are practiced. The materials of construction of the cases, seals, and terminals are quite inert and do not pose any significant chemical hazards to battery users. In the case of silver/cadmium cells, exposure to cadmium represents a serious health hazard.

8.6. MANGANESE/ZINC BATTERIES

We include a brief discussion of rechargeable alkaline manganese batteries. They are under development at the present time and hold some promise of becoming useful, low-cost, low-cycle-life rechargeable batteries for consumer applications. Basically, they are modified versions of the conventional primary alkaline manganese batteries. The active material of the positive electrodes is manganese dioxide or a complex oxide of manganese with other metals, for example, a mixed oxide of manganese and bismuth. The electrolyte is a concentrated solution of potassium hydroxide with added corrosion inhibitors in low concentrations. The negative electrode is made of zinc with the possible addition of small amounts of other metals as corrosion inhibitors, for example, lead. In some versions of the cells, the ratios of the electroactive materials have been changed from those of primary cells to prevent the formation of low-valency positive electrode species that may render the system nonrechargeable.

The prototype batteries for consumer applications are similar to the conventional cylindrical cells. They conform to the international standard cell sizes and have the same polarity. The interior configurations are either of the bobbin or spirally wound types, depending on the rate requirements of their intended applications. During their early life cycle, they deliver capacities that are about three-fourths of the capacities of comparable primary alkaline manganese cells. Prismatic, multicell batteries with flat-plate electrodes are under consideration for electric vehicle applications.

The active electrochemistry of the system is essentially the same as that of the primary alkaline manganese batteries, to which we refer, except for the presence of the mixed oxides in the positive electrodes. The batteries have voltage characteristics that are comparable with those of the primary versions. The recharge reactions are the reverse of the discharge reactions. Depending on the electrode structures employed, the batteries appear to be suitable for both high-rate and low-rate applications at ambient temperatures with well-regulated load voltages.

An intrinsic problem with the rechargeable manganese/zinc batteries is associated with the solubility of zinc oxide in the alkaline electrolyte and is the same as the problem that limits the usefulness of nickel/zinc batteries: the formation of zinc dendrites. An additional problem is caused by the soluble zinc species in the rechargeable manganese/zinc batteries: the soluble zinc species become incorporated in the positive electrode, where they may gradually destroy the rechargeability of the positive electrode on extended cycling. An essential requirement for the operational viability of the batteries is that the migration of soluble zinc species to the positive electrode be prevented. The development of complex multilayer separator structures that block the transfer of zincate ions holds some promise of progress in this direction without imposing undue impedance penalties on the batteries.

The chemical toxicity hazards of the rechargeable manganese/zinc batteries are the same as those discussed for the primary alkaline manganese batteries. The physical hazards are the generic hazards associated with all rechargeable, aqueous electrolyte battery systems with zinc electrodes, namely, the hazards caused by hydrogen formation on charge, overcharge, reversal, and hydrogen generation due to zinc corrosion. The reader is referred to Part II for a discussion of these processes and their related hazards.

8.7. ZINC/AIR BATTERIES

Rechargeable metal/air batteries constitute a class of batteries that has received considerable attention because of their high energy densities, which are comparable to those of rechargeable lithium batteries. The metals of principal interest have been and are still zinc, iron, and aluminum. Other metals have been considered, but most of them have more serious problems than the three preferred metals. The zinc and iron systems are electrically rechargeable, and the aluminum system is mechanically rechargeable. It may be noted that aluminum can be recharged in nonaqueous media, but this has not yet been implemented in any practical battery designs. The potential applications of metal/air systems range from portable consumer electronics to electric vehicles and high-capacity stationary power systems. The zinc/air system is of greatest interest for consumer applications, and we include a brief description of its properties as representative of the metal/air class of rechargeable batteries. The system is still in development and our description is based on the generic characteristics of prototype batteries. If the problems of short cycle lives and chemical instability can be overcome, the rechargeable zinc/air batteries may become important in both the consumer market and in special applications.

In the charged state, the system can be represented by the electrochemical cell

$$O_2(air)/catalyst/KOH, H_2O/Zn \qquad 1.6 \text{ V}$$

As in the case of the primary zinc/air cells, the high energy density of the system, about 50 Wh/lb for large prototypes, derives from the use of oxygen from an external source as the active material for the positive electrode. The operating voltages of the cells are close to 1.2 V and remain relatively flat during discharge. The rate capability depends strongly on the cell design. For high-rate applications, flat-pack, prismatic designs are employed with thin, large-surface-area electrodes. Thicker electrodes are used for low- to moderate-rate cell designs. It is difficult to make practical cylindrical cells with a high rate capability because of the need to prevent air access to the negative electrodes. This difficulty may limit the use of zinc/air batteries in consumer applications that require high-rate cylindrical cells, but not necessarily in high-capacity applications such as electric vehicles. In order to construct batteries with practically useful service lives, they must be provided with a means to control air access to the cells: free access during charge and discharge, and no access during idle stand.

The negative electrodes are porous zinc bodies similar to those used in rechargeable silver/zinc cells (Section 8.5), and the positive electrodes are highly porous, bifunctional, multilayer electrodes similar to the ones used in fuel cells, that is, large-surface-area catalytic current collectors partitioned into separate wettable and nonwettable segments for easy access of both air and electrolyte to the porous electrode structure. Bifunctionality is needed to achieve efficient discharge and charge. Some high-capacity batteries employ separate, third electrodes for the charging process. The electrolyte is a concentrated solution of potassium hydroxide, about 30% KOH by weight. Complicated separator structures are used to minimize the formation of zinc dendrites and short circuits (see Section 8.5 above). As in the case of rechargeable silver/zinc cells, they comprise regenerated cellulosic films, cellulosic absorbers, and various kinds of semipermeable membranes. The zinc electrode is the basic cycle life limiter in zinc/air cells.

The overall discharge reactions at the two electrodes are:

Negative electrode:

$$Zn + 2OH^- \rightarrow Zn(OH)_2 + 2e^-$$

Positive electrode:

$$\tfrac{1}{2}O_2 + H_2O + 2e^- \rightarrow 2OH^-$$

The reverse reactions occur on charge. Water is consumed at the positive electrode during discharge and regenerated during charge. The negative electrode contains a considerable excess of zinc relative to the rated cell capacity, of the order of 3 g/Ah of zinc. It may be noted that the positive electrode is not capacity-limited as long as air is available to the electrode. The cells may be discharged, therefore, until either all the zinc is consumed or until the water in the electrolyte is consumed. The cells require an amount of electrolyte of the order of 1 ml/Ah or more and are generally operated in the starved electrolyte condition. Deep discharges have an adverse effect on the subsequent cycling performance of the cells. The negative electrodes require the presence of corrosion inhibitors in either or both the electrode and the electrolyte to reduce the rate of spontaneous hydrogen evolution and zinc corrosion. The inhibitors may include lead, cadmium, and other unspecified metals in the zinc electrode and various proprietary inhibitors in the electrolyte. The air electrodes comprise fluorocarbon-bonded, porous carbon structures that contain wetting and wetproofing agents, manganese oxides, and possibly small amounts of heavy metal catalysts.

The stability of the system is affected by several reactions. As in all cells containing alkaline electrolytes, the zinc electrodes in zinc/air cells corrode and generate hydrogen whether the cells are in use or not. An additional factor that contributes to zinc corrosion is the oxygen that may reach the negative electrode from the air access ports. The severity of zinc corrosion due to oxygen can be reduced, but not eliminated, by incorporating structural elements in the cells that block direct access of air to the negative electrodes. Since air also contains carbon dioxide, contact between carbon dioxide and electrolyte is unavoidable during cell operation. This leads to the precipitation of insoluble potassium carbonate inside the porous positive electrode and reduces its rate capability, but it does not affect the safety of the cells except for the secondary effect of the lower rate capability lowering the charging rate tolerance of the positive electrodes. A more important consideration is the transfer of water to or from the cells, depending on the humidity of the ambient air. In very dry atmospheres, water escapes from the cells, and they tend to dry out and lose their effective capacity. In very humid air, water is likely to be absorbed by the cells, and, depending on their state of charge, it may cause cell flooding, a loss of rate capability, and electrolyte leakage, a potential safety problem. The most important side reactions from a safety point of view are the gas evolution reactions on charge and overcharge: the generation of hydrogen at the negative electrode and oxygen at the positive electrode. Depending on the cell designs, a greater or lesser fraction of the hydrogen may be absorbed by reaction with oxygen at the positive electrode. Since the cells are open to the atmosphere during charging, gas generation does not create an overpressure problem, except possibly inside the porous electrodes, where excessive local pressures may cause some deterioration of their structure. Gases that escape from the cells may form explosive gas mixtures in poorly ventilated battery compartments.

The physical hazards associated with zinc/air cells are related primarily to the possible generation of explosive gas mixtures in battery compartments. Since the cells cannot be pressurized, except when sealed during nonoperational stand, the likelihood of cell ruptures is low. The tendency of zinc to form dendrites on charge cannot be prevented, but their growth can be impeded by suitable separator structures to retard the formation of dendritic short circuits. If such short circuits do form, considerable cell heating may occur and lead to incipient thermal runaways. However, since the cells cannot contain pressures, heating is more likely to cause electrolyte expulsion in some form rather than ruptures. The chemical hazards are essentially limited to the possibility of contact with potassium hydroxide electrolyte that may escape from the cells due to abuse or defective seals. Extended contact with the electrolyte causes severe chemical burns. The likelihood of ingestion of any of the toxic cell materials is low as long as proper handling and cleanliness procedures are followed.

8.8. RECHARGEABLE BATTERY MATERIALS AND TOXICITY

Exposure to toxic solids in rechargeable batteries can occur only if the batteries rupture or explode (see Part II) or if the batteries are taken apart in some manner. Exposure to toxic electrolytes can occur under the same circumstances and, in addition, if the batteries leak. Since the battery solids are insoluble in the electrolytes, electrolyte leakage does not pose a threat of exposure to toxic battery solids. In the case of lead–acid batteries that contain antimony (and arsenic), users may be exposed to toxic stibine (SbH_3) and arsine (AsH_3) gases escaping from vented batteries during charging. Readers who need comprehensive information on material toxicity or who wish to learn about the medical aspects of poisoning may find it useful to consult one or more of the texts cited at the end of the chapter.[4–8]

8.8.1. Antimony

Chemical Species: Sb, SbH_3

Batteries: Lead–acid batteries. Note that antimony is more likely to be present in older type batteries. In the newer maintenance-free batteries, the antimony has been replaced with calcium, a nontoxic material.

Physical Form: When present in lead–acid batteries, the antimony is employed in low concentrations as an elemental alloying component of lead. The antimony reacts with hydrogen evolved at the negative electrode during charge and overcharge to form toxic stibine gas.

Toxicity: Although antimony is itself a toxic substance, the principal hazard to battery users is due to the possible inhalation of stibine gas, which is a central nervous system and cardiovascular toxicant and affects the healthy functioning of the liver and kidneys as well.

$$TWA: \ 0.5 \ mg/m^3$$

$$TLV: \ 0.5 \ mg/m^3$$

$$MCL: \ 0.006 \ mg/liter$$

Simple Symptoms: Ingestion of antimonials causes gastrointestinal damage manifested by abdominal pain, vomiting, and a burning esophageal sensation within half an hour to about two hours of exposure. Inhalation of stibine gas gives rise to headache, nausea, weakness, and a general sensation of being unwell, including abdominal pain and vomiting. The symptoms may appear only some time after exposure, generally within a day.

8.8.2. Arsenic

Chemical Species: As, AsH_3

Batteries: Lead–acid batteries. Arsenic is present, not as an additive, but as an impurity in the antimony used as an alloying constituent of lead. Whenever lead–acid batteries contain antimony (see Section 8.8.1), they contain small amounts of arsenic, and only then.

Physical Form: When present, arsenic occurs in its elemental form in very low concentrations in the lead metal. The volatile arsine gas is formed during charging, as in the case of stibine.

Toxicity: Arsenic and its compounds are very toxic. The only substance that battery users are likely to come in contact with is the arsine gas, unless they handle lead electrodes that contain antimony. Such handling is likely to cause lead poisoning rather than arsenic poisoning. The toxic effects of arsenic and arsine are essentially the same as described for antimony and stibine in Section 8.8.1.

$$TWA: \ 0.01 \ mg/m^3$$
$$TLV: \ 0.20 \ mg/m^3$$
$$MCL: \ 0.05 \ mg/liter$$

Simple Symptoms: The symptoms are the same as described for antimony poisoning in Section 8.8.1.

8.8.3. Cadmium

Chemical Species: Cd, CdO, $Cd(OH)_2$

Batteries: Nickel/cadmium batteries

Physical Form: Refer to Section 7.8.1.

Toxicity: Refer to Section 7.8.1.

Simple Symptoms: Refer to Section 7.8.1.

8.8.4. Cobalt

Chemical Species: Co, $Co(OH)_2$

Batteries: Cobalt and its hydroxides are present in many nickel/cadmium batteries as a minor constituent of the positive electrodes and may be present in other rechargeable batteries with nickel electrodes.

Physical Form: The metal and its hydroxide are present as essentially insoluble solids uniformly distributed in intimate contact with the nickel oxide electrode.

Toxicity: Cobalt is an essential nutritional element in low concentrations. Chronic exposures due to contact or inhalation may generate dermatitis and diseases of the respiratory tract. Ingestion of cobalt salts has been reported as a cause of cardiac failures; despite this, cobalt salts are used as a treatment for refractory anemia. Exposure to cobalt and its hydroxide should not be of concern to battery users. Long before they would become exposed to cobalt in sufficient concentrations to cause harm, they would have been exposed to far greater amounts of nickel and alkali that may cause harm.

$$TWA: \ 0.1 \ mg/m^3$$

Simple Symptoms: The most common symptoms resulting from chronic exposures to cobalt or from short-term exposure to excessive amounts of cobalt, most likely by inhalation in industrial environments, are coughing and a general feeling of malaise and, possibly, the gradual development of skin problems.

8.8.5. Lead

Chemical Species: Pb, $PbSO_4$, PbO_2

Batteries: Lead–acid batteries. Small amounts of lead may be present in batteries with zinc electrodes as a corrosion inhibitor.

Physical Form: The metal and the lead species are insoluble solids in lead–acid batteries. All the listed chemical species are present regardless of the state of charge of the batteries, with the relative amount of the lead sulfate increasing, the deeper the state of discharge of the batteries. Very little lead sulfate is present in fully charged batteries.

Toxicity: Lead and its compounds are all very toxic materials with pervasive, adverse effects on a wide variety of body systems, including the nervous system, the kidneys, and the organs responsible for the synthesis of vital blood components. Battery users can avoid lead poisoning by observing the elementary precautions of avoiding contact with lead and its compounds and by avoiding abusive treatments of lead–acid batteries.

TWA: 0.05 mg/m^3 (PbO$_2$)

TLV: 0.15 mg/m^3

MCL: 0.05 mg/liter

Simple Symptoms: The most common complaints resulting from lead poisoning are abdominal pain, fatigue, headache, irritability, and muscle and joint discomfort.

8.8.6. Manganese

Chemical Species: MnO$_2$, MnOOH, Mn(OH)$_2$

Batteries: Rechargeable alkaline manganese batteries. They are presently under development and may become important in the consumer market during coming years.

Physical Form: Refer to Section 7.8.2. Note that the rechargeable alkaline manganese batteries may contain some bismuth in the positive electrodes chemically combined with manganese to form a complex oxide. All the manganese species are present as insoluble solids.

Toxicity: Refer to Section 7.8.2. The toxicity of bismuth is low, and it is used as an ingredient in some therapeutic agents.

Simple Symptoms: Refer Section to 7.8.2.

8.8.7. Nickel

Chemical Species: Ni, Ni(OH)$_2$, NiOOH

Batteries: Nickel/cadmium batteries are the most important batteries in which nickel and its oxides and hydroxides are present. Small button cells, coin cells, and cylindrical cells are all available in the general consumer market, and they are widely used in portable electronic devices. Large Ni/Cd batteries are rarely used by the general consumer. Nickel/metal hydride batteries may enter the consumer market in significant numbers during coming years. Nickel/zinc batteries are of little importance at the present time, but may be developed to the point of practicality for general consumer applications in future years.

Physical Form: Metallic nickel occurs in appreciable amounts in batteries, mostly in the positive electrodes, but some nickel may be present in the negative electrodes. The hydroxides are confined to the positive electrodes, where they are present as reasonably cohesive, insoluble solids, distributed more or less uniformly in the porous structure of the electrodes.

Toxicity: Nickel is considered an essential nutritional trace element in low concentrations, and it has a low toxicity rating. The highly toxic nickel carbonyl is not present in any batteries. Metallic nickel is classified as an allergen and may cause dermatitis on contact. Only a small fraction of the general population is sensitive to this allergen. Continuous exposures to high concentrations of nickel in industrial environments may cause severe respiratory problems.

$$\text{TWA: } 1.0 \text{ mg/m}^3$$

$$\text{TLV: } 1.0 \text{ mg/m}^3$$

$$\text{MCL: } 0.10 \text{ mg/liter}$$

Simple Symptoms: The most commonly observed symptom of chronic exposure to nickel is dermatitis, a skin disorder of areas in contact with nickel. Ingestion of nickel compounds leads to almost immediate nausea, vomiting, abdominal pain, headache, and sensations of general discomfort that may last for several hours, followed by recovery in a few days with no lasting effects. Inhalation of low concentrations of nickel-bearing aerosols, primarily in industrial environments, may not generate any immediate symptoms, but chronic exposures lead to the gradual development of respiratory tract diseases.

8.8.8. Silver

Chemical Species: Ag, Ag_2O, AgO

Batteries: Silver/zinc batteries of various sizes. They are used to a limited extent by the general consumer.

Physical Form: Refer to Section 7.8.4.

Toxicity: Refer to Section 7.8.4.

Simple Symptoms: Refer to Section 7.8.4.

8.8.9. Zinc

Chemical Species: Zn, Zn(OH)$_2$, ZnO

Batteries: Silver/zinc, nickel/zinc, and zinc/air batteries. The former type is used to a limited extent by the general consumer, and the latter types have yet to enter the general consumer market.

Physical Form: Refer to Section 7.8.5.

Toxicity: Refer to Section 7.8.5.

Simple Symptoms: Refer to Section 7.8.5.

8.8.10. Metal Hydrides

The nickel/metal hydride batteries have yet to be produced in sufficient quantities to make a noticeable impact on the consumer market, but there is a possibility that they may do so during coming years. The likelihood of toxic exposures due to metal hydrides is low. A more likely hazard with this type of battery is exposure to the alkaline electrolyte. Some metal salts may be present in low concentrations in the electrolyte owing to their corrosion in the alkaline electrolyte. An additional hazard with metal hydrides is their flammability in air.

Chemical Species: The negative electrodes are of two types, which contain the following metals and their hydrides:

- AB$_2$ type: Ti, V, Zr, Ni
- AB$_5$ type: La, rare earths (mostly Ce), Ni, Co, Al, Mn

If the hydrides should be exposed to air and ignite, various oxides of the listed metals would be expected to form, possibly as inhalable fumes.

Batteries: Button type and cylindrical type cells are likely to be available in most of the standard sizes characteristic of rechargeable consumer batteries.

Physical Form: The metals and hydrides are all cohesive solids formed into relatively compact powder pellets or porous bodies. They are basically insoluble in the electrolyte, but some of the metals may corrode slowly to form oxides and/or hydroxides, and these species are likely to be present in aged cells, some of them in solution in low concentrations in the electrolyte.

Toxicity and Symptoms: We limit our discussion to a very brief summary of the properties of the metals present in the two types of electrodes.

1. AB$_2$ type electrodes

Titanium: Very low toxicity; oxide dusts may cause respiratory irritation.

Vanadium: Low oral toxicity, but high dosage may cause gastrointestinal disturbances. A pulmonary and skin irritant. Chronic exposures (industrial environments) may cause fibrosis and emphysema.

TWA: 1.0 mg/m^3 (V)

TLV: 0.05 mg/m^3 (V$_2$O$_5$)

Zirconium: Very low toxicity. Dusts may cause respiratory irritation, but without pulmonary complications. Some hypersensitivity reaction possible.

Nickel: Refer to Section 8.8.7.

2. AB$_5$ type electrodes

Aluminum: Very low toxicity. Ingestion and/or inhalation may have some adverse effects. Massive ingestion causes gastrointestinal irritation. Inhalation of oxide fumes causes weakness, fatigue, and respiratory distress, and possibly pulmonary fibrosis. None of the toxic aluminum fluorides are present in batteries.

TLV: 10 mg/m^3 (aluminum oxide dust)

Cobalt: Refer to Section 8.8.4.

Lanthanum, Cerium, and Rare Earth Metals: Little information is available on the toxicity of these metals (nonradioactive form), but since their chemical properties are quite similar, any information on any one of them may be expected to apply to them all to some degree. Cerium is classified as a hepatoxic agent that causes liver disease and lipid accumulation, and lanthanum is recognized as an inhibitor of calcium channels in cellular functioning. Until more specific information becomes available on the toxicity of these materials, they should be treated as heavy metals with corresponding toxicities, and inhalation and ingestion should be avoided to the fullest extent possible. Reference 7 contains much information on the animal toxicity of rare earth metals.

Manganese: Refer to Section 7.8.2.

Nickel: Refer to Section 8.8.7.

8.9. AQUEOUS ELECTROLYTE TOXICITY

8.9.1. Alkaline Electrolytes

Refer to Section 7.9.2.

8.9.2. Sulfuric Acid

Chemical Species: H_2SO_4, a concentrated solution of sulfuric acid in water.

Batteries: Rechargeable lead–acid batteries of all kinds.

Physical Form: Most of the batteries contain the acid as a liquid held in the porous electrodes and in the absorber/separator structures. In starved cells, there is little freely mobile electrolyte. Many of the smaller batteries contain the electrolyte in gelled form.

Toxicity: Acids destroy tissues chemically by virtue of their high hydrogen ion concentrations. Concentrated acids have the additional characteristics of dehydrating tissues and, in the process of being thus diluted, of creating excessive local heating that may cause thermal tissue damage. Acid exposures lead to coagulative necrosis of tissue proteins with subsequent scar formation.

Eye exposures may occur as a result of battery ruptures, explosions, or careless battery maintenance. They are very serious and may cause permanent eye damage by destroying corneal and other tissues and by ulceration and perforations. In contrast to alkalies, sulfuric acid penetrates the underlying tissues more slowly, but damage begins within seconds of exposure. Exposed eyes should be irrigated immediately after exposure with copious amounts of water for about 10–20 min, and medical attention should be obtained as soon as possible.

Inhalation may occur as a result of battery explosions or of exposure to fine mists escaping from batteries on charge, but it is a rare event. Depending on the quantities inhaled and the state of dispersal of the acid, the tissue damage may be limited to the upper respiratory tract or it may involve the entire respiratory system. In any event, serious tissue damage is likely to result from inhalation, and immediate medical attention should be obtained.

Ingestion has been reported in a few cases. It generally resulted from carelessness in battery maintenance operations. Serious tissue damage is likely to occur if more than minute quantities of acid have been ingested. The affected tissues are those of the entire gastrointestinal tract. Any ingested acid should be diluted with water or milk as soon as possible after exposure. No attempts should be made to neutralize the acid with alkaline media, and vomiting should not be induced. Immediate medical attention is required.

Skin exposures to sulfuric acid are a common occurrence in battery handling and maintenance. The hands, in particular, are likely to be involved, and the eyes may be involved if the operators are careless. The acid attacks the outer layers of the skin, and, unless the acid is rapidly removed, it may cause necrotic damage to the underlying tissues as well. Acid burns generally leave noticeable scars if tissue penetration has occurred. The extent of any potential injury may be considerably reduced by immediate irrigation of the affected parts to remove the acid and to extract acid that may have penetrated into underlying tissues.

$$TLV = TWA: \; 1 \; mg/m^3$$

Simple Symptoms: The most noticeable effect of sulfuric acid is a persistent burning sensation that is felt shortly after exposure. This is in contrast to alkalies, which do not cause any such early signs of exposure.

REFERENCES

1. D. Linden (ed.), *Handbook of Batteries and Fuel Cells,* McGraw-Hill, New York (1984).
2. H. Bode, *Lead–Acid Batteries,* John Wiley and Sons, New York (1977). [Translated from German original by R. J. Brodd and K. V. Kordesch.]
3. S. U. Falk and A. J. Salkind, *Alkaline Storage Batteries,* John Wiley and Sons, New York (1969).
4. Casarett and Doull's Toxicology, *The Basic Science of Poisons,* 3rd Ed., Macmillan, New York (1986).
5. M. J. Ellenhorn and D. G. Barceloux, *Medical Toxicology. Diagnosis and Treatment of Human Poisoning.* Elsevier, New York (1988).
6. F. W. Oehme (ed.), *Toxicity of Heavy Metals in the Environment,* Marcel Dekker, New York (1979).
7. N. I. Sax and R. J. Lewis, Jr., *Dangerous Properties of Industrial Materials,* Van Nostrand Reinhold, New York (1989).
8. J. B. Sullivan, Jr. and G. R. Krieger (eds.), *Hazardous Materials Toxicology. Clinical Principles of Environmental Health,* Williams and Wilkins, Baltimore (1992).

IV

LITHIUM BATTERIES

9

Solid Cathode Lithium Systems

A number of solid cathode lithium systems have been developed. The most commonly used are based on the lithium/manganese dioxide (Li/MnO_2) and the lithium/poly(carbonmonofluoride) [$Li/(CF_x)_n$] chemistries. They are available commercially in a variety of sizes and configurations for both consumer and military use. Most of the other chemistries have been developed for special applications and are not widely used. In many instances there is a paucity of safety information available.

9.1. LITHIUM/MANGANESE DIOXIDE

The Li/MnO_2 chemistry is a relatively high energy system, capable of delivering >220 Wh/kg and >400 Wh/liter. The most common commercially available cells use either lithium perchlorate or lithium trifluoromethanesulfonate (lithium triflate) dissolved in a mixture of propylene carbonate (PC) and dimethoxyethane (DME) as the electrolyte. $LiClO_4$ is a powerful oxidizing agent and an irritant to the skin on contact. It forms explosive mixtures with carbonaceous materials or finely divided metals.[1] Lithium triflate (CF_3SO_3Li) is a hygroscopic solid that is moderately irritating to the eyes and skin. It is slightly toxic upon ingestion, having an LD_{50} for rats between 1250 and 5000 mg/kg of body weight.[2] Propylene carbonate is a colorless liquid having a boiling point of 242°C. PC is a skin, eye, and mucous membrane irritant and has a low order of toxicity in large oral doses.[3] Dimethoxyethane is a flammable liquid with a boiling point of 82–83°C and a flash point of 4.5°C. It is a very dangerous fire hazard when exposed to heat, flame, or oxidizers and can readily form an explosive peroxide. When heated to decomposition, it emits acrid smoke and fumes.[1] Manganese dioxide is a strong oxidizing agent and is both a skin and an eye irritant. It is a neurotoxin and, upon ingestion, can result in metal fume fever, an influenza-like illness.[4]

9.1.1. General Safety Considerations

The lithium/manganese dioxide system has a high degree of safety and has been recognized as one of the safest lithium systems.[5,6] During discharge or on open circuit, Li/MnO_2 cells have a minimal rise in internal pressure, making them suitable for consumer use. If two cells are connected in parallel, one having a lower capacity, the sloping voltage curve allows the discharge to be balanced between the cells with no safety hazards.[7] Also, in a series string with one low cell, no abnormal heating or gassing occurs. Gassing on discharge can be minimized by discharging cells ~ 2–10% after manufacture.[8]

9.1.2. Chemistry

The only significant safety-related chemistry in the Li/MnO_2 system concerns reactions involving the electrolyte and reactions of metallic lithium, which may plate on the cathode during cell reversal. Electrolyte reactions result in the formation of various gaseous products. The gas composition (in mole percent) found in cells that have vented after abuse includes $\sim 55\%$ nitrogen, $\sim 18\%$ oxygen, $\sim 10\%$ carbon dioxide, $\sim 8\%$ hydrogen, $\sim 4\%$ carbon monoxide, $\sim 3\%$ methane, $\sim 1\%$ various hydrocarbons, and $\sim 0.7\%$ argon.[9]

During cell reversal, the morphology of the plated lithium on the cathode is highly dependent on the electrolyte.[10] The use of nonreactive electrolytes, for example, $PC/DME/LiClO_4$, enhances the Li plating process, allowing a coherent metallic deposit to form on the cathode. This deposit is in intimate contact with the cathode and its discharge products. If the cell is heated, for example, by short circuit, this intimate mixture can react violently.

A number of modifications have been made to the electrolyte by various scientists, to enhance the safety of this system. Alterations to the solvent include precomplexing a volatile ether with the $LiClO_4$ prior to adding propylene carbonate,[11] use of a ternary solvent (PC/4-methyldioxolane/dioxolane) containing a variety of solutes,[12–19] use of the ternary solvent PC/DME/tetrahydrofuran (THF) with lithium triflate,[20] and use of the 1:1 binary solvent 1,3-dimethyl-2-imidazolidinone/dioxolane containing lithium triflate.[21]

Changes to the solute have focused on either eliminating $LiClO_4$ or reducing its oxidizing power. Addition of a second solute was found to reduce the oxidizing power of $LiClO_4$. Mixing $LiCF_3SO_3$[22] or $Li_2B_{10}Cl_{10}$[23] with $LiClO_4$ as the solute in PC/DME solvent results in cells that are safe even at high temperatures. A $LiCF_3CO_2/LiClO_4$ mixture in PC/DME/THF also resulted in a safer cell than one using $LiClO_4$ by itself as the solute.[24] Substitution of either $LiCF_3SO_3$, $LiPF_6$, $LiBF_4$, or $LiAsF_6$ for $LiClO_4$ in a solvent containing diethylene glycol dimethyl ether and/or triethylene glycol has resulted in a safer battery.[25] Although complete elimination of $LiClO_4$ from the electrolyte is desired for improved safety, some have found that a small amount ($\sim 1\%$) is needed to prevent corrosion of metallic parts.[26] This small amount does not seem to impact the improved safety of the system.

Explosions can occur in Li/MnO_2 cells due to the reaction of lithium with water. The water is usually introduced into the cell by improperly dried MnO_2 cathodes. This problem was completely eliminated by one manufacturer by using an infrared (IR) oven to dry the cathode pellets prior to cell assembly.[27]

9.1.3. Safety-Related Design Features

For enhanced safety, Li/MnO_2 cells should contain a safety vent.[28,29] The operation of the vent may be improved by utilizing an electrolyte containing a significant proportion of a volatile component having a boiling point of 30–130°C, such as, DME.[30,31] Another approach is to add a component to the electrolyte which evolves gas when the temperature is increased. Such a material is $LiPF_6$. It has been found that the addition of $0.1M$ $LiPF_6$ to an electrolyte containing $0.7M$ $LiClO_4$ in PC/THF/DME prevents discharged cells from exploding when subsequently charged to 6 V.[32]

The separator plays an important role in Li/MnO_2 cell safety. It was found that cells with a wound configuration, in which a porous polypropylene separator was used, had less puncturing by the cathode active mass than control cells.[33] Coating the separator on both sides with a polymer, for example, polyacrylamide, polyethylene, or polypropylene, decreases the likelihood of internal shorts and only slightly increases the cell internal resistance.[34] A safety shutdown separator has been developed for use in Li/MnO_2 cells for 9-V batteries.[35] It consists of a microporous polypropylene film backed by a layer of nonwoven polypropylene fiber. The fiber flows readily at 91°C, which causes the resistance of the separator to increase markedly, shutting down the cell. It has been found to work effectively over the ambient temperature range −20°C to +70°C under a variety of abusive conditions, including short circuit, forced overdischarge, charge, nail penetration, crush, ignition, and sawing in half followed by water immersion. Internal short circuits have also been prevented by coating the cathode on its separator-facing side, with an ion-permeable film, such as, polyacrylamide, sodium polyacrylate, polyethylene, or polypropylene. This coating also improves the uniformity of the electrode thickness.[36]

A design feature to prevent plating of lithium metal onto the cathode during forced overdischarge, and to provide a shunt for the current without generating excess heat, comprises an inert metal coupled mechanically and electrically to the cathode, facing a dendrite target coupled mechanically and electrically to the anode. A separator is located between the two metal members. During voltage reversal of Li/MnO$_2$ cells, dendrites grow from the metal on the cathode to the anode target, creating a low resistance path and enhancing cell safety.[37] A similar means employs an anode tab, located on a section of the anode that is not fully utilized during discharge. An electrically conductive member is located opposite the face of the anode section that does not contain the tab and is insulated from the anode by a separator. During voltage reversal, lithium metal is preferentially deposited on the member.[10]

In the design of a high-rate, spirally wound, Li/MnO$_2$ cell, consideration must be given to a number of features to ensure safe performance, even under abusive conditions.[38] A vent must be included in the case or lid and must have tight tolerances (should vent at $\sim$110°C), and the electrode assembly should be rigid to survive shock and vibration. All connections should be welded or riveted. The cathode grid should have selvaged edges on both sides, so sharp points cannot occur when the cathodes are cut or handled. Placing the lithium on a large-area current collector will help to ensure that it will all be consumed during discharge.

9.1.4. Abuse Tests

A number of abuse tests have been conducted on Li/MnO$_2$ cells of different sizes. The following paragraphs summarize the reported results.

Short Circuit. Shorting three DL1/3N cells in series yielded a peak current of $\sim$300 mA and a peak temperature of $\sim$92°C. No ventings, explosions, leaks, or disfiguration were noted.[9] Short circuit of 200-mAh button cells resulted in a small temperature increase, with no ventings or case bulge.[6] Shorting of "$\frac{1}{2}$AA" cells led to a benign venting through the safety vent mechanism in the can bottom.[39] "A" cells did not vent when shorted and reached a temperature of 100–110°C in $\sim$4 min. "C" cells delivered a short-circuit current of 6.5 A, and the case temperature rose to <200°C with no ventings. When "D" cells were shorted, an explosive condition was reached at the melting point of lithium. Internal pressure approached 400 psi, and the case temperature was $\sim$140°C. Initially, the current peaked at >50 A, and the case temperature rose to $\sim$100°C in 2 min.[40]

Forced Overdischarge. Discharge and 150% reversal of DL1/3N cells at 80 mA to a final voltage of -10 V did not cause venting, explosion, leakage, or disfiguration. The maximum temperature reached was $\sim$57°C.[9] "C" cells having a balanced design were driven into reversal at 800 mA and 20°C. The maximum temperature attained was 27°C with no vents or leaks.[39]

Charge. DL1/3N cells that were 50% discharged at 10 mA and then charged at 100 mA exhibited a peak temperature of 84°C with no hazardous behavior. Similar cells discharged 80% at 10 mA and then charged at 100 mA vented with some flame after 30 min. A repeat of the latter test did not result in cells venting. A peak temperature of 82°C was reached after 30 min, and then the voltage dropped.[9] Two small spiral wound cells connected in series were force overdischarged at 1 A and then charged at 1 A with a 12-V power supply. The battery reached the 12-V limit in both cases and was stable for three hours with no hazardous behavior noted.[41] Charging 200-mAh button cells at 5 V for 100 hours did not cause any vents or explosions. The initial current was 40 mA, and the current dropped to 10 mA in one hour and then asymptotically to zero. A slight bulge in the case was noted.[6] "D" cells were charged at 12 V through a 3.6-Ω current-limiting resistor. They vented safely at 5 V with a case temperature of 40°C in 30 minutes.[40]

Heat. Heat taping DL1/3N cells causes them to vent at temperatures above 200°C. At temperatures between 350 and 435°C, cells emitted flame for less than 10 s.[9] External heating of "D" cells at a rate of 20°C/min resulted in venting after 7 min, at a temperature of 150°C.[40]

Puncture. Nail penetration of 200-mAh button cells caused the open-circuit voltage to drop to zero with no safety hazards noted, except leakage of electrolyte.[6] Shrapnel penetration of "$\frac{2}{3}$A" cells was simulated by shooting the cells with an M16 rifle. Cells were either 100%, 50%, or 0% charged, in a dry environment or within a water spray. Fully charged cells were warmer to the touch after penetration than either half or fully discharged cells. These tests indicated that partly or fully discharged cells present a slightly lower risk upon penetration than do fully charged cells, especially in a wet environment.[42]

Crush. Button cells, with 200-mAh capacity, were crushed at 3800 pounds of force. The open-circuit voltage dropped to zero, and electrolyte leaked at the weld. No other hazardous behavior was observed.[6]

It can be concluded from the safety abuse tests described above and tests performed on Li/MnO$_2$ cells by the U.S. Navy, according to NAVSEA 9310,[28] that the smaller sizes, both button and cylindrical designs, are safe. However, the larger cells (size C and above) should be tested for each specific application.

9.2. LITHIUM/POLY(CARBONMONOFLUORIDE)

The Li/(CF$_x$)$_n$ chemistry is considered a safe system, capable of delivering energy densities of 280 Wh/kg and 550 Wh/liter. It is not capable of exploding, even under short-circuit conditions.[43] Commercial cells normally contain an electrolyte of either $LiBF_4$ in γ-butyrolactone or $LiAsF_6$ in dimethyl sulfite. $LiBF_4$ is a white powder that is irritating to the mucous membranes and upper respiratory tract, as well as to the eyes and skin. The chemical, physical, and toxicological properties have not yet been thoroughly investigated.[44] γ-Butyrolactone is a liquid with a boiling point of 204°C. Upon heating to decomposition, it emits acrid smoke and fumes.[1] It has an oral LD_{50} for rats of 17.2 ml/kg.[45]

$LiAsF_6$ may be fatal if inhaled, swallowed, or absorbed through the skin. It is also a carcinogen, so care must be taken when handling this material.[46] Dimethyl sulfite is a flammable liquid with a boiling point of 126–127°C. It is an irritant to the skin, eyes, mucous membranes, and upper respiratory tract. Dimethyl sulfite may be absorbed through intact skin in toxic amounts.[47]

Poly(carbonmonofluoride) is an off-white powder which will ignite in a hydrogen atmosphere at 400°C and explode at 500°C in an inert atmosphere. It decomposes at 500–600°C, giving off toxic fumes.[1]

9.2.1. Safety-Related Designs

Large $Li/(CF_x)_n$ cells incorporate a pressure relief vent in the can bottom. One "DD" design utilizes a 10-atm vent plus a tapered nickel tab on the cathode which fuses at ~ 10 A.[48] Smaller cells have been built with a crimped plastic cover, made from a low-melting-point thermoplastic resin, which deforms at high temperature. This allows the electrolyte to escape before a significant pressure builds up inside the cell.[28]

A large multicell battery was designed with shunt diodes placed across each cell.[49] The diodes prevent any cell that may have a low capacity from being driven into deep reversal.

9.2.2. Safety/Abuse Tests

Lithium poly(carbonmonofluoride) cells of various sizes have been put through a variety of safety abuse tests. Reported results are summarized below.

Short Circuit. Short-circuit tests of "$\frac{1}{2}$A", "$\frac{2}{3}$A", and "DD" cells resulted in no safety hazards, only a slight rise in the skin temperature.[48,50,51]

Forced Overdischarge. Driving "$\frac{1}{2}$A" cells into reversal at 500 mA[50] or "$\frac{2}{3}$A" cells at 250 mA did not cause any hazardous behavior. Forced overdischarge of "$\frac{2}{3}$A" cells at 1 A resulted in a maximum case temperature of 84°C, but again no safety concerns.[51] Insulated "DD" cells, reversed at -10°C, showed a 15°C temperature rise with 20% of the cells venting.[48]

Charging. Charging fresh or 50% discharged "$\frac{1}{2}$A" cells at 1 A did not cause any hazardous behavior.[50] However, in another test program,[51] charging of "$\frac{2}{3}$A" cells at 250 mA and at 1 A caused cells to vent with smoke.

Heat. External heating of Li/$(CF_x)_n$ cells results in case rupture, via a built-in vent mechanism or otherwise. Tests on "$\frac{1}{2}$A" cells resulted in venting at 200–250°C,[50] while "$\frac{2}{3}$A" cells vented at 200–275°C with some smoke.[51] Heat taping "DD" cells to 186°C caused them to either rupture or explode.[48] Thermal shocking of "$\frac{2}{3}$A" and "C" cells (+70 to −60 to +70°C) did not have any effect on the safety or discharge performance of the cells.[52]

Puncture. Simulated shrapnel penetration of "$\frac{2}{3}$A" cells was carried out using a projectile from an M16 rifle.[42] Cells were either fully charged, half discharged, or fully discharged. No ignition occurred on any cells tested, either in the dry or the wet state (i.e., within mist from hose). Puncture of "DD" cells results in a hydraulic vent at the puncture site. Skin temperatures may rise to >100°C, and some lithium fires may occur.[48]

Mechanical Abuse. A series of tests were conducted on "DD" cells.[48] A 1-m drop onto concrete caused no safety hazards with either fresh or discharged cells. Explosive shock initiated internal reactions that resulted in case rupture at the shock point. Crushing cells caused a hydraulic vent but no fires.

One manufacturer of Li/$(CF_x)_n$ cells reported on a series of mechanical, electrical, and special abuse tests.[53] The mechanical tests were performed on "$\frac{2}{3}$A" and "C" cells and consisted of vibration, shock, thermal shock, and drop tests. No safety problems arose during any of these tests. The electrical tests were conducted on "$\frac{2}{3}$A" cells and consisted of short circuit, forced discharge, heavy-duty discharge, and charge. No safety problems were noted during any of these tests. The special tests were also performed on "$\frac{2}{3}$A" cells and included high-temperature storage, heating on a hot plate to 180°C, drilling, crushing, burning, water immersion, and salt water spray. No safety concerns were raised during these tests. However, in the burning test, in which cells were subjected to a 1000°C fire from a gas burner, all cells vented and caught fire within 3 min. No explosions occurred.

These tests indicate, as with the Li/MnO_2 cells, that the small cells are safe. However, larger cells ("C" and above) should be abuse tested for each application.

9.3. LITHIUM/COPPER OXIDE (Li/CuO)

The lithium/copper oxide chemistry is a low-voltage system, capable of delivering 250 Wh/kg and 470 Wh/liter. It is considered to be intrinsically safe at temperatures up to 150°C.[54] Cells contain an electrolyte of $LiClO_4$ in dioxolane. 1,3-Dioxolane is a flammable liquid with a flash point of 35.6°C. It is a very dangerous fire hazard when exposed to heat and/or flames. It is moderately toxic by ingestion, skin contact, and inhalation. The LD_{50} for inhalation by rats is 20.73 g/m^3 in 4 h.[1] Dioxolane is a skin and severe eye irritant and emits acrid smoke and irritating fumes when heated to decomposition. The toxicological properties of $LiClO_4$ were described earlier. Cupric oxide is a fine black powder, which is used as a fungicide. Cupric oxide inhalation may be responsible for one form of metal fume fever.[1]

9.3.1. Safety-Related Design Features

In bobbin type cells, the lithium anode is centrally located, surrounded by the CuO/graphite cathode. Upon penetration or crushing of the cells, the cathode acts as an energy absorber. Thus, the kinetic energy of impact is dissipated before it can lead to ignition of the lithium.[54]

Spirally wound cells have been designed to have limited power, eliminating the short-circuit hazard.[55] This is accomplished by limiting the electrode surface area and using a nonwoven polypropylene separator.

Button cells have been designed to prevent separator breakage during assembly by utilizing a matrix construction.[56] The thickness of the lithium anode is made thinner than the depth of the anode case. A porous electrolyte matrix of either polyethylene, polypropylene, or PTFE (Teflon) is coated on the anode-facing side of the cathode, which fits into the space provided on the anode side of the cell. In a test build, 20 of 100 cells using a standard design had broken separators, while no cells utilizing the matrix construction had broken separators.

9.3.2. Safety/Abuse Tests

A number of electrical and physical abuse tests have been performed on Li/CuO cells. The following paragraphs summarize the reported results.

Short Circuit. Short circuit of fresh "AA" cells does not pose a safety hazard.[54,57,58] However, shorting of partially discharged cells resulted in venting.[59] The worst condition for these cells has been found to be when cells are connected in a series/parallel arrangement and shorted at the end of life.[58] Therefore, blocking diodes should be incorporated into batteries wired in this manner.

Forced Overdischarge. Driving Li/CuO cells into reversal will normally lead to a violent rupture of the cell case.[54,60] In one study, "AA" cells were force overdischarged between 20 mA and 1 A over the temperature range −30 to +70°C.[57] All cells failed, but one cell discharged at 1 A and −30°C ruptured violently. In other studies, "AA" cells force overdischarged at 25 mA were found to be safe in most instances[61] and to meet the requirements of NAVSEA 9310 testing.[59]

Charging. Charging of Li/CuO cells will result in a violent rupture in most cases.[60] Prior to violent rupture, the cell voltage rises rapidly.[54] A charging matrix was carried out on lithium-limited, crimp-sealed "AA" bobbin cells.[57] Cells were either fresh, half discharged, or fully discharged, charging currents ranged from 20 mA to 1 A, and the temperature varied between −30 and +70°C. At 20°C, all the fresh cells failed, with the plastic grommet and cap being forced from the case. All the partially and fully discharged cells survived an 8-Ah charge up to 200 mA. All cells failed the 1-A charge; the greater the depth of discharge of the cell, the longer was the time to failure. Another study indicated that partially discharged cells are more hazardous on charge than fresh cells.[61]

Heating. Both crimp-sealed and hermetic Li/CuO "AA" cells met the heat tape requirements of NAVSEA 9310.[59] The heating of crimp-sealed "AA" cells at <250°C does not cause violent rupture,[54] but incineration at 500°C leads to violent rupture of fresh cells after ∼2 min and of discharged cells after ∼4 min.[57]

Shock and Vibration. Typical transportation shock and vibration environments do not cause any hazardous behavior in Li/CuO "AA" cells.[57,59]

Puncture and Crushing. Neither puncturing nor crushing "AA" cells results in any hazardous behavior.[54,57] Crushing an "AA" cell to half its diameter distorted the crimp seal, allowing electrolyte to escape. The temperature rose to 38°C after 12 min owing to internal shorting, but the voltage returned to normal after the pressure was relaxed.[61]

9.4. LITHIUM/COPPER OXYPHOSPHATE

The $Li/Cu_4O(PO_4)_2$ system is similar to the Li/CuO system, but has a higher operating voltage (2.5 V vs. 1.5 V) and can operate safely at elevated temperatures (up to 200°C). Two versions are available: the standard, which utilizes a $LiClO_4$ in dioxolane electrolyte, and a high-temperature design, containing a $LiCF_3SO_3$ in propylene carbonate/tetraglyme electrolyte. Tetraglyme is a high-boiling glycol ether, 2-, 5-, 8-, 11-, 14-pentaoxapentadecane (bp = 275.3°C), and is mildly toxic by ingestion, the oral LD_{50} for rats being 5.14 g/kg.[1] It is an eye irritant and may possibly have dangerous effects on the human reproductive system. Upon heating to decomposition, it emits acrid smoke and irritating fumes. Little has been written about the toxicological properties of copper oxyphosphate, but they are probably similar to those of other copper salts. The health effects of $LiClO_4$, $LiCF_3SO_3$, dioxolane, and propylene carbonate have been described earlier.

9.4.1. Safety Features

The high-temperature version of this cell is hermetically sealed in a nickel-plated steel case. The high-boiling electrolyte results in low internal pressure, even at elevated temperatures. Cells have been discharged at 200°C, which is ~20°C above the melting point of lithium, with no adverse effects. Short circuit at 200°C did not result in cells leaking, venting, or exploding.[62]

9.5. LITHIUM/COPPER SULFIDE

The Li/CuS chemistry is a high-energy system that has been utilized in button cells for consumer applications and in flat "D-shaped" cells for use in implantable cardiac pacemakers. Cells contain an electrolyte of $LiClO_4$ in a mixture of dimethoxyethane (ethylene glycol dimethyl ether, glyme) and dioxolane. Some designs contain dimethylisoxazole as a stabilizer. Dimethoxyethane is a flammable liquid, having a boiling point of 82–83°C and a flash point of 4.5°C. It is a very dangerous fire hazard and forms explosive peroxides. Upon heating to decomposition, acrid smoke and fumes occur. As with other glycol ethers, exposure may cause reproductive problems including sterility. The lowest published lethal dose is 2–16 g/kg, orally in mice.[1] Cupric sulfide is a black powder that has mutagenic properties. DNA damage has been noted in hamsters at doses of 10 mg/liter.[1] The toxicological properties of $LiClO_4$ and dioxolane have been described previously in this chapter.

9.5.1. Chemistry

The chemical reactions leading to cell rupture during forced overdischarge have been postulated.[63] They involve the presence of strongly oxidizing perchlorate and breakdown of the ether solvents.

The normal discharge reactions are:

At the anode:

$$Li \rightarrow Li^+ + e^- \tag{9.1}$$

At the cathode:

$$CuS + nLi^+ + ne^- \rightarrow Li_nCuS \qquad (9.2)$$

The overall reactions during forced overdischarge are:

At the cathode:

$$2Li_nCuS + nCH_3OCH_2CH_2OCH_3 + 4Li^+ + 4e^- \rightarrow$$

$$2Li_2S(s) + 2nLiOCH_3 + nCH_2CH_2 + 2Cu \quad (9.3)$$

At the anode:

$$2nLiOCH_3 \rightarrow 2nLi^+ + nH_2O + (-CH_2CH_2-O-)n + 2ne^- \quad (9.4)$$

$$nH_2O + 2nLi^+ + 2nClO_4^- \rightarrow 2nH^+ + 2nClO_4^- + nLi_2O \qquad (9.5)$$

The final reactions leading to cell rupture involve the production of large quantities of gaseous products:

$$6H^+ + 3Li_2S \rightarrow 3H_2S + 6Li^+ \qquad (9.6)$$

$$6ClO_4^- + 4CH_2CH_2 \rightarrow 8CO_2 + 6Cl^- + 8H_2O \qquad (9.7)$$

$$nH_2O + 2Li_nCu \rightarrow nLi_2O + nH_2 + 2Cu \qquad (9.8)$$

Cell safety can be improved by substituting $LiCF_3SO_3$ for $LiClO_4$ and/or replacing the ether solvents with propylene carbonate.

9.5.2. Design Features

To protect cells from rupturing on short circuit, they should be designed with a high surface area-to-volume ratio and limited rate capability.[64] A proprietary design has been developed which eliminates cell rupture, even during forced overdischarge into reversal.[63]

9.5.3. Safety/Abuse Tests

A limited number of safety/abuse tests have been reported for Li/CuS cells.

Short Circuit. Shorting of a four-cell, high-rate 9-V battery caused a rise in the skin temperature of ~20°C but no safety problems.[63] Short circuiting of pacemaker cells was also found to be safe.[65]

Forced Overdischarge and Charge. Both forcing into reversal and charging of a four-cell, 9-V battery at 3-mA constant current resulted in case rupture and/or explosion.[63] Newer designs have minimized this problem.

Physical/Mechanical Abuse. Puncturing of Li/CuS cells with nails or bullets or drilling through the case did not cause any safety problems. Crushing or dropping cells likewise did not result in any hazardous behavior.[63,65]

Heat. Boiling pacemaker cells in water was not hazardous;[65] however, exposing 9-V batteries to fire resulted in an explosion when the cell temperatures reached 180°C.[63]

9.6. LITHIUM/VANADIUM PENTOXIDE

The Li/V_2O_5 chemistry is an energetic system capable of delivering 250 Wh/kg and 650 Wh/liter. An electrolyte of $LiAsF_6$ plus $LiBF_4$ in methyl formate is most commonly used. Methyl formate is a flammable liquid with a boiling point of 31.5°C. It is a very dangerous fire hazard, and its vapors are explosive when exposed to heat or flames. Inhalation of methyl formate vapors may cause irritation to nasal passages and conjunctiva. In high concentrations, death may result from pulmonary irritation.[1] Vanadium pentoxide dust is a respiratory irritant and is poisonous by either ingestion or inhalation. Absorption by the body through inhalation is nearly 100%. The lowest published lethal concentration of V_2O_5 is 346 mg/m^3, or 1 mg/m^3 for 8 h.[1] Toxicological information for $LiAsF_6$ and $LiBF_4$ has been given previously.

9.6.1. Chemistry

The only safety-related chemistry involves decomposition (hydrolysis) of the methyl formate to produce carbon monoxide gas, which could result in sufficient pressure to rupture the cell. If any water is present in the cell, hydrolysis occurs, yielding formic acid and methanol:

$$HCOOCH_3 + H_2O \rightarrow HCOOH + CH_3OH \qquad (9.9)$$

In acid solutions, the formic acid will dehydrate to give carbon monoxide:

$$HCOOH \rightarrow H_2O + CO \qquad (9.10)$$

Methanol will also dehydrate to form additional water plus dimethyl ether:

$$2CH_3OH \rightarrow H_2O + (CH_3)_2O \qquad (9.11)$$

These reactions indicate that for every mole of water consumed in the hydrolysis of methyl formate, 1.5 moles of water are produced. Thus, the water concentration continually increases, promoting the hydrolysis of methyl formate and the formation of CO gas.[66]

To improve the safety of the Li/V_2O_5 system in fire environments, cells were built with an electrolyte of $LiAsF_6$ in organosilicon heterocyclic solvents (e.g., 3-trimethylsilyl-γ-butyrolactone or 1,1-dimethylsilacylopentane-2-one). These solvents are much less flammable than methyl formate.[67,68]

9.6.2. Abuse Conditions

Abuse testing of lithium/vanadium pentoxide cells, such as crushing, penetration, shock, vibration, or forced overdischarge, does not usually cause violent explosions.[69] Cells built with a pile-type construction were shock tested to 17,000 g with no adverse safety effects.[70] The most dangerous condition occurs upon leakage or venting, if methyl formate fumes are trapped in a small area and form an explosive mixture with air.

9.7. LITHIUM/CHROMIUM OXIDE

The lithium/chromium oxide system can deliver energy densities in excess of 300 Wh/kg and 700 Wh/liter. The preferred cathode material is a mixed oxide of chromium, Cr_3O_8, and the electrolyte used is $LiClO_4$ in propylene carbonate/dimethoxyethane. Chromium oxides are strong oxidizing agents and severe eye, skin, and mucous membrane irritants. The hexavalent state is more toxic than the lower oxidation states and is carcinogenic by inhalation.[1] Toxicological properties of the electrolyte components have been discussed earlier.

9.7.1. Safety/Abuse Tests

Li/Cr_3O_8 cells in the "$\frac{1}{2}$AA" size have been subjected to the following series of abuse tests.

Short Circuit. Upon short circuit at room temperature, a maximum current of 1.4 A was observed; at 70°C, the maximum current was 1.2 A. In both cases, there was a slight swelling at the case bottom but no electrolyte leakage or explosion.[71]

Forced Overdischarge. Cells were discharged at 2 kΩ to 2 V. They were then forced into reversal at 50-mA constant current. Maximum cell temperature was 45°C. After ~2 h, with the cell voltage at −30 to −50 V, cells will burst.[71]

Heat. Cells were heated to 150°C for 2 h. Some swelling of the cover and bottom was noted. No leakage of electrolyte occurred.[71]

Compression. Cells were compressed to ~60% of their original height. Internal shorting occurred, but no explosion or leakage of electrolyte was noted.[71]

Puncture. Cells were fully penetrated by a 3-mm-diameter nail perpendicular to their longitudinal axes. Cell temperatures did not increase, no spray of electrolyte was observed, and there was no swelling or explosion.[71]

Underwriters Laboratory (UL) Tests. These cells have passed the UL safety tests and have been granted UL component recognition.[72]

9.8. LITHIUM/IRON SULFIDE

The Li/FeS$_2$ system is a 1.5-V chemistry that is utilized in small cylindrical and button cells and has been used as a replacement for Zn/Ag$_2$O cells in watches, calculators, etc. Typical electrolytes used with this system include LiClO$_4$ in propylene carbonate/dimethoxyethane and LiCF$_3$SO$_3$ in dimethoxyethane/dioxolane/3-methyl-2-oxazolidone. No specific data are available on the health effects of FeS$_2$, but it may cause irritation to the eyes, mucous membranes, and respiratory tract, mainly due to mechanical action.[73] The solvent 3-methyl-2-oxazolidone is mildly toxic by ingestion and skin contact. The oral LD$_{50}$ for rats is 7.13 g/kg, and the LD$_{50}$ for skin contact in rabbits is 10.8 g/kg. Upon heating to decomposition, toxic fumes of NO$_x$ are released.[1] The other electrolyte components have been discussed earlier in this chapter.

9.8.1. Design Features

The lithium/iron sulfide system is considered a "safe" chemistry and is available to consumers in both "AA" and small button cells. These cells do not contain $LiClO_4$ in the electrolyte. Cells do contain a positive temperature coefficient (PTC) resistor that limits the current when the battery reaches 85–90°C. A vent is designed into the case of "AA" cells which activates if the cells reach 120–130°C.

Internal design features to protect against shorting due to separator penetration include a foil cathode collector and a foil anode with no backing. Cathode particles are small, providing a smooth surface, and a split-resistant separator is used. To protect against edge shorts, the cathode foil edge extends beyond the anode foil edge. Also, the anode tab-to-can position is isolated to prevent shorting.[74]

9.8.2. Safety/Abuse Tests

A series of electrical, environmental, and physical abuse tests have been conducted on Li/FeS_2 "AA" cells.[74] No vents, fires, or explosions occurred during the electrical tests, which included short circuit, charge, and forced overdischarge. The environmental tests comprised 50,000-ft altitude simulation, vibration, shock to 150 g, and thermal stability. No vents, leakage, bulge, or significant weight loss was noted. The physical abuse consisted of water submersion, crush, puncture, laceration, and heat, both by propane torch and in a bonfire. No vents, fires, or explosions occurred during the submersion, crush, puncture, or laceration tests. Cells did catch fire during the incineration tests, but no projectiles or thermal or blast effects, sufficient to hinder fire fighters, occurred.

9.9. LITHIUM/SILVER VANADIUM OXIDE

The lithium/silver vanadium oxide chemistry provides a high-energy system capable of delivering greater than 200 Wh/kg and 600 Wh/liter. It contains an electrolyte of $LiBF_4$ in either propylene carbonate or a mixture of propylene carbonate and dimethoxyethane. This system has been developed for high-power implantable medical applications, for example, cardiac defibrillators, but has also been used in nonmedical applications.

9.9.1. Safety Tests

The $Li/AgV_2O_{5.5}$ system is manufactured in both prismatic and cylindrical cells. Both designs have undergone extensive safety testing to qualify the chemistry for human implants. These tests are summarized below.

Short Circuit. Short-circuit tests of prismatic cells did not result in any safety hazards. Peak currents reached ~ 17 A at room temperature and 20–25 A when shorted at 37°C. Skin temperature rose to 85–140°C.[75-78] Short-circuit tests of "AA" cells at room temperature yielded a maximum current of 34 A and peak temperatures of 80 to 130°C. The cells vented mildly through the glass-to-metal seal.[75]

Forced Overdischarge. Prismatic cells were driven into reversal at either the $C/10$ rate or through a low resistive load (two cells in series, one fresh and one discharged). A slight case warming was noted, but no safety hazards were observed. Forced overdischarge at higher rates (2 A) caused skin temperatures to reach 130 to 140°C, and one of eight cells exploded.[76,77]

Charging. Charging of prismatic $Li/AgV_2O_{5.5}$ cells in either the fresh, half discharged, or fully discharged condition did not result in any safety hazards; only a slight bulge in the case was observed.[76,77,79]

Crush. Prismatic cells were crushed until an internal short circuit was generated, as noted by a sharp drop in voltage. The skin temperature reached 60–80°C, and no violent cell behavior was observed.[75-77]

Puncture. Prismatic cells punctured with a sharp metal rod leaked electrolyte at the puncture site. Skin temperatures rose to 90–130°C, but there was no violent behavior noted.[76,77]

Shock and Vibration. Prismatic cells were shocked at 1000 g for 0.5 ms in two directions for each of three orthogonal axes and vibrated for two 15-min sweeps at 5–500 Hz in three orthogonal axes for a total of 1.5 h. Peak acceleration was 5.0 g. No safety hazards were noted in any of these tests.[76,77]

Heat. Prismatic lithium/silver vanadium oxide cells were autoclaved at 125°C, with no safety problems occurring.[79]

9.10. LITHIUM/IODINE

The lithium/iodine couple is an all-solid-state system that was developed to power implantable cardiac pacemakers. Button cells have also been developed for memory backup applications. The cathode consists of a charge transfer complex of iodine with poly(2-vinylpyridine). The electrolyte is lithium iodide, which forms *in situ* from the reaction of I_2 with Li. Iodine is a crystalline material which has a vapor pressure of 1 mm at 38.7°C. It is poisonous by ingestion and moderately toxic by inhalation. Its effects are similar to those of Cl_2 and Br_2, but it is more irritating to the lungs. The lowest published lethal dose for humans orally is 28 mg/kg.[1] Lithium iodide is similar in toxicity to LiBr. Prolonged absorption may produce a condition called "iodism," which causes skin rash, running nose, headache, and irritation of mucous membranes. Upon heating to decomposition, iodine fumes may be released.[1]

The poor ionic conductivity of the LiI electrolyte limits the current capability of the system, making it very safe under most conditions. In addition, the electrolyte also acts as the separator and is self-healing. Any cracks that may occur in this film are immediately filled in by reaction of the Li with cathode material, thus preventing internal short circuits from occurring.[80]

Cells have been subjected to 2000 g shocks, 16.9 g RMS random vibration, 36 g repetitive bumps (4000 repetitions), and thermal cycles between −40 and +60°C. All cells tested survived without catastrophic failure or deviation in electrical output.[81]

As with all lithium systems, the temperature of the lithium/iodine system should not be allowed to exceed the melting point of lithium (180°C). If it does, extremely violent reactions can occur, resulting in case rupture.

9.11. OTHER SYSTEMS

A number of other primary lithium systems have been built, either commercially or experimentally. Included among them are the systems described briefly below. These systems have not found widespread use, and there are no published safety data. If one should ever find the need to use one of them in an application, we highly recommend that the manufacturer be contacted for any safety data that might be on file. In addition, some safety/abuse tests should be conducted by the user to verify the cell's safety in the intended application.

9.11.1. Lithium/Silver Chromate

The Li/Ag_2CrO_4 system was developed for powering implantable cardiac pacemakers. It uses a $LiClO_4$ in propylene carbonate electrolyte and can deliver over 200 Wh/kg and 700 Wh/liter.

9.11.2. Lithium/Silver–Bismuth Chromate

The $Li/AgBi(CrO_4)_2$ system was studied as a pacemaker battery. The electrolyte used is $LiClO_4$ in propylene carbonate, and the cells are capable of delivering 650 Wh/liter.

9.11.3. Lithium/Bismuth Oxychromate

The $Li/Bi_2O(CrO_4)_2$ chemistry was also investigated as a power source for pacemakers. The electrolyte of choice is $LiClO_4$ in propylene carbonate.

9.11.4. Lithium/Bismuth Oxide

The Li/Bi_2O_3 cell is manufactured in a button configuration for use in watches. The manufacturer does not give the composition of the electrolyte, but it probably consists of $LiClO_4$ in a mixture of propylene carbonate and dimethoxyethane.

9.11.5. Lithium/Lead Iodide

The Li/PbI_2 system was developed for use as a pacemaker battery. It is a solid-state system, with an electrolyte of lithium iodide, and has an energy density of 660 Wh/liter.

9.11.6. Lithium/Bismuth–Lead Oxide

The $Li/Bi_2Pb_2O_5$ ($Bi_2O_3 \cdot 2PbO$) chemistry has been utilized in button cells manufactured for consumer applications, such as wristwatches. It contains an electrolyte of $LiClO_4$ in dioxolane.

9.11.7. Lithium/Cobalt Polysulfide

The Li/Co_2S_7 and Li/Co_2S_9 systems have been built into experimental cells. The electrolyte used was propylene carbonate/dimethoxyethane with any of several lithium salts. A limited number of safety tests have been performed on these chemistries, with no hazardous behavior observed.[82]

REFERENCES

1. N. I. Sax and R. J. Lewis, Sr., *Dangerous Properties of Industrial Materials,* Van Nostrand Reinhold, New York (1989).
2. Material Safety Data Sheet, "Methanesulfonic Acid, Trifluoro-Lithium Salt," 3M, St. Paul, Minnesota (1990).
3. Material Safety Data Sheet, "Propylene Carbonate," Occupational Health Services, Inc., New York (1989).
4. Material Safety Data Sheet, "Manganese Dioxide," Occupational Health Services, Inc., New York (1991).
5. H. Ikeda, M. Hara, S. Narukawa, and S. Nakaido, Characteristics of lithium/manganese dioxide cells, in *Proceedings of the 2nd Manganese Dioxide Symposium,* pp. 395–413 (1980).

6. T. B. Reddy, J. R. Sullivan, and A. Shakked, Performance and safety characteristics of the MDX-200A, an hermetically sealed Li/MnO₂ primary cell, in *Proceedings of the Symposium on Lithium Batteries*, Proc. Vol. 84-1, The Electrochemical Society, Pennington, New Jersey (1984).

7. T. Umeda and C. Kawamura, *Prog. Batteries Sol. Cells* **5**, 81 (1984).

8. Matsushita Electric Industrial Co. Ltd., Batteries, Jpn. Kokai Tokkyo Koho, 80 80,276 (June 17, 1980).

9. J. W. Baker, J. A. Barnes, R. F. Bis, P. B. Davis, and L. A. Kowalchik, Safety Evaluation of the Expendable Sound Velocity Probe (SSXSV), Report NSWC TR 86-390, Naval Surface Weapons Center, Silver Spring, Maryland (1986).

10. P. R. Moses, F. J. Berkowitz, and A. H. Taylor, Electrochemical cell, U.S. Patent 4,937,154 (June 26, 1990).

11. W. L. Bowden and P. R. Moses, Improving the reliability of nonaqueous batteries, Belg. BE 890,071, 16 Dec 1981; U.S. Patent Appl. 182,897 (September 2, 1980).

12. K. Shinoda, K. Yamamoto, and Y. Harada, Batteries with organic electrolytes, Jpn. Kokai Tokkyo Koho JP 63,148,565 (June 21, 1988).

13. K. Shinoda, K. Yamamoto, and Y. Harada, Batteries with organic electrolytes, Jpn. Kokai Tokkyo Koho JP 63,148,566 (June 21, 1988).

14. K. Shinoda, K. Yamamoto, and Y. Harada. Batteries with organic electrolytes, Jpn. Kokai Tokkyo Koho JP 63,148,567 (June 21, 1988).

15. K. Shinoda, K. Yamamoto, and Y. Harada. Batteries with organic electrolytes, Jpn. Kokai Tokkyo Koho JP 63,148,568 (June 21, 1988).

16. K. Shinoda, K. Yamamoto, and Y. Harada. Batteries with organic electrolytes, Jpn. Kokai Tokkyo Koho JP 63,148,569 (June 21, 1988).

17. K. Shinoda, K. Yamamoto, and Y. Harada. Batteries with organic electrolytes, Jpn. Kokai Tokkyo Koho JP 63,148,570 (June 21, 1988).

18. K. Shinoda, K. Yamamoto, and Y. Harada. Batteries with organic electrolytes, Jpn. Kokai Tokkyo Koho JP 63,148,571 (June 21, 1988).

19. K. Shinoda, K. Yamamoto, and Y. Harada. Batteries with organic electrolytes, Jpn. Kokai Tokkyo Koho JP 63,148,572 (June 21, 1988).

20. S. Sadakuni, M. Higuchi, J. Komatsu, M. Yoshida, and F. Ooo, Nonaqueous lithium batteries, Jpn. Kokai Tokkyo Koho JP 02 21,567 (January 24, 1990).

21. K. Shinoda, K. Yamamoto, Y. Harada, and M. Kitakata, Nonaqueous batteries with improved safety, Jpn. Kokai Tokkyo Koho JP 01 281,678 (November 13, 1989).

22. Sanyo Electric Co., Ltd., Nonaqueous electrolyte battery, Jpn. Kokai Tokkyo Koho JP 58,220,367 (December 21, 1983).

23. Sanyo Electric Co., Ltd., Nonaqueous electrolyte battery, Jpn. Kokai Tokkyo Koho JP 58,220,366 (December 21, 1983).

24. F. Kita and K. Kajita, Organic electrolyte batteries, Jpn. Kokai Tokkyo Koho JP 02 239,571 (September 21, 1990).

25. K. Shinoda, K. Yamamoto, Y. Harada, and M. Kitakata, Nonaqueous batteries with improved safety, Jpn. Kokai Tokkyo Koho JP 01 281,677 (November 13, 1989).

26. K. Shinoda, K. Yamamoto, Y. Harada, and M. Kitakata, Nonaqueous batteries, Jpn. Kokai Tokkyo Koho JP 01 281,676 (November 13, 1989).

27. A. Sadamura and M. Nakai, Manufacture of nonaqueous batteries, Jpn. Kokai Tokkyo Koho JP 61,240,571 (October 25, 1986).

28. R. F. Bis, J. A. Barnes, W. V. Zajac, P. B. Davis, and R. M. Murphy, Safety Characteristics of Lithium Primary and Secondary Battery Systems, Report NSWC TR 86-296, Naval Surface Weapons Center, Silver Spring, Maryland (1986).

29. H. Ikeda and N. Furukawa, Manganese dioxide–lithium cells, in *Practical Lithium Batteries* (Y. Matsuda and C. Schlaikjer, eds.), JEC Press, Cleveland, Ohio (1988).

30. A. H. Taylor, M. J. Turchan, and S. E. Bascom, Batteries and methods for improving their safety, Belg. BE 898,928 (June 18, 1984); U.S. Patent Appl. 466,187 (February 16, 1983).

31. P. R. Moses, A. H. Taylor, and M. J. Turchan, Electrochemical cells and methods for improving safety, Ger. Offen. DE 3,405,610 (August 16, 1984); U.S. Patent Appl. 466,817 (February 16, 1983).

32. F. Kita, K. Kajita, and T. Manabe, Organic-electrolyte batteries with safety means, Jpn. Kokai Tokkyo Koho JP 01 272,053 (October 13, 1989).

33. K. Yamamoto, T. Mizuno, and H. Hamada, Batteries with coiled electrode stacks, Jpn. Kokai Tokkyo Koho JP 01 213,963 (August 28, 1989).

34. K. Yamamoto, Y. Tanaka, Y. Sasaki, and Y. Abe, Lithium–manganese batteries with coated separators, Jpn. Kokai Tokkyo Koho JP 01 319,250 (December 25, 1989).

35. P. S. Clark, Plastic cased lithium batteries, the challenge to achieve hermeticity and safety, in *Proceedings of the 35th International Power Sources Symposium*, pp. 4–9, IEEE, Piscataway, New Jersey (1992).

36. K. Yamamoto, Y. Tanaka, Y. Sasaki, and Y. Abe, Lithium batteries with coated cathodes, Jpn. Kokai Tokkyo Koho JP 01 319,253 (December 25, 1989).

37. P. L. Bedder, P. R. Moses, B. Patel, T. F. Reise, and A. H. Taylor, Electrochemical cells, U.S. Patent 4,622,277 (November 11, 1986).

38. B. Becker-Kaiser, P. Schmode, and J. R. Welsh, Li–MnO$_2$ cells for high rate applications, in *Proceedings of the 34th International Power Sources Symposium*, pp. 53–56, IEEE, Piscataway, New Jersey (1990).

39. M. P. Hart, E. P. Thurston, W. Andruk, and T. B. Reddy, Performance and safety characteristics of primary lithium manganese dioxide systems, in *Proceedings of the 35th International Power Sources Symposium*, pp. 129–132, IEEE, Piscataway, New Jersey (1992).

40. A. Jeffery, Recent developments in Dowty primary high rate Li/MnO$_2$ batteries, in *Proceedings of the 35th International Power Sources Symposium*, pp. 133–137, IEEE, Piscataway, New Jersey (1992).

41. T. Iwamuru and O. Kajii, Li/MnO$_2$ 2CR5 battery for automatic cameras, in *Practical Lithium Batteries* (Y. Matsuda and C. Schlaikjer, eds.), JEC Press, Cleveland, Ohio (1988).

42. W. N. C. Garrand, Effect of Shrapnel Penetration on Lithium–Carbon Monofluoride and Lithium–Manganese Dioxide Batteries, MRL Technical Note MRL-TN-579, Material Research Labs, Ascot Vale, Australia (1990).

43. S. Kawauchi, E. Kawakubo, K. Aoki, and M. Eza, *Prog. Batteries Sol. Cells* **4**, 87 (1982).

44. Material Safety Data Sheet, "Lithium Tetrafluoroborate," Sigma-Aldrich Corp., Milwaukee, Wisconsin (1992).

45. S. Budavari (ed.), *The Merck Index*, Merck & Co., Inc., Rahway, New Jersey (1989).

46. Material Safety Data Sheet, "Lithium Hexafluoroarsenate (V)," Sigma-Aldrich Corp., Milwaukee, Wisconsin (1992).

47. Material Safety Data Sheet, "Methyl Sulfite," Pfaltz & Bauer, Inc., Waterbury, Connecticut (1985).

48. T. M. Potts, Safety of double "D" Li/CF cells, in *Proceedings of the 30th Power Sources Symposium*, pp. 183–184, The Electrochemical Society, Pennington, New Jersey (1983).

49. D. Miller and R. Higgins, Shunt diode designs in Li/CF shuttle batteries, in *The 1983 Goddard Space Flight Center Battery Workshop*, NASA Conference Publication 2331, pp. 171–182 (1984).

50. S. E. Bucholz, J. W. Baker, J. A. Barnes, R. F. Bis, P. B. Davis, F. C. DeBold, and L. A. Kowalchik, Safety Evaluation of the NC-106 Multi-Order Gradient, Noise-Cancelling Mi-

crophone, Report NSWC-TR-86-388, Naval Surface Weapons Center, Silver Spring, Maryland (1986).

51. P. B. Davis, R. F. Bis, J. A. Barnes, S. E. Bucholz, F. C. DeBold, and L. A. Kowalchik, Safety Evaluation of an Electronic Totalizer Containing a Li/(CF)$_x$ Cell, Report NSWC-TR-83-508, Naval Surface Weapons Center, Silver Spring, Maryland (1983).

52. J. W. Baker and L. A. Kowalchik, Thermal Shock Abuse of Panasonic Li/(CF)$_x$ 2/3A and C Size Cells, Report NSWC-TR-85-398, Naval Surface Weapons Center, Silver Spring, Maryland (1985).

53. Panasonic Industrial Co., *Lithium Batteries Technical Handbook,* Panasonic Industrial Co., Secaucus, New Jersey (1984).

54. R. Bates and Y. Jumel, Lithium–cupric oxide cells, in *Lithium Batteries* (J. P. Gabano, ed.), Academic Press, London (1983).

55. Y. Jumel, O. Lamblin, and M. Broussely, Spirally wound lithium cupric oxide organic electrolyte cells, in *Proceedings of the Symposium on Lithium Batteries,* Proc. Vol. 84-1, pp. 293–300, The Electrochemical Society, Pennington, New Jersey (1984).

56. K. Murakami, K. Sato, S. Nishino, K. Miura, T. Izumikawa, and K. Momose, Nonaqueous batteries, Jpn. Kokai Tokkyo Koho JP 61,273,856 (December 4, 1986).

57. R. E. Bates, B. P. Murphy, G. D. White, and A. Attewell, The performance of Li–CuO cells and batteries under abusive conditions, in *Proceedings of the 30th Power Sources Symposium* pp. 175–178, The Electrochemical Society, Pennington, New Jersey (1983).

58. J. P. Arzur, *Prog. Batteries Sol. Cells* **4,** 49 (1982).

59. J. D. Jensen, D. L. Warburton, and G. F. Hoff, Safety and Performance Characteristics of SAFT Lithium Copper Oxide Cells and Batteries, Report NSWC TR 86-478, Naval Surface Weapons Center, Silver Spring, Maryland (1986).

60. R. F. Bis, J. A. Barnes, W. V. Zajac, P. B. Davis, and R. M. Murphy, Safety Characteristics of Lithium Primary and Secondary Battery Systems, Report NSWC TR 86-296, Naval Surface Weapons Center, Silver Spring, Maryland (1986).

61. Y. Jumel, M. Broussely, A. Thunder, and G. W. Allvey, Properties of the Li–CuO Couple, SAFT SOGEA Report, Hampton, England.

62. Lithium Copper Oxyphosphate Data Sheet, SAFT Leclanche, Poitiers, France.

63. A. M. Bredland, T. G. Messing, J. W. Paulson, Performance and safety characteristics of a Li/CuS NEDA 1604 battery, in *Proceedings of the 29th Power Sources Symposium,* pp. 82–85, The Electrochemical Society, Pennington, New Jersey (1981).

64. N. Margalit, Lithium–copper(II) sulfide cells, in *Lithium Batteries* (J. P. Gabano, ed.), Academic Press, London (1983).

65. A. J. Cuesta and D. D. Bump, The lithium cupric sulfide cell, in *Proceedings of the Symposium on Power Sources for Biomedical Implantable Applications and Ambient Temperature Lithium Batteries,* Proc. Vol. 80-4, pp. 95–101, The Electrochemical Society, Pennington, New Jersey (1980).

66. W. B. Ebner and C. R. Walk, Stability of LiAsF$_6$–methyl formate electrolyte solutions, in *Proceedings of the 27th Power Sources Symposium,* pp. 48–52, PSC Publications Committee, Red Bank, New Jersey (1976).

67. S. Sugihara, J. Yamamoto, M. Arakawa, I. Yoshimatsu, H. Otsuka, and S. Okada, Fire-resistant secondary batteries containing heterocyclic electrolyte solvents, Jpn. Kokai Tokkyo Koho JP 03,236,170 (October 22, 1991).

68. S. Sugihara, J. Yamaki, M. Arakawa, and I. Yoshimatsu, Fire-resistant batteries with silicon-containing heterocyclic electrolyte solvents, Jpn. Kokai Tokkyo Koho JP 03,236,171 (October, 22, 1991).

69. C. R. Walk, Lithium–vanadium pentoxide cells, in *Lithium Batteries* (J. P. Gabano, ed.), Academic Press, London (1983).

70. S. C. Levy, Electrical and environmental testing of lithium V_2O_5 cells, in *Proceedings of the 27th Power Sources Symposium*, pp. 52–56, PSC Publications Committee, Red Bank, New Jersey (1976).
71. Technical Data Sheet, "VARTAlith Li/CrO_x-Cells, Type ER," VARTA Batterie AG, Kelkheim, Germany.
72. F. J. Kruger, Characteristics of Low Drain Li/CrO_x Cells, in *Practical Lithium Batteries* (Y. Matsuda and C. Schlaikjer, eds.), pp. 42–45, JEC Press, Cleveland, Ohio (1988).
73. Material Safety Data Sheet, "Iron Disulfide," Occupational Health Services, Inc., New York (1991).
74. R. A. Langan, A safe, high performance 1.5 volt lithium/iron disulfide primary battery, *Wescon/92 Conference Record*, Western Periodicals Co., Ventura, California (1992).
75. E. S. Takeuchi, D. R. Tuhovak, and C. J. Post, Low temperature performance of lithium/silver vanadium oxide cells, in *Proceedings of the 34th International Power Sources Symposium*, pp. 355–358, IEEE, Piscataway, New Jersey (1990).
76. P. A. Cashmore, Safety and Qualification Testing Program and Results for Implantable Grade Model 8615 Lithium/Silver Vanadium Pentoxide Primary Cell, Wilson Greatbatch Ltd., Clarence, New York (1987).
77. P. A. Cashmore, Safety and Qualification Testing Program and Results for Implantable Grade Model 8830 Lithium/Silver Vanadium Oxide Primary Cell, Wilson Greatbatch Ltd., Clarence, New York (1989).
78. C. F. Holmes, P. Keister, and E. Takeuchi, *Prog. Batteries Sol. Cells* **6**, 64 (1987).
79. P. Keister, R. T. Mead, S. J. Ebel, and W. R. Fairchild, Performance and safety characteristics of a lithium/silver vanadium pentoxide battery for low to moderate rate applications, in *Proceedings of the 31st Power Sources Symposium*, pp. 331–338, The Electrochemical Society, Pennington, New Jersey (1985).
80. C. F. Holmes, Lithium/halogen batteries, in *Batteries for Implantable Biomedical Devices* (B. B. Owens, ed.), Plenum Press, New York (1986).
81. M. V. Tyler, Environmental Test Report for Model 802/23 Lithium Iodine Cell, Catalyst Research Corp., Baltimore, Maryland (1976).
82. W. L. Bowden, L. H. Barnette, and D. L. DeMuth, *J. Electrochem. Soc.* **136**, 1614 (1989).

10

Lithium/Thionyl Chloride Batteries

The lithium/thionyl chloride battery is one of the highest energy systems available, delivering up to 480 Wh/kg (950 Wh/liter). Due to its high energy content, care must be taken to ensure that cells and batteries are properly designed for each application and used in a safe manner. In addition to their high energy content, these batteries contain liquid thionyl chloride, which is toxic by inhalation and corrosive to the skin, eyes, and mucous membranes on contact. Continuous inhalation of the fumes may cause lung damage. Thionyl chloride has an LC_{50} of 500 ppm for 1 h exposure.[1] On contact with water (moist air), thionyl chloride reacts violently to give off corrosive fumes of HCl and SO_2.

10.1. DOCUMENTED SAFETY INCIDENTS

Over the years, a number of safety incidents involving lithium/thionyl chloride batteries have been documented.[2,3] Several are discussed below to give the reader a feel for the hazards associated with this system if proper precautions are not taken.

(a) A single cell in a large (10,000 Ah) Minuteman battery exploded, resulting in a fatality. The cell was being drained of electrolyte by a technician, prior to disposal, when an internal short occurred. The high short-circuit current coupled with the reduced electrolyte volume resulted in hot spots. This caused the lithium to melt, leading to thermal runaway.

(b) Other failure modes for Minuteman batteries that have been identified include:

1. An internal short followed by a pressure buildup, leading to an explosion.
2. Failure of a diode, allowing cells connected in parallel to be charged, leading to an explosion of the cells receiving charge.
3. Failure of the vent gas scrubbing system.

(c) A battery containing 28 "D" cells, used for oil well logging, exploded. Postmortem of other cells from the same lot showed $\sim$900 ppm of hydrolysis product when the cells were less than 50% discharged. The exact cause of the explosion was not determined, but the possibility of hydrogen generation from insufficiently dried components remains a possibility.

(d) A 50-pound lithium/thionyl chloride battery failed and exploded while in an explosive vault at the manufacturer. The explosion was contained within the vault, except for smoke.

(e) Three explosions occurred in the "AA" memory backup batteries in portable computers. It was concluded that, upon servicing, the new batteries were installed backward.

10.2. HAZARDOUS REACTIONS

Explosion hazards of $Li/SOCl_2$ cells include spontaneous reaction between the discharge products and cell constituents; a reaction between oxidation products at the anode after reversal; a reaction triggered by reactive lithium on the cathode after reversal; or a reaction triggered by polarization heat during cell reversal.[4]

Numerous studies have been conducted to characterize the chemical and electrochemical reactions that occur in $Li/SOCl_2$ cells during normal discharge, forced overdischarge into reversal, and charge and to identify the products of these reactions.

10.2.1. Normal Discharge

Analyses of discharged cells have found a variety of reaction products in the electrolyte. Typical species identified are $LiAlCl_4 \cdot 3SO_2$, $LiAlCl_4 \cdot 2SOCl_2$, $LiAlCl_4 \cdot SOCl_2 \cdot SO_2$ plus trace SO_3,[5] Cl_2 and SO_2,[6] SO_2, S_2Cl_2, Cl_2, and proposed OClS,[7] and OClS.[8] OClS has a short lifetime at room temperature so the concentration does not get high enough to cause a safety problem. However, at low temperatures the concentration may build to a point that, upon warming, it could initiate hazardous behavior.

In the gas phase HCl, CS_2, SO_2, S_2O, SCl_2, and $SOCl_2$ have been identified by mass spectroscopy.[9] Five hours after removal of the load, there was a large increase in the amount of HCl and SO_2 in the gas phase, plus the presence of CF_2^+ was observed.

X-ray diffraction studies of cathodes from discharged cells have ruled out the presence of Li_2O_2 and rhombohedral sulfur. All the lines can be accounted for by $LiCl \cdot H_2O$.[10]

Reactive intermediates that have been identified during the normal discharge of $Li/SOCl_2$ cells are SO, $(SO)_n$, OCIS, and Li_2O_2,[11] Cl_2, SCl_2, and OCIS having a half-life of seconds to minutes depending on the temperature,[12] and S_2O.[13] During low-temperature discharge, the concentration of S_2O can become fairly high. Upon warming, S_2O will decompose to SO_2 with the generation of heat and pressure.

10.2.2. Forced Overdischarge into Reversal

A number of studies have been carried out to determine the products formed when $Li/SOCl_2$ cells are driven into reversal. If the cells are anode (lithium) limited, the main reaction is the oxidation of thionyl chloride at the anode grid. The oxidation products found include SO_2Cl_2, $SOCl^+AlCl_4^-$, SCl_2, and Cl_2,[14] SO^{2+} and Cl_2O,[15] Cl_2, SO_2, SCl_2, and S_2Cl_2, but no ClO_2 or Cl_2O,[16] SO_3 and S_2Cl_2,[17] SO_2Cl_2, SO_2, SO^+, and Cl_2O,[18] and $SOCl^+AlCl_4^-$, $AlCl_3$, Cl_2, and SO_2Cl_2.[19] In studies of laboratory cells, the products in each compartment were identified.[20] Gas chromatographic analysis of the anode compartment found Cl_2, SCl_2, and SO_2 while Cl_2, SO_2, and S_2Cl_2 were found in the cathode compartment. Infrared analysis of the electrolyte compartment indicated the presence of SO_2Cl_2 and S_2Cl_2.

In cathode (carbon) limited cells, the reaction products found include LiCl, Li_2O_2,[21] and $LiAlSCl_2$.[22] The following reactions have been proposed to occur during the forced overdischarge of carbon-limited cells[14]:

$$2Li^+ + S + 2e^- \rightarrow Li_2S \qquad (4.1)$$

$$Li_2S + LiAlCl_4 \rightarrow LiAlSCl_2 + 2LiCl \qquad (4.2)$$

10.2.3. Charging

Gas analysis has been performed on $Li/SOCl_2$ cells which had vented after abusive charging.[23] The species identified were CO, CO_2, COS, SO_2, and CS_2, indicating that the carbon cathode takes part in the chemical reactions that occur in the cells during charging.

10.2.4. Other Studies

Studies on individual cell components indicate that the exothermic reaction with the lowest initiation temperature is between lithium and sulfur, in the dry state. However, the presence of electrolyte raises the initiation temperature to >300°C.[24] Evidence was found for a reaction between S and $SOCl_2$, which would tend to inhibit the Li + S reaction.

The hazards associated with sulfur in discharged $Li/SOCl_2$ cells can be further diminished with the addition of S_2Cl_2, which solubilizes the sulfur.[6,25] Two moles of S_2Cl_2 per mole of sulfur will keep all free sulfur in solution.[26]

The presence of water in $Li/SOCl_2$ cells has been found to increase the safety hazards. Water not only reacts exothermally with $SOCl_2$ but also catalyzes the explosive reaction between Li_2S and $SOCl_2$.[27,4]

Lithium nitride, Li_3N, has also been shown to react violently with thionyl chloride.[28] Therefore, care should be taken to minimize the exposure of lithium to ambient air prior to assembly into cells.

Lithium metal itself is very reactive with most of the chemical species found in $Li/SOCl_2$ cells, but the presence of a passive film of LiCl inhibits its reactivity. However, when the lithium is mixed with carbon, its reactivity increases markedly.[28-32] This becomes a serious consideration during reversal of cathode-limited cells, when Li deposits on the carbon surface of the cathode.

Exothermic chemical reactions involving the reduction products of $SOCl_2$ have been found to occur after cell discharge has been completed.[33] Spontaneous explosions of partially discharged cells on storage are believed to be initiated by a series of reactions as indicated below.

$$SOCl_2 + e^- \rightarrow SOCl + Cl^- \qquad (4.3)$$

$$SOCl + e^- \rightarrow SO + Cl^- \qquad (4.4)$$

$$2SO \rightarrow (SO)_2 \qquad (4.5)$$

$$(SO)_2 \rightarrow S + SO_2 \qquad (4.6)$$

10.2.5. Increased Pressure

There are four categories of gas that can be found in $Li/SOCl_2$ cells[34]:

1. SO_2 generated electrochemically during discharge
2. H_2 from the reduction of hydrolysis products or other protic species
3. Inert gases trapped by ad/absorption on the cathode and released during discharge
4. N_2 released from the Li anode during discharge

There are a number of reactions that produce pressure in Li/SOCl$_2$ cells. Sulfur dioxide is produced as a result of the normal discharge of thionyl chloride. However, more SO$_2$ is produced in intermittently discharged or low rate discharged cells than in high rate discharged cells. This is due to a higher concentration of intermediates that are present under these conditions. Also, intermittent or low rate discharged cells produce hydrogen more readily from the reaction of lithium with protic species. Hydrogen is not generated in fresh cells; some discharge is required.

Gases such as N$_2$, He, or Ar that may be trapped in the porous carbon cathode may be released during discharge. They can be eliminated by an SO$_2$ purge prior to filling. Nitrogen can also be introduced into the cell via the lithium, either as a dissolved gas or as Li$_3$N.

10.3. CELL DESIGNS

There are a number of design parameters for Li/SOCl$_2$ cells, each of which strongly influences safety. External configuration (cylindrical vs. prismatic vs. button), internal configuration (spirally wound vs. flat plate vs. bobbin), balance (anode limited vs. cathode limited vs. electrolyte limited), whether or not to use vents (type of vent and operating pressure), separator (ceramic vs. glass), and can (positive vs. negative) are among the design features that must be considered when either designing or choosing a Li/SOCl$_2$ cell for a specific application.

10.3.1. Configuration

Cells designed to deliver high rates are usually either spirally wound or flat plate, having thin electrodes. Low-rate cells are normally bobbin or flat plate, having thicker electrodes. High-rate cells are more hazardous because of their inherently higher short-circuit current and because of heat transfer problems not encountered in cells with thicker cathodes.[35]

Small bobbin Li/SOCl$_2$ cells have been shown to be safe when short circuited[35] and charged.[11] In fact, vents are not needed on bobbin cells if the electrode area is limited and the lithium is on the outside, in contact with the cell case.[36] This design limits the short-circuit current and provides good heat transfer. Cells of this type have survived insulated shorting at 25 and 72°C, forced overdischarge at the 3C rate, charging of both fresh and discharged units, heating to 250°C on a hot plate, compression to 30% of original height, and puncturing. Even large cells of this design have been shown to be safe. A 45-A h cell, equipped with a polyswitch for overcurrent protection, was short-circuited.[37] A maximum current of 5 A was obtained which decreased to 200 mA within 30 seconds. No hazards were observed.

A small prismatic cell for use in a 9-V battery has been designed with special safety features.[38] The cell uses a ceramic felt separator and hermetic ceramic seal, contains reverse voltage protection, and has a 3-A fuse in the terminal board. No hazards were observed during charging or forced over-discharge. A safe, controlled vent was obtained during incineration and crush. Short circuit yielded a maximum current of 7 A and a maximum temperature of 100°C with no ventings.

Spirally wound cells, even in very small sizes such as those for memory retention applications, are a safety hazard if short-circuited.[39] There have been no reports of hazards on shorting of bobbin cells of this type, and they are believed to be safe.

10.3.2. Balance

In designing a Li/SOCl$_2$ cell, one has the option of a balanced design (all active materials in equivalent amounts), an anode-limited design (end of life due to lithium depletion), a cathode-limited design (end of life due to plugging of the carbon cathode collector with reaction products), or an electrolyte-limited design (end of life due to depletion of SOCl$_2$). The electrolyte-limited design is extremely hazardous and should never be used under any circumstances. As the cell discharges, dry areas form, increasing the current density throughout the rest of the cell. Eventually, enough heat will be generated locally to melt the lithium, initiating exothermic reactions that will lead to thermal runaway and catastrophic failure. A balanced design is effectively cathode limited during normal use[40] and is seldom used.

Most Li/SOCl$_2$ cell designs are either lithium or carbon limited. A cell can be anode limited even though it is not lithium limited if the lithium loses contact with the anode grid.[41,42,19] This can be an extremely hazardous situation since electric arcs between the lithium and the grid may initiate an explosion.[43]

There has been considerable debate as to whether the lithium-limited or the carbon-limited design is safer when cells are overdischarged into reversal. Each configuration will be discussed separately.

Lithium-Limited Design. During reversal, the electrolyte is oxidized at the anode grid (usually a Ni screen).[44] A number of studies have been carried out to define the products of this reaction. Species identified include Li$_2$O$_2$,[45] SO^{2+} and Cl$_2$O,[15] Cl$_2$,[46] SO$_2$Cl$_2$, SOCl$^+$AlCl$_4^-$, SCl$_2$, and Cl$_2$,[14] Cl$_2$O,[47] Cl$_2$ and AlCl$_3$,[48] Cl$_2$,[22] and SO$_3$ and S$_2$Cl$_2$.[17] Studies in laboratory cells have identified Cl$_2$, SCl$_2$, and SO$_2$ in the anode compartment and Cl$_2$, SO$_2$, and S$_2$Cl$_2$ in the cathode compartment.[20]

Overdischarging lithium-limited cells at low rates (<1 mA/cm^2) can lead to electrolyte depletion due to oxidation of the SOCl$_2$ at the anode grid. As the lithium is depleted, higher current densities are forced through the remaining lithium, resulting in local heating.[45] This can eventually lead to thermal runaway.

Anode-limited cells that are driven into reversal usually experience wide voltage fluctuations, compared to cathode-limited cells, which have smooth voltage traces. This is caused by intermittent contact between residual lithium and the anode substrate.[46]

Researchers in the field are divided as to the safety of anode-limited Li/SOCl$_2$ cells. Some feel this design is safer because there is a minimum amount of lithium remaining at the end of life, resulting in less energy within the cells.[49,50,43,48] Several believe it is a safe design during low-rate overdischarge,[11,42,51] while others have found that anode-limited cells can sustain high current densities in reversal with no ventings or explosions.[52] Both lower temperature and pressure maxima[53] and higher temperature and pressure maxima[54] have been reported, with no ventings or explosions noted. One study found no evidence of any hazardous species generated during oxidation of SOCl$_2$ during forced overdischarge.[16] Several researchers state that the lithium-limited design is safe.[55,56] It is preferred if cells are to experience reversal[35] and is less likely to result in thermal runaway than other designs.[57]

Conversely, a number of people in the field believe that anode-limited designs are more hazardous.[58] It is believed by some that the oxidation of SOCl$_2$ during reversal leads to reactive species and/or intermediates that may result in an explosion, for example, SO^{2+} and Cl$_2$O.[15,19,22,47] Others have found that forcing anode-limited cells into reversal will result in an explosion.[14,41,45,59] Tests have shown that driving a lithium-limited cell into reversal at >10 mA/cm^2 will generate pressures inside the cell in excess of 300 psi.[60]

It should be noted that lithium-limited cells, when discharged at high rates (~20 mA/cm^2) and/or low temperatures, can become carbon limited.[17,35,45,61] This is due to the carbon collector becoming saturated with discharge product (LiCl) at the surface and has been verified by the presence of lithium dendrites on the cathode.

Carbon-Limited Design. During reversal, lithium dendrites will plate on the cathode (Refs. 14, 17, 19, 35, 40, 41, 43, 45, 48, 52, 57–62, 63). There are two schools of thought as to the effect of deposited lithium on the cathode in terms of safety. The first considers it a hazardous situation due either to the reactive nature of the high-surface-area deposits[45,57,61] or to the fact that internal shorts may develop which could lead to localized heating followed by venting or explosions.[35,45,40] The second believes that the cells are safer, since the dendrites will shunt the internal current and prevent hazardous reactions from occurring (Refs. 14, 19, 41, 43, 48, 51, 58, and 63).

Dendrites formed during reversal of cathode-limited cells at low temperature ($-40°C$) have been found to be more reactive, and thus more hazardous, than those formed at higher temperatures. Several reasons have been postulated for this behavior: (1) more SO_2 is in the electrolyte because of higher solubility and slower reaction with $SOCl_2$, making the Li more reactive toward the electrolyte[57]; (2) the Li deposits are finely divided and pyrophoric and react violently when the cell is warmed[48,64]; however, if the cell is left at $-40°C$ for ~5 days, the lithium slowly reacts and the cell is safe when warmed; (3) many of the dendrites detach from the cathode; when warmed, they react with the discharge products resulting in a thermal runaway[61]; (4) the low temperature allows an intermediate to either decompose or be converted to a second intermediate which can react with the Li dendrites; as the cell warms, thermal runaway may occur. Also, since carbon catalyzes the reactivity of lithium, the lithium deposited on the cathode will be more reactive than lithium in other parts of the cell.

A number of potentially hazardous products have been identified in cathode-limited cells that have been driven into reversal. At high currents, Cl_2O has been observed.[45] It is postulated that decomposition of Cl_2O may lead to explosions. Several researchers have found Li_2O_2 in the cathodes of overdischarged cells.[21,42,45] This is a strong oxidizing agent and may initiate hazardous behavior under certain conditions. Sulfur is a product of the normal reduction of $SOCl_2$ and will precipitate near the end of life, as the $SOCl_2$ is depleted. Lithium will react with sulfur to form Li_2S. The reaction is accelerated when lithium is in intimate contact with carbon, as occurs in cathode-limited cells in reversal. This mixture (Li, C, S) has been reported to be shock sensitive and can lead to thermal runaway.[35] It has also been reported that the Li, C combination is unstable in the presence of $SOCl_2$[35] and may even be explosive.[16]

A Taguchi analysis was carried out to study a number of variables affecting Li/$SOCl_2$ cell safety.[65] Specially designed cells, having a pressure transducer built into the header, were used. It was found that cathode-limited cells developed lower pressures during reversal and that the maximum pressure is inversely proportional to the amount of lithium in the cell at the start of reversal.

There is a difference of opinion among workers in the field concerning the safety benefits of using a cathode-limited design. Some believe that this design generally enhances the safety of Li/$SOCl_2$ cells,[40,42,60] and others feel it is safe in situations where cells are forced into reversal.[14,22,56] On the other hand, there are those who believe that the cathode-limited design is more hazardous in reversal.[46,53] This design has been reported to become shock sensitive in reversal[11,43] and capable of venting or exploding.[52,54,66]

One can conclude from the above discussions that the hazardous behavior of Li/SOCl$_2$ cells is very dependent on their design and on the way in which they are used/abused. Therefore, to ensure a safe battery, one must design and test each cell and battery pack for every specific application. If the battery is to be used in another application, it must be tested to the new requirements before being put into service. This is true even if the electrical requirements are the same, since the location of the battery or the device may be such that the heat transfer properties will be different, resulting in a more severe thermal environment.

10.3.3. Vents

Vents are normally incorporated into the case of Li/SOCl$_2$ cells, except in certain low-rate bobbin designs. The purpose of the vent is to prevent an explosion in a hazardous cell by allowing the electrolyte to be ejected prior to the melting of the lithium. Ideal properties of a safety vent are a low operating pressure [$\sim$200 psi (gauge)], a consistent operating pressure, sufficient expulsion of electrolyte through the open vent, no degradation with age, small size, simple design for mass production, and low cost.[67] Several different vent designs have been used in Li/SOCl$_2$ cells.[68–71]

When designing a cell to be used with a vent, consideration must be given to the void volume within the can. If the void volume is too low, the cell will vent prematurely. If it is too high, the internal temperature may reach $>180°C$ before the cell vents, which can lead to severe safety problems. Void volume should be in the range of 4–8%, which will allow for electrolyte expansion up to temperatures of 90–110°C.[72]

A model has been developed to predict the internal pressure of Li/SOCl$_2$ cells as a function of temperature, state of charge, void volume, and inert gas pressure initially in the cell.[73] On the basis of this model, a suitable design for a 250-Ah cell includes a vent pressure of $\sim$200 psi, a void volume of 112 cm^3, and an inert gas pressure of 40 psi.

A gas is sometimes added to a cell to improve its venting ability.[74] Cells containing 50% SO$_2$ in a 1.2M LiAlCl$_4$/SOCl$_2$ electrolyte were tested in a fire. The cells vented normally while control cells, without the SO$_2$ additive, exploded.

In one study, cells with low-pressure vents were driven into reversal. All exploded, although the case temperature only reached 40°C. Thus, low-pressure vents do not appear to be effective for protecting cells against reversal.[75]

10.3.4. Separators

Glass separators have traditionally been used in $Li/SOCl_2$ cells, but recent studies have shown that other materials may be suitable.[76] The development of separators specifically designed to improve safety has concentrated on preventing molten lithium from migrating through the cell.

It has been reported that an Al_2O_3/SiO_2 ceramic separator, porous enough to allow normal cell operation, can act as an ion-permeable mechanical barrier and prevent most of the hazards associated with voltage reversal.[77]

An "explosion-proof" bobbin type $Li/SOCl_2$ cell has been designed which contains a microporous film of ethylene–tetrafluoroethylene copolymer together with a nonwoven glass separator. The micropores are uniform and tortuous and retard the migration of molten lithium through the separator to the cathode. This allows the vent to operate before molten lithium can react with the positive electrode. The nonwoven glass is used to retain electrolyte.[78]

A novel method has been developed to prevent the violent reaction of molten lithium with other cell components or discharge products. An inorganic solid (95% Al_2O_3, 5% SiO_2) is used as the separator and is placed in intimate contact with the lithium anode. At a predetermined temperature, the lithium reacts with the separator endothermically, or mildly exothermically, to form new solid phases that are chemically nonreactive.[79] Below the critical temperature, the separator does not interfere with the normal operation of the cell.

10.3.5. Other Safety Innovations

Nickel powder, either dispersed in the carbon cathode to catalyze the reaction of unstable discharge products[80,81] or as the cathode collector,[82] has been shown to improve the safety of $Li/SOCl_2$ cells under abusive conditions.

An electrochemical switch has been devised which closes upon the cell entering reversal.[46] In a cathode-limited, case-positive design, copper is present on the case (e.g., as a braze in the ceramic-to-metal seal). Opposite the copper is placed a small anode grid, which favors dendritic growth of lithium from the copper to the grid during reversal. The current then passes harmlessly through the dendrite bridge, and no hazardous reactions occur. Alternatively, a porous metal member can be placed between the lithium and carbon, and in electrical contact with the carbon.[83] Upon voltage reversal, lithium will plate on the porous metal, and dendrites will bridge to the anode. This prevents the lithium from mixing with the carbon and allows the cell to act as a resistor, shunting the current.

Use of a copper foil sandwiched between the lithium in the anode minimizes local "hot spots" during forced overdischarge. This results in a more even distribution of current density over the anode, and because of the high thermal conductivity of the copper foil, distributes heat more evenly throughout the anode. Upon consumption of the lithium, copper is deposited at the cathode. The copper then forms dendrites which short to the anode, forming a low-resistance path. It also inhibits the reactivity of dendritic lithium on the cathode.[84]

Since one of the basic causes of safety-related problems in Li/SOCl$_2$ cells is the high reactivity of lithium coupled with its low melting point, alloying of the lithium can improve the safety characteristics of these cells. Alloys of lithium with low concentrations (4–6%) of Al, Mg, Ca, Si, and Cd have been studied.[85,86] Initially, the short-circuit current in cells using alloy anodes was similar to that in cells with pure lithium anodes. However, after a brief predischarge and storage, the short-circuit current was reduced. This is due to the formation of a barrier layer upon storage which is rich in the alloying element, inhibiting the transport of lithium ions through the layer.

The concentration of electrolyte has an impact on the safety of Li/SOCl$_2$ cells. One study has shown that if the salt concentration is too low, a buildup of sulfur may occur, increasing the risk of hazardous behavior. A concentration of 1.5–1.6M is preferred.[55] However, other studies have found that a lower electrolyte concentration reduces the risk of hazardous behavior. In one case, similar cells were built with either 0.3M or 1.8M LiAlCl$_4$ electrolyte. In short-circuit tests of the cells, the cells with 1.8M electrolyte exploded in 2–3 min, while the cells with 0.3M electrolyte did not explode.[86] In a Taguchi matrix designed to study the effects of various parameters on the safety of Li/SOCl$_2$ cells, it was found that a lower electrolyte concentration (1.0M vs. 1.8M) resulted in lower temperatures and pressures during forced overdischarge into reversal.[87]

A number of additives have been proposed to increase the safety of Li/SOCl$_2$ cells. The addition of S$_2$Cl$_2$ keeps free sulfur in solution and avoids the exothermic reaction of sulfur with lithium.[26] It has been claimed that 0.9M NbCl$_5$ added to the electrolyte absorbs electrons released by lithium under load until the cell is completely dead, making the cells safe under reverse voltage conditions.[88] Adding poly(vinyl chloride) in the form of a gel to the electrolyte protects the cells at high temperature, for example, during short circuit. As the internal temperature rises to ~90°C, the gel dissolves, limiting the current flow.[89] Finally, PCl$_5$ has been used as an additive, either to the electrolyte or to the cathode, to prevent overpressurization due to SO$_2$ buildup during high-rate discharge.[90]

In multicell battery packs, use of a low-melting, high-heat-capacity material between the cells can mitigate a hazardous situation in the event one cell is short-circuited. The material melts and transfers heat to all the cells in the battery, increasing the available heat capacity and avoiding thermal runaway. Heat-absorbing materials such as paraffin, wax, and tar having a melting point in the range of 40–150°C may be used.[91]

10.4. THERMAL STUDIES

A number of thermal studies have been carried out for Li/SOCl$_2$ cells to help identify and characterize hazardous exothermic reactions that occur in this system. Some studies looked at cell components using either differential scanning calorimetry (DSC) or differential thermal analysis (DTA). It was found that lithium reacts violently with sulfur in the dry state but in the presence of electrolyte either does not react due to the presence of a passive film on the lithium[31] or must be heated to >300°C for the reaction to initiate.[24] Lithium mixed with carbon was found to be very reactive with the thionyl chloride electrolyte.[31,28] As the temperature was raised above the melting point of lithium, several hazardous reactions were found to occur.[92] At 209°C lithium reacts exothermally with the glass separator, and at 223°C lithium reacts with thionyl chloride in a violent manner.

Lithium sulfide was shown by DTA to react explosively with thionyl chloride,[27] the reaction being catalyzed by water. DTA of fresh cells has indicated there are no exothermic reactions occurring below the melting point of lithium.[35]

Several studies have looked at heat dissipation in Li/SOCl$_2$ cells. It was found that ~90% of the heat is dissipated in the radial direction.[93,94] If a cell were designed with fins on the surface, three times more heat would be removed during discharge.

The thermal masses of Li/SOCl$_2$ cell components are given in Table 10.1, along with the approximate percentage of each component in a typical cell.[93]

The specific heat of 1.8M LiAlCl$_4$ in SOCl$_2$ electrolyte is ~1215 J/(kg·K). Thus, a flooded cell is safer from a heat transfer point of view also, since the percentage loss of thermal mass through electrolyte consumption is less than in an electrolyte-starved cell.

TABLE 10.1. Thermal masses of cell components

Component	J/°C	Percent in cell
Lithium	19.5	20
Nickel	6.7	6
Stainless steel	19.2	20
Electrolyte	53.0	54

Rectangular cells are safer than cylindrical cells during short circuit, from a thermal viewpoint, because of their better heat radiation characteristics.[95] Thermal analysis of prismatic cells indicated that six cells in series could operate safely at rates up to $C/5$. At higher rates, free space is needed between the cells for convection, in order to maintain the temperature below $100°C$.[96]

Studies of discharged $Li/SOCl_2$ cells by microcalorimetry have shown that the cells continue to produce heat after discharge.[97,98] Thus, controlled thermal treatment of discharged and partially discharged cells may prevent explosions during casual storage. Similar studies have found that when the load voltage drops below 2.7 V, the rate of heat generation becomes significant.[99]

10.5. CATALYSTS

The incorporation of a catalyst into the cathodes of $Li/SOCl_2$ cells has been found to improve their safety characteristics. Cobalt tetraazoannulene (Co-TAA) significantly reduces hazardous behavior at high rates[100] as well as at low temperatures.[101]

The catalysts promote less hazardous behavior by modifying the cathode reaction mechanism to reduce or eliminate unstable intermediates or reaction by-products.[102-104] Specific reaction mechanisms include the direct reduction of $SOCl_2$ to sulfur with no SO-type intermediates or a two-electron reduction, eliminating the reactive intermediates formed during a normal one-electron reduction, by incorporating Co-phthalocyanine as the catalyst.[105]

10.6. MODELING

Models have been developed over the years to explain/predict unsafe behavior in $Li/SOCl_2$ cells. A model of how catastrophic failure occurs starts with cracks that develop in the LiCl passive film on the surface of the lithium anode. These cracks are repaired by reaction with the solvent via an exothermic chemical reaction. As the current increases above some critical value, the crack formation rate increases markedly, resulting in a high internal heating. When the local temperature reaches the melting point of lithium, the film breaks down, and venting or explosion occurs.[106] A model developed to explain thermal runaway from a single heat point source includes an induction period, followed by ignition, propagation of the reaction zone, venting, and flame extinction.[107] Thermal runaway has been found to be associated with ignition and burning rather than explosion.

A mathematical model has been developed which indicates that the initial rise in cell temperature is due to electrode polarization. At the end of discharge, localized reactant depletion results in severe polarization and an increase in the rate of temperature rise. Therefore, for a cell to remain safe, it must operate below some critical current density, limits on the ambient temperature must be imposed, and the current should be cut off before the cell is depleted.[108]

A computer program has been developed for thermal modeling of Li/ $SOCl_2$ cells which calculates the instantaneous heat generation rates and the instantaneous wall temperature.[109] During normal discharge rates, the primary heat dissipation modes are convection and radiation. At high discharge rates, convection and radiation only dissipate 40% of the heat generated. During charging, vigorous chemical reactions occur in the cell as the point of explosion is neared.

An equation has been developed to estimate the heat generation in Li/ $SOCl_2$ cells[66]:

$$Q = I(E_{oc} - E_{op}) - IT \, dE_{oc}/dT \qquad (4.7)$$

where Q is heat, I is current, E_{oc} is the open-circuit voltage, E_{op} is the operating voltage, and T is temperature. The electrochemical heat generated at 20°C is

$$Q = I(3.65 - E_{op}) - I(0.316) \qquad (4.8)$$

An empirical equation has also been generated which gives a quick estimate of heat generation:

$$Q = (3.7 - E_{op})I \qquad (4.9)$$

10.7. SAFETY/ABUSE TESTS

A significant number of safety/abuse tests have been carried out for Li/ $SOCl_2$ cells of various sizes and configurations. Results vary, depending on the cell design and the type of test performed.

10.7.1. Wound Cells

Wound cells are high-rate designs and can be hazardous when abused. Severe abuse, for example, short circuit, charging, penetration with a nail or bullet, or compression, will result in cells venting or exploding.[110-114] Casual abuse, for example, reversal or shock, may result in a mild vent or leakage.

10.7.2. Bobbin Cells

The low-rate-design bobbin cells appear to be safe under abusive conditions and for most applications do not require a vent.[95,92] Overheating or short circuit results in a swelling of the case or benign venting. Other abusive conditions, for example, high-rate discharge, charging to 10 mA/cm^2, forced overdischarge, puncture, and compression, have been shown not to be safety hazards. Bobbin cells can be hazardous if partially discharged cells are charged at high current.[115]

10.7.3. Prismatic Cells

Safety/abuse tests conducted on prismatic cells have shown them to be relatively safe.[95,116,117] During short circuit, this design allows for more efficient heat transfer from the cell, with only bulged cans and mild vents or leaks resulting. Overdischarge and reversal have caused slight bulging of cans with no vents or leaks. When prismatic cells were subjected to nail penetration, some electrolyte leaked from the nail hole, and a fire test caused normal venting to occur. Shock and vibration, as well as thermal shock of large (2000 and 10,000 Ah) cells, presented no hazardous behavior.[117]

10.7.4. Flat Circular Cells

Flat circular cells have been built by several manufacturers in a range of sizes. A number of safety/abuse test programs have shown these cells to pose no safety problems with either nail or bullet penetration, crushing, short circuit, or reverse voltage.[25,118,119] Heating with a propane torch or on a hot plate resulted in a mild vent, while incineration caused normal venting. Driving a cell into reversal while cold, and allowing it to warm on open circuit while insulated, caused a violent venting. Also, driving a cell into reversal at ambient temperature and then storing it cold resulted in an explosion after ~2 h.[120]

Other tests with a 3-in.-diameter, 0.9-in.-high cell found that it will vent during a 100-A discharge and will vent with a "Roman candle" effect on short circuit.[111] Tests with a high-rate four-cell battery pack containing a diode and slow-blow fuse in series gave the following results: all cells vented during 150°C storage, one cell in one battery (of seven) vented during forced overdischarge at 290 mA, and a benign vent occurred during incineration after ~2 min.[121]

10.7.5. Small and Large Cells

Low-rate "AA" cells, crimp-sealed with a ~400-psi vent, were put through a series of electrical, environmental, and mechanical abuse tests.[122]

It was found that charging at >100 mA for a discharged cell and >1 A for a fresh cell resulted in venting. Immersing a cell in fresh or salt water, placing a cell on a hot plate, or exposing an opened cell to fire or water did not result in any hazardous behavior. Finally, cutting with a saw, crushing, puncturing, dropping, shock, and vibration did not cause any safety hazards, except for fumes in confined areas.

Large 16,500-Ah cells were electrically and mechanically abused.[50] Fresh, half-capacity, and fully discharged units were short-circuited. Worst case temperature and pressure rise was found in fresh cells. Glass-to-metal seals fractured in ~ 8 min, and the temperature rose to $\sim 316°C$. Similar results were obtained in partially discharged cells. Reversal at 0.6 mA/cm^2 for >45 h showed no safety hazards. Charging of fresh, half-capacity, and fully discharged cells did not result in any hazardous behavior. Puncture tests of a fully charged cell resulted in a violent venting. Puncture of a cell discharged at 10°C caused an explosion. Ramming a fully charged cell with a forklift tine dented it 2–3 in.; it then exploded after ~ 2 min. Similar tests were repeated on 400-Ah cells, and there was no hazardous behavior, indicating that size does play a role in cell safety.

10.7.6. Alloy Anode Cells

Cells using a Li–Mg alloy anode were subjected to short circuit, heating, and short-circuit/heating tests. The hot short-circuit test resulted in a skin temperature of 245°C and a violent explosion. The other tests were safe, since the temperature remained below the melting point of the alloy.[123] Short-circuit tests of a "D" cell having a 3% Mg alloy anode yielded a peak current of 16 A, and the cell cover flew off violently. Similar tests with a cell having a 5% Ca alloy anode bulged the cell case, but there was no rupture or explosion.[85]

The sometimes conflicting results reported for safety/abuse testing of Li/ SOCl$_2$ cells again point out the fact that hazardous behavior is related to cell design and the conditions to which the cell is exposed. Thus, before using any Li/SOCl$_2$ cell, it should be tested to the worst case conditions the cell might experience in that application.

REFERENCES

1. N. I. Sax and R. J. Lewis, Sr., *Dangerous Properties of Industrial Materials,* Van Nostrand Reinhold, New York (1989).
2. N. Marincic, Hazardous behavior of lithium batteries, case histories, in *The 1982 Goddard Space Flight Center Battery Workshop,* NASA Conference Publication 2263, pp. 15–22 (1983).

3. P. K. Raj, Safety Issues Related to the Storage of Chemicals in Advanced Missile Bases for Power Generation and Life Support Systems, Technical Report: Volume 2, Technology and Management Systems, Inc., Burlington, Massachusetts (1989).

4. A. N. Dey, Primary Li/SOCl₂ cells VI. Identification of the possible explosion causing cell constituents by DTA studies, in *Proceedings of the 28th Power Sources Symposium*, pp. 251–255, PSC Publications Committee, Red Bank, New Jersey (1978).

5. R. C. McDonald, F. W. Dampier, P. Wang, and J. M. Bennett. Investigations of Lithium Thionyl Chloride Battery Safety Hazards, *Technical Report No. 60921-81-C-0229*. Final Report for Period Sept. 1981 to Nov. 1982, Naval Surface Weapons Center, White Oak, Silver Spring, Maryland.

6. H. V. Venkatasetty and D. J. Saathoff, *J. Electrochem. Soc.* **128**, 773 (1981).

7. B. Carter, R. Williams, F. Tsay, A. Rodriguez, and H. Frank, Lithium–thionyl chloride battery safety, in *Proceedings of the 17th Intersociety Energy Conversion Engineering Conference*, Vol. 2), pp. 638–641, IEEE, Piscataway, NJ (1982).

8. B. J. Carter, R. M. Williams, F. D. Tsay, A. Rodriguez, S. Kim, M. Evans, and H. Frank, *J. Electrochem. Soc.* **132**, 525 (1985).

9. A. I. Attia, K. A. Gabriel, and R. P. Burns, Investigation of Lithium–Thionyl Chloride Battery Safety Hazards, Report No. 838-012, Naval Surface Weapons Center, Silver Spring, Maryland (1983).

10. R. M. Williams, S. Surampudi, and C. P. Bankston, *J. Electrochem. Soc.* **136**, 1287 (1989).

11. S. Subbarao and G. Halpert, Safety of Li–SOCl₂ cells, in *The 1984 Goddard Space Flight Center Battery Workshop*, NASA Conference Publication 2382, pp. 153–173 (1985).

12. B. J. Carter, S. Subbarao, R. Williams, M. Evans, S. Kim, and F. D. Tsay, The chemistry of lithium–thionyl chloride cells, in *Proceedings of the 31st Power Sources Symposium*, pp. 400–405, The Electrochemical Society, Pennington, New Jersey (1984).

13. W. K. Istone and R. J. Brodd, *J. Electrochem. Soc.* **131**, 2467 (1984).

14. K. M. Abraham, R. M. Mank, and G. L. Holleck, Investigation of the safety of Li/SOCl₂ cells, in *Proceedings of the Symposium on Power Sources for Biomedical Implantable Applications and Ambient Temperature Lithium Batteries*, Proc. Vol. 80-4, The Electrochemical Society, Pennington, New Jersey (1980).

15. D. J. Salmon, M. E. Peterson, L. L. Henricks, L. L. Abels, and J. C. Hall, *J. Electrochem. Soc.* **129**(11), 2496 (1982).

16. B. J. Carter, H. A. Frank, and S. Szpak, *J. Power Sources* **13**, 287 (1984).

17. R. C. McDonald and F. W. Dampier, Investigation of Lithium Thionyl Chloride Battery Safety Hazard, Technical Report N60921-81-C-0229, Final Report for Period September 1981–November 1982, Naval Surface Weapons Center, Silver Spring, Maryland (1982).

18. D. J. Salmon, M. E. Adamczyk, L. L. Hendricks, L. L. Abels, and J. C. Hall, Spectroscopic studies of the hazards of Li/SOCl₂ batteries during anode limited cell reversal, in *Proceedings of the Symposium on Lithium Batteries*, Proc. Vol. 81-4, pp. 64–77, The Electrochemical Society, Pennington, New Jersey (1981).

19. K. M. Abraham, G. L. Holleck, and S. B. Brummer, Studies of explosion hazards of Li/SOCl₂ cells on forced overdischarge, in *Proceedings of the Symposium on Battery Design and Optimization*, Proc. Vol. 79-1, pp. 356–364, The Electrochemical Society, Princeton, New Jersey (1979).

20. B. J. Carter, R. M. Williams, M. Evans, Q. Kim, S. Kim, F. D. Tsay, H. Frank, and I. Stein, The chemistry of Li–SOCl₂ cells during anode-limited voltage reversal, in *Proceedings of the Symposium on Lithium Batteries*, Proc. Vol. 84-1, pp. 162, The Electrochemical Society, Pennington, New Jersey (1984).

21. R. C. McDonald, Investigation of Lithium Thionyl Chloride Battery Safety Hazards, Contract No. N60921-81-C-0229, Quarterly Technical Progress Report Period 1 April 1982–30 June 1982, Naval Surface Weapons Center, Silver Spring, Maryland (1982).

22. K. M. Abraham, R. M. Mank, and G. L. Holleck, Investigations of the Safety of Li/SOCl$_2$ Batteries, Report DELET-TR-78-0564-F, U.S. Army Electronics Research and Development Command, Fort Monmouth, New Jersey (1980).

23. J. C. Bailey, Analysis of gaseous products produced by the reaction of lithium–thionyl chloride cell components, in *Proceedings of the Symposium on Lithium Batteries,* Proc. Vol. 87-1, pp. 121–128, The Electrochemical Society, Pennington, New Jersey (1987).

24. D. L. Chua, S. L. Deshpande, and H. V. Venkatasetty, Lithium–thionyl chloride system studies on safety aspects, in *Proceedings of the Symposium on Battery Design and Optimization,* Proc. Vol. 79-1, pp. 365–376, The Electrochemical Society, Princeton, New Jersey (1979).

25. J. F. McCartney, T. J. Lund, and W. J. Sturgeon, Lithium–thionyl chloride batteries—past, present and future, in *Proceedings of the 1979 Near Surface Ocean Experimental Technology NORDA Workshop,* pp. 205–217, National Space Technology Laboratories, (1979).

26. W. H. Shipman and J. F. McCartney, Nonexplosive lithium battery, U.S. Patent Appl. 767,606 (February 10, 1977).

27. A. N. Dey, The design and optimization of Li/SOCl$_2$ cells with respect to energy density, storability and safety, in *Proceedings of the Symposium on Battery Design and Optimization,* Proc. Vol. 79-1, pp. 336–347, The Electrochemical Society, Princeton, New Jersey (1979).

28. S. Dallek, S. D. James, and W. P. Kilroy, Exothermic reactions among components of lithium–sulfur dioxide and lithium–thionyl chloride cells, in *Proceedings of the Symposium on Lithium Batteries,* Proc. Vol. 81-4, pp. 90–97, The Electrochemical Society, Pennington, New Jersey (1981).

29. S. D. James, P. H. Smith, and W. P. Kilroy, *J. Electrochem. Soc.* **130,** 2037 (1983).

30. S. D. James and W. P. Kilroy, Explosive and spontaneously flammable mixtures among components of Li–SOCl$_2$ and Li–SO$_2$ cells, Electrochemical Society Fall Meeting, Hollywood, Florida, 1980, Paper No. 706 RNP.

31. S. Dallek, S. D. James, and W. P. Kilroy, *J. Electrochem. Soc.* **128,** 508 (1981).

32. W. P. Kilroy and S. D. James, *J. Electrochem. Soc.* **128,** 934 (1981).

33. W. L. Bowden and A. N. Dey, Primary Li/SOCl$_2$ cells X. SOCl$_2$ reduction mechanism in a supporting electrolyte, in *Proceedings of the Symposium on Power Sources for Biomedical Implantable Applications and Ambient Temperature Lithium Batteries,* Proc. Vol. 80-4, pp. 471–485, The Electrochemical Society, Pennington, New Jersey (1980).

34. R. C. McDonald, *J. Electrochem. Soc.,* **129,** 2453 (1982).

35. F. W. Dampier, Lithium–Thionyl Chloride Cell System Safety Hazard Analysis, AFWAL-TR-84-2112, Final Report for Period December 1981–December 1984, Aero Propulsion Laboratory, Wright Patterson Air Force Base, Ohio (1985).

36. M. Babai, *Prog. Batteries Sol. Cells* **3,** 110 (1980).

37. M. Mizutani and Y. Okamura, *Prog. Batteries Sol. Cells* **6,** 48 (1987).

38. J. C. Hall, Performance and safety of Li/SOCl$_2$ batteries for the 9V application, in *Proceedings of the 31st Power Sources Symposium,* pp. 443–452, The Electrochemical Society, Pennington, New Jersey (1985).

39. T. Iwamaru and Y. Uetani, *Prog. Batteries Sol. Cells* **5,** 36 (1984).

40. J. C. Hall and D. J. DeBiccari, Design, performance and safety of lithium–thionyl chloride man pack radio batteries, in *Proceedings of the 32nd International Power Sources Symposium,* pp. 566–573, The Electrochemical Society, Pennington, New Jersey (1986).

41. K. M. Abraham, P. G. Gudrais, G. L. Holleck, and S. B. Brummer, Safety aspects of Li/ SOCl₂ batteries, in *Proceedings of the 28th Power Sources Symposium*, pp. 255–257, PSC Publications Committee, Red Bank, New Jersey (1978).

42. S. Surampudi, G. Halpert, and I. Stein, Safety Considerations of Lithium–Thionyl Chloride Cells, JPL Publication 86-15, Jet Propulsion Laboratory, California Institute of Technology, Pasadena, California (1986).

43. C. R. Schlaikjer, Lithium-oxyhalide cells, in *Lithium Batteries* (J. P. Gabano, ed.), Academic Press, London (1983).

44. Duracell International Inc., Electrochemical cell resistant to cell abuse, Indian Patent No. 290/BOM/80 (June 30, 1984).

45. R. C. McDonald and F. W. Dampeir, Investigation of Lithium Thionyl Chloride Battery Safety Hazards, Contract No. N60921-81-C-0229, Final Report for Period September 1981– November 1982, Naval Surface Weapons Center, Silver Spring, Maryland (1983).

46. M. Domeniconi, Controlling hazardous reactions during voltage reversal of high energy lithium cells, in *The 1982 Goddard Space Flight Center Battery Workshop*, NASA Conference Publication 2263, pp. 67–74 (1983).

47. D. J. Salmon, M. E. Peterson, L. L. Henricks, L. L. Abels, and J. C. Hall, *J. Electrochem. Soc.* **129**, 2496 (1982).

48. D. Vallin and M. Broussely, *J. Power Sources* **26**, 201 (1989).

49. D. L. Chua, J. O. Crabb, and S. L. Deshpande, Large lithium–thionyl chloride cells: Cell performance and safety, in *Proceedings of the 28th Power Sources Symposium*, pp. 247– 251, The Electrochemical Society, Princeton, New Jersey (1979).

50. K. F. Garoutte and D. L. Chua, Safety performance of large Li/SOCl₂ cells, in *Proceedings of the 29th Power Sources Symposium*, pp. 153–157, The Electrochemical Society, Pennington, New Jersey (1981).

51. W. P. Kilroy, L. Pitts, and K. M. Abraham, High Rate Li/SOCl₂ Cells I. Effect of Design Variables in Uncatalyzed Cells, Report NSWC-TR-85-98, Naval Surface Weapons Center, Silver Spring, Maryland (1985).

52. N. Doddapaneni and G. F. Hoff, The Effect of cell design on the safety of the Li/SOCl₂ battery, in *Proceedings of the Symposium on Lithium Batteries*, Proc. Vol. 84-1, pp. 146, The Electrochemical Society, Pennington, New Jersey (1984).

53. K. M. Abraham and L. Pitts, *J. Electrochem. Soc.* **132**, 2301 (1985).

54. N. Doddapaneni, Safety Studies on Li/SOCl₂ Reserve Battery, Contract No. N60921-81-C-0305, Final Report for Period 30 September 1981–30 April 1983, Naval Surface Weapons Center, Silver Spring, Maryland (1983).

55. D. H. Johnson, Design of a safe cylindrical lithium/thionyl chloride cell, in *The 1982 Goddard Space Flight Center Battery Workshop*, NASA Conference Publication 2263, pp. 75–84, (1983).

56. K. M. Abraham, Forced overdischarge related safety aspects of Li/SO₂ and Li/SOCl₂ cells, in *The 1982 Goddard Space Flight Center Battery Workshop*, NASA Conference Publication 2263, pp. 47–59 (1983).

57. W. Clark, F. Dampier, R. McDonald, A. Lombardi, D. Batson, and T. Cole, Lithium Cell Reactions, AFWAL-TR-85-2003, Final Report for Period December 1981–December 1984, Aero Propulsion Laboratory, Wright Patterson Air Force Base, Ohio (1985).

58. K. M. Abraham and R. M. Mank, Some safety related chemistry of Li/SOCl₂ cells, in *Proceedings of the 29th Power Sources Symposium*, pp. 135–137, The Electrochemical Society, Pennington, New Jersey (1981).

59. R. C. McDonald, Investigation of Lithium Thionyl Chloride Battery Safety Hazards, Contract No. N60921-81-C-0229, Quarterly Technical Progress Report Period 29 September 1981– 31 December 1981, Naval Surface Weapons Center, Silver Spring, Maryland (1982).

60. D. J. DeBiccari, High current Li/SOCl$_2$ batteries, in *Proceedings of the 32nd International Power Sources Symposium*, pp. 574–579, The Electrochemical Society, Pennington, New Jersey (1986).

61. F. W. Dampier and R. C. McDonald, Lithium deposition during overdischarge in Li/SOCl$_2$ cells, in *Proceedings of the Symposium on Lithium Batteries*, Proc. Vol. 84-1, pp. 154–161, The Electrochemical Society, Pennington, New Jersey (1984).

62. R. C. McDonald, Investigation of Lithium Thionyl Chloride Battery Safety Hazards, Contract No. N60921-81-C-0229, Quarterly Technical Progress Report Period 1 January 1982–31 March 1982, Naval Surface Weapons Center, Silver Spring, Maryland (1982).

63. A. N. Dey, Abuse resistant cells containing fluid depolarizers, U.S. Patent 4,184,014 (December 9, 1980).

64. S. Subbarao, F. Deligiannis, D. H. Shen, S. Dawson, and G. Halpert, Low temperature safety of lithium–thionyl chloride cells, in *Proceedings of the Symposium on Primary and Secondary Ambient Temperature Lithium Batteries*, Proc. Vol. 88-6, pp. 187–200, The Electrochemical Society, Pennington, New Jersey (1988).

65. S. C. Levy, Safety and reliability studies of primary lithium batteries, *J. Power Sources*, **43**, 247–251 (1993).

66. H. Frank, Heat generation rates in lithium thionyl chloride cells, in *The 1981 Goddard Space Flight Center Battery Workshop*, NASA Conference Publication 2217, pp. 91–99 (1982).

67. R. J. Staniewicz, R. A. Dixon, and J. A. Willson, The safety and performance of spiral wound Li/SOCl$_2$ cells for military applications, in *Proceedings of the 32nd International Power Sources Symposium*, pp. 559–565, The Electrochemical Society, Pennington, New Jersey (1986).

68. B. C. Navel, An innovative rupture disk vent for lithium batteries, in *Proceedings of the Symposium on Primary and Secondary Ambient Temperature Lithium Batteries*, Proc. Vol. 88-6, pp. 169–174, The Electrochemical Society, Pennington, New Jersey (1988).

69. E. J. Chaney, Jr., Hermetically sealed galvanic cell having safety vent construction, European Patent Appl. EP 158,703 (October 23, 1985); U.S. Patent Appl. 595,365 (March 30, 1984).

70. S. Sato, H. Yoshikowa, S. Ikenari, K. Yokoyama, and Y. Uetani, Batteries with nonaqueous liquid active mass, Jpn. Kokai Tokkyo Koho JP 63,285,861 (November 22, 1988).

71. S. Sato, H. Yoshikowa, S. Ikenari, K. Yokoyama, and Y. Uetani, Batteries with nonaqueous liquid active mass, Jpn. Kokai Tokkyo Koho JP 63,285,859 (November 22, 1988).

72. G. R. Tucholski, G. R. Drengler, and J. C. Bailey, Cells with pressure sensitive vents, European Patent Appl. EP 284,444 A2 (September 28, 1988).

73. H. Frank and A. Attia, Factors affecting pressure in Li–SOCl$_2$ cells, in *Proceedings of the 25th Intersociety Energy Conversion Engineering Conference*, American Institute of Chemical Engineers, New York, Vol. 3, pp. 117–121 (1990).

74. M. Babai and J. R. Goldstein, Safety device for a high energy density battery, Ger. Offen. DE 3,222,294 (January 20, 1983).

75. A. N. Dey, Primary Li/SOCl$_2$ cells II. Thermal runaways and their prevention in hermetic D cells, in *Proceedings of the 27th Power Sources Symposium*, pp. 42–44, PSC Publications Committee, Red Bank, New Jersey (1976).

76. W. R. Cieslak, Compatibility and performance of separators in Li/SOCl$_2$ cells, in *Proceedings of the 33rd International Power Sources Symposium*, pp. 233–239, The Electrochemical Society, Pennington, New Jersey (1988).

77. N. Doddapaneni and D. L. Chua, Reversal resistant non-aqueous electrochemical cell, U.S. Patent 4,450,029 (July 1, 1986).

78. K. Hisatome, H. Sasama, K. Ishida, and S. Sekido, Inorganic nonaqueous electrolytic solution type cell, U.S. Patent 5,030,525 (July 9, 1991).

79. V. O. Catanzarite, Anode neutralization, U.S. Patent 4,407,910 (October 4, 1983).
80. A. N. Dey and W. L. Bowden, Fluid depolarized cell, U.S. Patent 4,264,687 (April 28, 1981).
81. W. L. Bowden, Batteries with liquid depolarizers, Ger. Offen. 3,034,783 (April 2, 1981); U.S. Patent Appl. 78,120 (September 24, 1979).
82. T. M. Watson, B. Codd, J. D. Jolson, and M. J. Cole, Recent improvements in Li/SOCl₂ cell design, in *Proceedings of the 32nd International Power Sources Symposium,* pp. 498–507, The Electrochemical Society, Pennington, New Jersey (1986).
83. B. C. Bergum, Lithium cell anti-reversal protection, U.S. Patent 4,830,935 (May 16, 1989).
84. A. N. Dey and R. W. Holmes, Cell having improved abuse resistance, U.S. Patent 4,450,213 (May 22, 1984).
85. E. Peled, A. Lombardi, and C. R. Schlaikjer, *J. Electrochem. Soc.* **130,** 1365 (1983).
86. E. Peled, Lithium cells, Israeli Patent 52,555 (October 26, 1980).
87. C. C. Crafts, S. C. Levy, and W. R. Cieslak, Results of FY90 Lithium Battery Safety Testing, Unpublished memo, Sandia National Laboratories, Albuquerque, New Mexico (1990).
88. W. H. Shipman and J. F. McCartney, Lithium–thionyl chloride battery with niobium pentachloride electrolyte, U.S. Patent 4,307,160 (December 22, 1981).
89. F. W. Dampier and R. T. Kalivas, Electrochemical battery cell, U.S. Patent 4,351,888 (September 28, 1982).
90. G. R. Ramsay and D. J. Salmon, Prevention of overpressurization of lithium thionyl chloride battery cells, U.S. Patent 4,490,446 (December 25, 1984).
91. E. Peled, M. Brand, E. Elster, J. Kimel, and R. Cohen, Multi-cell battery, U.S. Patent 4,833,726 (November 28, 1989).
92. M. Babai and U. Zak, Safety aspects of low rate Li/SOCl₂ batteries, in *Proceedings of the 29th Power Sources Symposium,* pp. 150–153, The Electrochemical Society, Pennington, New Jersey (1981).
93. Y. I. Cho and G. Halpert, *J. Power Sources* **18**(2/3), 109 (1986).
94. Y. I. Cho, *J. Electrochem. Soc.* **134,** 771 (1987).
95. D. Reshef, J. Bineth, and M. Babai, Performance characteristics of Li/SOCl₂ packs for military communications equipment, in *Proceedings of the 30th Power Sources Symposium,* pp. 218–221, The Electrochemical Society, Pennington, New Jersey (1982).
96. Y. I. Cho and G. Halpert, Thermal analysis of prismatic Li–SOCl₂ primary cells, in *Proceedings of the 32nd International Power Sources Symposium,* pp. 547–552, The Electrochemical Society, Pennington, New Jersey (1986).
97. A. N. Dey, *J. Power Sources* **5,** 57 (1980).
98. L. D. Hansen and H. Frank, *J. Electrochem. Soc.* **134,** 1 (1987).
99. K. Y. Kim and D. L. Chua, *Prog. Batteries Sol. Cells* **4,** 58 (1982).
100. F. Walsh, A Safe, High-Power-Density Lithium Battery, Office of Naval Research Contract No. N00014-84-C-0724, Final Report for Period September 1984–February 1985, Office of Naval Research, Arlington, Virginia (1985).
101. W. P. Kilroy, L. Pitts, and K. M. Abraham, High Rate Li/SOCl₂ Cells II. Effect of Catalyst on Cell Performance, Report NSWC-TR-85-104, Naval Surface Weapons Center, Silver Spring, Maryland (1985).
102. N. Doddapaneni, Safety Studies on Li/SOCl₂ Reserve Batteries, Contract No. N60921-81-C-0305, Third Quarter, Honeywell Power Sources Center, Horsham, Pennsylvania (1982).
103. F. Walsh and M. Yaniv, Improved safety Li/SOCl₂ battery performance, in *Proceedings of the 31st Power Sources Symposium,* pp. 420–426, The Electrochemical Society, Pennington, New Jersey (1985).

104. F. Walsh and J. Hopewood, A New Bonded Catalyst for Safe Lithium–Thionyl Chloride Batteries, Contract No. DAAK20-82-C-0379, Final Technical Report, U.S. Army Electronics Research and Development Command, Fort Monmouth, New Jersey (1982).

105. P. A. Bernstein and A. B. P. Lever, *Inorg. Chem.* **29,** 608 (1990).

106. H. Frank, Catastrophic event modeling, in *The 1980 Goddard Space Flight Center Battery Workshop,* NASA Conference Publication 2177, pp. 99–103 (1981).

107. S. Szpak, C. J. Gabriel, and J. R. Driscoll, *Electrochim. Acta* **32,** 239 (1987).

108. T. I. Evans, T. V. Nguyen, and R. E. White, *J. Electrochem. Soc.* **136,** 328 (1989).

109. Y. I. Cho, H. Frank, and G. Halpert, *J. Power Sources* **21,** 183 (1987).

110. I. McVey, Batteries, Lithium, Thionyl Chloride, Reverse Current, GIDEP SAFE-ALERT, G1-S-87-01, Rockwell International, Canoga Park, California (1987).

111. A. N. Dey and N. Hamilton, *J. Appl. Electrochem.* **12,** 33 (1982).

112. M. J. Kaduboski and R. E. Horvath, 'F' size Li/SOCl₂ cells—service/abuse, *Proceedings of the 32nd International Power Sources Symposium,* pp. 553–558, The Electrochemical Society, Pennington, New Jersey (1986).

113. J. F. McCartney, W. H. Shipman, C. R. Gunderson, and C. W. Koehler, Development of Lithium Inorganic Electrolyte Batteries for Navy Applications, Report NUC TP 564, Naval Undersea Center, San Diego, California (1977).

114. K. Uno and M. Mizutani, Safety tests of spiral-type lithium–thionyl chloride D-cells, *GS News Tech. Rep.* **47**(2), 29 (1988).

115. Y. Okamura and M. Mizutani, Safety tests of bobbin-type lithium–thionyl chloride D-cells, *GS News Tech. Rep.* **46**(2), 26 (1987).

116. G. H. Boyle and F. Goebel, High rate lithium/thionyl chloride batteries, in *Proceedings of the 30th Power Sources Symposium,* pp. 215–217, The Electrochemical Society, Pennington, New Jersey (1982).

117. N. Marincic and F. Goebel *J. Power Sources* **5,** 73 (1980).

118. A. E. Zolla, R. R. Waterhouse, D. J. DeBiccari, and G. L. Griffin, Primary Lithium Thionyl–Chloride Cell Evaluation, Report AFWAL-TR-80-2076, Air Force Wright Aeronautical Laboratories, Wright Patterson Air Force Base, Ohio (1980).

119. J. Suprenant and D. Snuggerud, Abuse resistant high rate lithium/thionyl chloride cells, in *Proceedings of the 17th Intersociety Energy Conversion Engineering Conference,* IEEE, Piscataway, New Jersey, Vol. 2, p. 635 (1982).

120. J. Bene, Safety testing of lithium cells, in *The 1980 Goddard Space Flight Center Battery Workshop,* NASA Conference Publication 2177, pp. 111–117 (1981).

121. U. Zak, D. Reshef, and H. Kreinin, High rate military lithium inorganic (Li/SOCl₂) battery, *Proceedings of the 31st Power Sources Symposium,* pp. 427–432, The Electrochemical Society, Pennington, New Jersey (1984).

122. R. L. Zupancic, L. F. Urry, and V. S. Alberto, Performance and safety characteristics of small cylindrical Li/SOCl₂ cells, in *Proceedings of the 29th Power Sources Symposium,* pp. 157–159, The Electrochemical Society, Pennington, New Jersey (1981).

123. P. Kane, N. Marincic, J. Epstein, and A. Lindsey, Lithium based alloy–thionyl chloride cells, in *Proceedings of the Symposium on Lithium Batteries,* Proc. Vol. 87-1, pp. 171–184, The Electrochemical Society, Pennington, New Jersey (1987).

Lithium/Sulfur Dioxide Batteries

The lithium/sulfur dioxide battery is a high-energy system capable of delivering 320 Wh/kg (520 Wh/liter). The active cathode material is a gas, SO_2, which is present at a pressure of ~ 30 psi at room temperature. All cells, therefore, contain a vent mechanism in the case. Sulfur dioxide is a poisonous gas with an Occupational Safety and Health Administration (OSHA) permissible exposure limit of 5 ppm in an 8-h day. Exposure of 100–500 ppm can pose immediate danger to life and health.[1,2] The electrolyte solvent, acetonitrile, is a moderately toxic, flammable liquid having a boiling point of 81°C. The lowest published toxic concentration for inhalation by humans is 160 ppm for 4 h.[1]

11.1. HAZARDOUS REACTIONS

A number of chemical reactions may occur in Li/SO_2 cells that can lead to unsafe behavior. The nature of these reactions is a function of cell balance, type and extent of electrochemical abuse, and temperature.

11.1.1. Normal Discharge

The discharge chemistry of the Li/SO_2 system involves the electrochemical oxidation of lithium by sulfur dioxide to form lithium dithionite [Eq. (11.1)]. The direct chemical reaction of these two materials results in the formation of a passive film of lithium dithionite on the surface of the lithium. This film is normally broken down physically as the cell discharges.

$$2Li + 2SO_2 \rightarrow Li_2S_2O_4 \qquad (11.1)$$

Lithium dithionite is unstable, and its thermal decomposition is a major factor in the hazardous behavior of Li/SO_2 cells.[3] At temperatures near the melting point of lithium, lithium dithionite has been reported to decompose exothermally as follows[4,5]:

$$2Li_2S_2O_4 \rightarrow SO_2 + S + 2Li_2SO_3 \qquad (11.2)$$

Impurities, for example, H_2O, enhance the decomposition of $Li_2S_2O_4$, and the heat output is increased by as much as ten times.[6]

Analysis of cells discharged at 72°C and high rates showed that these cells had less $S_2O_4^{2-}$ than calculated ($\sim 50\%$ less than at room temperature). Other species found in place of the missing $Li_2S_2O_4$ were lithium dithionate ($Li_2S_2O_6$), lithium pyrosulfite ($Li_2S_2O_5$), and sulfur.[7]

Near the end of life, as the sulfur dioxide is becoming depleted, the passive film of $Li_2S_2O_4$ is not repaired as rapidly and/or completely. This allows lithium to come in contact with the acetonitrile electrolyte solvent. Lithium catalyzes the polymerization of acetonitrile, with methane gas generated as a by-product. Methane is highly flammable, and any ventings that occur at or near the end of life could result in a fire.

The lithium–acetonitrile reaction has been identified as an initiator of thermal runaway in Li/SO_2 cells. This reaction is catalyzed by the electrolyte salt LiBr.[8]

Analyses of cells that have undergone pulse discharge have indicated the presence of CO_2 in the gas phase. Similar cells undergoing steady-state discharge have shown no evidence of CO_2. The cathodes of some pulsed cells have $Li_2S_2O_4$ on one side and $Li_2S_nO_6$ on the opposite side. This may be due to secondary reactions of $Li_2S_2O_4$ involving carbon.[9]

Cells were analyzed that had been 50% discharged and stored at room temperature for one year. Only SO_2 and acetonitrile were found in the vapor phase. At the anode, $Li_2S_2O_4$, lithium thionates, and Li_2SO_4 were observed.[10] Other studies found $Li_2S_2O_4$ and Li polythionates at the anode and only $Li_2S_2O_4$ at the cathode.[11]

Li/SO_2 cells can become shock sensitive after discharge if the resistive load is left on the cell or if a noncorrosion-resistant glass is used in the glass-to-metal seal.[12] This is due to the formation of Li–Al alloy in the cathode grid by a process known as spontaneous electrochemical alloying. It has been shown that the violent reaction between Li–Al alloy and the $Li_2S_2O_4$ found in discharged cathodes is initiated by shock.

11.1.2. Forced Overdischarge

When Li/SO_2 cells are force overdischarged into reversal, major safety problems can result from the plating of lithium metal on the cathode. This lithium has 100–200 times the surface area of the anode and is highly reactive, since there is no SO_2 present to passivate its surface. A series of reactions may occur that can lead to thermal runaway.[6] First, lithium can react exothermally with acetonitrile. Second, the heat generated from this reaction can lead to the thermal decomposition of $Li_2S_2O_4$. Finally, lithium can react exothermally with the sulfur resulting from the decomposition of the dithionite.

Analysis of Li/SO_2 cells force overdischarged at room temperature showed the presence of di- and triacetonitriles, LiCN, Li_2SO_3, and Li_2SO_4 at the anode; Li, Li–Al, LiF, and LiH at the cathode; and CH_4 in the gas phase. In cells force overdischarged at $-10°C$, very little or no CH_4 was found.[11]

The chemistry of Li/SO_2 cell reversal was studied in glass laboratory cells.[13] The major reactions found were the reduction of Li^+ to Li^0, the alloying of lithium with aluminum, and reaction of Li–Al with acetonitrile (even in the presence of excess SO_2). The following reaction scheme was postulated:

$$Li(Al) + CH_3CN \rightarrow LiCN + CH_3 \cdot \quad \text{(slow)} \tag{11.3}$$

$$CH_3 \cdot + CH_3CN \rightarrow CH_4 + \cdot CH_2CN \tag{11.4}$$

$$2CH_3 \cdot \rightarrow C_2H_6 \tag{11.5}$$

$$2 \cdot CH_2CN \rightarrow (CH_2CN)_2 \tag{11.6}$$

$$Li(Al) + CH_3CN \rightarrow {}^+Li \cdot {}^-CH_2CN + \tfrac{1}{2}H_2 \tag{11.7}$$

$${}^+Li \cdot {}^-CH_2CN \rightarrow LiCN + (CH_2) \tag{11.8}$$

$$2(CH_2) \rightarrow C_2H_4 \tag{11.9}$$

On high-rate discharge, there is more excess Li and SO_2 when cells go into reversal, due to the inefficiency of the carbon collector. Also, the resistivity of the cathode increases near the tab (possibly due to Li–Al alloy formation). I^2R heating in this area can initiate the reactions between Li and acetonitrile, SO_2, and/or CH_4.[14] At low temperatures, forced overdischarge is hazardous even at low currents. The plated lithium at the cathode can react with SO_2 absorbed in the carbon. The heat generated can initiate reactions between Li and acetonitrile or methane, or it can initiate the thermal decomposition of $Li_2S_2O_4$.

11.1.3. Charging

Li/SO$_2$ cells have been found to explode in a violent manner when charged at high rates ($>C/7$). The time to explosion varies inversely with the charging current in an exponential manner.[15] At low rates ($<C/70$), charging of these cells is not hazardous.

Anodes from Li/SO$_2$ cells that have been discharged and then charged have been found to be covered with fine filaments or dendrites of lithium. These anodes are very reactive when dry, they sputter when cut or hit, and they burn when exposed to dry air. The carbon cathodes are benign.[16]

A number of reactions have been postulated to occur during charging of Li/SO$_2$ cells.

At the anode:

$$Li^+ + e^- \rightarrow Li^0 \tag{11.10}$$

$$2Li^0 + Br_2 \rightarrow 2LiBr \tag{11.11}$$

$$2Li^0 + 2SO_2 \rightarrow Li_2S_2O_4 \tag{11.12}$$

$$Li_2S_2O_4 + Br_2 \rightarrow 2LiBr + 2SO_2 \tag{11.13}$$

At the cathode:

$$2Br^- \rightarrow Br_2 + 2e^- \tag{11.14}$$

$$Br_2 + Li_2S_2O_4 \rightarrow 2LiBr + 2SO_2 \tag{11.15}$$

At the Al grid in the cathode:

$$2Al + 3Br_2 \rightarrow 2AlBr_3 \tag{11.16}$$

$$AlBr_3 + Br^- \rightarrow AlBr_4^- \tag{11.17}$$

Charging a fully discharged cell can be extremely hazardous, resulting in an explosion, but charging a fresh cell has not been found to be a safety hazard.[6]

11.1.4. Reaction Enhancement

Carbon has been found to catalyze the reactions of lithium metal in Li/SO$_2$ cells.[17,18] Lithium is inert to the CH$_3$CN–LiBr–SO$_2$ electrolyte at room temperature. However, grinding Li with carbon causes spontaneous ignition on mixing with the electrolyte. It has been found that increasing the Li/C ratio lowers the spontaneous flammability and increases the shock sensitivity.

LiBr has been found to have a catalytic effect in promoting the reaction between lithium and acetonitrile.[8]

11.2. CELL DESIGNS

Since the Li/SO$_2$ system is pressurized, cells are normally built in a cylindrical shape. A wound or jelly-roll electrode assembly is the most common internal configuration. The steel case is protected from corrosion by maintaining it at the lithium potential, that is, making it the negative terminal. The positive terminal must be a valve metal (i.e., having a protective oxide film on its surface), for example, Ta or Mo, to protect it from corrosion by the electrolyte. A polypropylene separator is normally used between the lithium anode and the acetylene black cathode collector. Design features that impact safety are the cell balance, vent design and location, and additives to the electrolyte to inhibit exothermic reactions.

11.2.1. Balance

Cell balance is the most critical aspect of safety in Li/SO$_2$ cells. Three components of the cell may be limiting: the anode (Li), the cathode (SO$_2$), or the cathode collector (C). Hazardous behavior has, for the most part, been related to the deposition of Li dendrites on the cathode collector. Therefore, cells containing excess lithium tend to pose more of a safety risk than those that are lithium limited or balanced.

A note of caution needs to be mentioned at this point. Cells may be designed to be lithium limited or balanced, but excess lithium may still be present at the end of life under certain discharge and/or storage conditions. High-rate and/or low-temperature discharge can result in severe polarization of the cathode collector due to plugging of the pores near the surface with solid discharge product (Li$_2$S$_2$O$_4$). As the cell is then driven into reversal, Li dendrites will plate on the carbon. Reaction with SO$_2$ will form a coating of Li$_2$S$_2$O$_4$ on their surface, but hazardous reactions may occur as the cell is driven further into reversal. Under certain storage conditions, SO$_2$ may react parasitically with other cell components, resulting in a cell with excess lithium. Cells built with the standard wrap, that is, Li on the outside, cannot be Li limited because some of the Li is unavailable for electrochemical reaction.[19] The cathode outer wrap (COW) configuration leads to a more efficient use of the lithium.[20] Therefore, in choosing or designing a Li/SO$_2$ cell, one must consider the worst case condition and perform safety tests at that level before putting the cell into use.

A number of studies have looked at the effect of cell balance on safety. Both the Li-limited and balanced designs (Li/SO$_2$ ratio $\approx$ 1) have been shown to be safer on forced overdischarge.[6,4,14,20–30] One study found Li/SO$_2$ cells to be abuse resistant even if they contain up to 15% excess lithium.[31]

The Li/SO$_2$ ratio is related to the safety of Li/SO$_2$ cells only when the materials are utilized efficiently.[22] Balancing the Li:SO$_2$:C ratio results in optimum performance (i.e., optimum capacity) at some optimum current I_0.[28] Discharge at I_0 or lower currents is safe, but at currents greater than I_0, cells become cathode limited and hazardous on reversal. Since cathode utilization is proportional to temperature, I_0 decreases as the cell temperature is lowered. At higher temperatures and lower current densities, the Li/SO$_2$ ratio can be increased and the cells will still be safe during forced overdischarge.[32]

Although the lithium-limited or balanced design cells are safer under certain abusive conditions, a slight performance penalty is associated with these designs. Also, the standard deviation of their capacities, within a lot, are greater than for similar cells with excess lithium.[33] Cathode-limited Li/SO$_2$ cells have been used in low-rate applications, providing a high reliability (>0.9995) with no safety problems.[34]

11.2.2. Vents

Two types of safety vent are commonly used in Li/SO$_2$ cells: a coined groove in the bottom or along the side of the can or a convolution vent in the can bottom. At a specified internal pressure, corresponding to the vapor pressure of the CH$_3$CN–LiBr–SO$_2$ electrolyte at $\sim$125°C (400 psi), the vent opens, allowing the rapid escape of electrolyte.[33] Since vent actuation is associated with a slight bulging of the cell case, an external clearance must be provided to allow the vents to operate properly. Special care must be taken when using cells with a side vent, since the vent is covered by the label.

A vent has been designed for crimp-sealed Li/SO$_2$ "D" cells, consisting of a two-piece aluminum cover welded together in a way that produces a rupture at high internal pressures.[35] The rupture pressure can be varied by changing the location and size of the welds. Tests of cells having vents designed to operate between 300 and 400 psi showed them to be safe under short circuit, reversal, external heating, shock, immersion in salt water, puncture, and deformation.

11.2.3. Other Safety-Related Features

A number of innovations have been devised to improve the safety of Li/SO$_2$ cells. A copper foil has been sandwiched between two Li layers in the anode to minimize local hot spots during forced overdischarge. This leads to an even distribution of current density and heat throughout the anode. After the Li is depleted, the Cu deposits at the cathode. Dendrites form and short to the anode to provide a low-resistance and low-heat path. They also inhibit the reactivity of dendritic Li on the cathode.[36] A Cu-alloy strip the length of the anode has been found to prevent detrimental effects when cells are subjected to forced discharge[37] and to suppress erratic voltages during reversal.[38]

A means of preventing Li from depositing on the cathode during forced overdischarge involves coupling an exposed inert metal to the cathode and a dendrite target to the anode.[39] The two metals should face each other and have a separator between them. This can be accomplished by scraping the carbon from one side of the cathode, exposing the Al grid, and leaving an exposed Ni tab at the anode or cold welding Cu metal to the Li. During voltage reversal, dendrites grow from the metal on the cathode to the anode target, creating a low-resistance path. This prevents the Li from contacting the carbon and provides a shunt for the current without generating excess heat.

Improved safety during reversal of Li/SO_2 cells is associated with higher cathode efficiency, which has been shown to be related to carbon loading.[40] Higher carbon loadings can be accomplished by increasing the density of the cathode and lowering the binder content.

A Li/SO_2 cell was designed to prevent extreme pressure buildup by connecting the cathode contact to the cell bottom. As the pressure inside the cell rises, the can bottom distends, causing a loss of contact to the cathode. This opens the circuit, preventing further reaction.[41]

The addition of $\sim 10\%$ propylene carbonate to the electrolyte acts as a "protective" solvent to inhibit thermal runaway.[8,42] The propylene carbonate acts by raising the initiation temperature of the Li–acetonitrile reaction from ~ 50 to $\sim 110°C$.

11.3. THERMAL STUDIES

The components of Li/SO_2 cells have been studied by a number of researchers using differential thermal analysis (DTA). The Li–acetonitrile reaction has been found to be the most exothermic, generating heat at room temperature.[43,44] $Li_2S_2O_4$ decomposes exothermally to form sulfur at $>100°C$. Lithium in discharged cathodes containing $Li_2S_2O_4$ exhibits a stronger exothermic reaction, probably due to reaction with sulfur from the decomposition of $Li_2S_2O_4$. The presence of moisture in the cathode yields additional heat, due to reaction with both lithium and $Li_2S_2O_4$. Li and Al alloy exothermally at the melting point of Li. Lithium does not react exothermally with SO_2 at temperatures up to 370°C.

DTA was performed on miniature Li/SO_2 cells. Thermal runaway occurred in discharged cells owing to the reaction of Li with CH_3CN, the decomposition of $Li_2S_2O_4$, and the reaction of Li with sulfur from the decomposition of $Li_2S_2O_4$.[45] During cell reversal, stronger exotherms were observed due to Li in the cathode containing $Li_2S_2O_4$.

Differential scanning calorimetry (DSC) has also been used to study the exothermic reactions of Li/SO_2 cell components. Li and CH_3CN react at room temperature, generating a large amount of heat.[46] The presence of SO_2 increases the activation temperature to near the melting point of Li at the battery concentration of 14M. In contrast to the findings of the DTA studies, trace concentrations of water have been shown to lower the exothermicity of the reaction. The presence of water was also found to reduce the exothermicity of the $Li-Li_2S_2O_4$ reaction.[47]

A discharged cathode, washed with CH_3CN, exhibited a small exotherm. When the cathode was mixed with lithium, a sharp exotherm was observed at 251°C.[48]

Accelerating rate calorimetry (ARC) has been used in a number of studies involving overdischarge of Li/SO_2 cells. Four designs were looked at in one study: Li limited, balanced, excess C, and excess Li. Exotherms during overdischarge were found only in cells containing excess Li.[49] Another study showed no exothermic reactions in any cell discharged at room temperature to a 2.0-V cutoff. However, cells discharged to 2.0 V at −35°C and cells force overdischarged exhibited exothermic reactions on ARC at <100°C.[50] The overdischarged cell had an explosive reaction at 108°C. Most reactions occur between 140 and 170°C and between 195 and 220°C.

Cells having excess Li exhibited an exothermic reaction shortly after entering reversal. This reaction can occur even in the presence of up to 25 mg SO_2/cm^2 Li.[51]

ARC was used to study the effect of storage on the safety of partially discharged cells.[52] The cells were discharged to either 50% or 100% of capacity and stored for up to 6 months. After storage, the cells were either pulsed, shorted, or force overdischarged into reversal. The only safety hazard observed was with cells that were force overdischarged at −20°C. The violent reactions that occur appear to be triggered by the depletion of Li. At −35°C, forced overdischarge presents a significant explosion hazard.[53] The $Li-CH_3CN$ reaction triggers the hazardous behavior and can be prevented only in severely Li limited cells. Such cells, however, are not practical from a performance perspective. The heat of reaction for the $Li-CH_3CN$ reaction was determined by microcalorimetry to be 54 ± 1 kcal/mol Li.

11.4. MODELING

A computer program has been written to thermally model Li/SO$_2$ cells.[54] It calculates the instantaneous heat generation rates and the instantaneous wall temperature. At normal discharge rates, the dominant heat dissipation modes are convection and radiation. At high discharge rates, convection and radiation only dissipate 40% of the heat, allowing the cell temperature to rise rapidly. On charging, as the point of explosion is approached, vigorous chemical reactions have been found to occur in the cell.

A model to predict a safe envelope for Li/SO$_2$ cells has been developed.[55]

Fault Tree Analysis. Although fault tree analysis is not normally considered modeling, the fault tree itself may be thought of as a graphic model of various combinations of faults showing logical interrelationships of events that will lead to the top event, that is, a cell/battery hazard.[56] It may be used as an aid in designing cells and batteries that will operate safely under specified conditions and also to determine the cause of safety incidents.

Fault tree analysis was used to identify the underlying faults that resulted in a safety incident in which a string of Li/SO$_2$ cells that had been on test for several years suddenly vented violently.[57] The analysis also suggested a means for preventing this hazardous behavior.

11.5. SAFETY/ABUSE TESTS

In one series of abuse tests, Li/SO$_2$ "D" cells were force discharged to 100% reversal at several temperatures ranging from −40 to 65°C, while in an insulated container.[58] No problems occurred during discharge with SO$_2$-limited cells (Li:SO$_2$ ratio 1.5:1) at 250 mA to 4 A from −30 to 65°C. However, many cells vented in reversal, some violently. In the Li-limited design (10–30% excess SO$_2$), no problems were observed during normal discharge or on forced overdischarge. However, ~20% of the cells vented under load at −30°C, but not as violently as the SO$_2$-limited cells.

Another abuse test series was conducted on Li/SO$_2$ "D" cells in the fresh, stored, nondischarged, and fully discharged conditions.[59] On short circuit at ambient temperature, only those cells predischarged at −30 and −40°C vented. At 72°C, cells predischarged at −40, −30, and 24°C vented. Cells predischarged at 52°C did not vent at temperatures below 82°C. In reversal, fresh cells did not vent at a constant current of 1 A but did vent at 2 A. Cells stored at 87°C for 102 days did not vent at 1 A in reversal but vented mildly at 2 A. Charging fresh cells did not pose a safety hazard, and the voltage stabilized at 3.5 V owing to the generation of Br$_2$. Discharged cells charged at 0.5 A vented violently. Discharged cells stored for 30 days at 72°C exploded at a charging rate of 0.5 A.

Several charging studies have been conducted on Li/SO_2 "D" cells.[60,61] At 10 A and room temperature, fresh cells will vent violently, and discharged cells will explode. Partially depleted and aged cells will also explode on 10-A charge. Similar behavior has been noted at 1 A and 5 A, but the time to explosion increases as the charging current decreases. Similar results were observed at −20 and 70°C.

Abuse tests were carried out on low-rate "$\frac{1}{2}$AA" cells.[62] Short circuit at room temperature yielded a maximum current of 1.1 A, and no venting occurred. At 70°C, the maximum current was 1.25 A, and the cells vented. Charging at 20 mA for >500 h did not cause the cells to vent. Charging at 500 mA resulted in the cells venting in ∼1 h, while charging at 1 or 2 A caused the cells to vent in a few minutes. Charging at 6.5 V resulted in cells venting in <2 min. Forced overdischarge at 500 mA caused cells to vent after 10–20 min.

11.6. MULTICELL BATTERY PACKS

The assembly of individual cells into a multicell battery pack presents safety problems over and above those related to single cells. Thus, the use of cells that have successfully passed all safety/abuse testing does not guarantee a safe battery pack. Each battery pack must be designed and tested for safety during the anticipated worst case condition the battery will see during its life (i.e., shipping, handling, storage, use, and disposal).

11.6.1. Battery Design

All Li/SO_2 cells have a vent designed into the case to prevent the excessive buildup of pressure that can lead to case rupture and/or explosion. In designing battery packs,[63–66] room must be left for the vents to operate normally. A volume for the gas to vent into and/or a path for the gas to exit the battery must also be provided. If possible, all cell vents should be facing in the same direction. Thermal management is another important consideration. In large batteries where one or more cells is surrounded by other cells, provision must be made for the removal of excess heat that may occur during normal operation and/or under abusive conditions.

Fuses, current-limiting resistors, and/or thermal switches should be considered, depending on the battery design and application. A nonreplaceable fuse will eliminate the possibility of defective fuse holders, which can have a high resistance, leading to I^2R heating and venting of cells.[67] Thermal switches will react faster than electrical fuses during a short circuit plus provide added reliability to the system. Blocking diodes should be used if cells, or strings of cells, are connected in parallel to prevent charging of one string by another. Reverse-bias or bypass diodes may be used on each cell to limit cell reversal currents.[68,69]

Intercell connections should be welded, rather than soldered, and all internal wires and leads must be insulated. Tapping of strings to obtain multiple voltage outputs is not recommended.

The U.S. Army had experienced a safety incident rate of 0.011% with early Li/SO_2 battery packs. After implementing a safety modification program, the incident rate dropped to 0.0006%.[67] Also, with the incorporation of a manually activated circuit to discharge the balanced design batteries completely, the Environmental Protection Agency has declared them nonhazardous. The Army can now dispose of their used batteries in ordinary sanitary landfills.[70]

Sometimes, a simple mechanical change can eliminate a hazardous condition. In one reported case, venting of a cell inside a battery box resulted in rupture of the box.[71] This situation was eliminated by simply adjusting the catches on the box lid so that it can move up slightly. A spring tension in the catches returns the lid to its normal position when the pressure is released.

11.6.2. SO_2 Getter

An SO_2 getter has been developed for shipping Li/SO_2 batteries on passenger aircraft.[72] It consists of a coal base, granular activated carbon impregnated with various metallic oxides (Calgon Corporation, type ASC) contained in a porous paper bag. It can also be used within a battery pack to prevent emission of SO_2 gas upon cell venting. Batteries containing getter material have been built for use in equipment containing expensive electronics that would be damaged by acid gas vapors. The getter could also be used in confined areas containing Li/SO_2 batteries to protect personnel.

REFERENCES

1. N. I. Sax and R. J. Lewis, Sr., *Dangerous Properties of Industrial Materials,* Van Nostrand Reinhold, New York (1989).
2. A. M. Ducatman, B. S. Ducatman, and J. A. Barnes, *J. Occup. Med.* **30**(4), 309 (1988).
3. W. P. Kilroy, S. A. Chmielewski, and D. W. Bennett, Investigation of Li/SO_2 Cell Hazards, III. Raman Spectroscopy of Lithium Dithionite, Report NSWC TR 88-76, Naval Surface Warfare Center, Silver Spring, Maryland (1988).
4. K. M. Abraham and L. Pitts, Investigation of the Safety Related Chemistry of the Lithium Sulfur Dioxide (Li/SO_2) Battery, Report NSWC TR 83 478, Naval Surface Weapons Center, Silver Spring, Maryland (1983).

5. W. P. Kilroy, *J. Electrochem. Soc.* **132**, 998 (1985).
6. M. W. Rupich and K. M. Abraham, Investigation of Lithium Sulfur Dioxide (Li/SO$_2$) Battery Safety Hazard, Contract No. N60921-81-C-0084, Naval Surface Weapons Center, Silver Spring, Maryland (1981).
7. W. L. Bowden, L. Chow, D. L. Demuth, and R. W. Holmes, *J. Electrochem. Soc.* **131**, 229 (1984).
8. A. N. Dey and R. W. Holmes, *J. Electrochem. Soc.* **127**, 1877 (1980).
9. K. M. Abraham and S. M. Chaudhri, Performance and Safety of the Li/SO$_2$ Batteries Used in Sonobuoys, Naval Surface Weapons Center Contract No. N60921-83-C-0126, EIC Laboratories, Norwood, Massachusetts (1984).
10. K. M. Abraham and L. Pitts, *J. Electrochem. Soc.* **130**, 1618 (1983).
11. K. M. Abraham and L. Pitts, The chemistry of the discharge, forced overdischarge, and partial discharge and storage of the Li/SO$_2$ cell, in *Proceedings of the Symposium on Lithium Batteries,* Proc. Vol. 84-1, pp. 265–275, The Electrochemical Society, Pennington, New Jersey (1984).
12. C. C. Crafts and S. C. Levy, Safety Investigations of Discharged Li/SO$_2$ Cells, Report SAND82-0644, Sandia National Laboratories, Albuquerque, New Mexico (1982).
13. H. Taylor, W. Bowden, and J. Barrella, Li/SO$_2$ Cells of Improved Stability, in *Proceedings of the 28th Power Sources Symposium,* pp. 183–188, The Electrochemical Society, Princeton, New Jersey (1979).
14. K. M. Abraham, M. W. Rupich, and L. Pitts, Investigation of Lithium Sulfur Dioxide (Li/SO$_2$) Battery Safety Hazards—Chemical Studies, Contract No. N60921-81-C-0084, Final Report, Naval Surface Weapons Center, Silver Spring, Maryland (1982).
15. H. Frank and D. Lawson, Safety considerations of inadvertent charging of Li/SO$_2$ cells, in *Proceedings Lithium '87,* pp. 22–35, Waste Resource Associates, Niagara Falls, New York (1987).
16. S. Subbarao, D. Lawson, H. Frank, and G. Halpert, *J. Power Sources* **21**, 227 (1987).
17. S. D. James and W. P. Kilroy, Explosive and Spontaneously Flammable Mixtures among components of Li–SOCl$_2$ and Li–SO$_2$ Cells, The Electrochemical Society, Fall Meeting, Hollywood, Florida, 1980, Paper No. 706 RNP.
18. W. P. Kilroy and S. D. James, *J. Electrochem. Soc.* **128**, 934 (1981).
19. W. P. Kilroy, W. Ebner, D. L. Chua, and H. V. Venkatasetty, *J. Electrochem. Soc.* **132**, 274 (1985).
20. J. Barrella and M. Kumbhani, Primary Lithium Organic Electrolyte Battery BA-5090 (   )/ U, BA-5585 (   )/U, BA-5598 (   )/U, Report DELET-TR-76-1735-F, U.S. Army Electronics Research and Development Command, Fort Monmouth, New Jersey (1978).
21. L. J. Blagdon and B. Randall, Safety Studies of Lithium Sulfur Dioxide Cells, Report DELET-TR-78-0530-F, Final Report for Period 27 April 1978–27 October 1978, U.S. Army Electronics Research and Development Command, Fort Monmouth, New Jersey (1979).
22. A. N. Dey, *J. Electrochem. Soc.* **127**, 1886 (1980).
23. G. DiMasi, Behavior of Li/SO$_2$ cells under forced discharge, in *Proceedings of the 24th Power Sources Symposium,* pp. 75–77, PSC Publications Committee, Red Bank, New Jersey (1977).
24. G. DiMasi and J. A. Christopulos, The effects of the electrochemical design upon the safety and performance of the lithium–sulfur dioxide cells, in *Proceedings of the 28th Power Sources Symposium,* pp. 179–182, The Electrochemical Society, Princeton, New Jersey (1979).
25. K. M. Abraham, M. W. Rupich, and L. Pitts, Studies of the safety of Li/SO$_2$ cells, in *Proceedings of the 30th Power Sources Symposium,* pp. 124–127, The Electrochemical Society, Pennington, New Jersey (1983).
26. H. V. Venkatasetty, W. Ebner, D. L. Chua, and W. P. Kilroy, Studies on lithium–sulfur dioxide (Li/SO$_2$) battery safety hazards, in *Proceedings of the Symposium on Lithium Batteries,* Proc. Vol. 84-1, pp. 276–292, The Electrochemical Society, Pennington, New Jersey (1984).

27. A. N. Dey, Safety studies on Li/SO$_2$ cells IV. Effect of design variables on the abuse resistance of hermetic D cells, in *Proceedings of the Symposium on Power Sources for Biomedical Implantable Applications and Ambient Temperature Lithium Batteries,* Proc. Vol. 80-4, pp. 589–604, The Electrochemical Society, Pennington, New Jersey (1980).

28. V. Manev, A. Nassalevska, and R. Moshtev, *J. Power Sources* **6**, 347 (1981).

29. J. N. Barrella, Abuse resistant active metal anode/fluid cathode depolarized cells, U.S. Patent 4,238,554 (December 9, 1980).

30. A. N. Dey and P. Witalis, Safety Studies of Lithium–Sulfur Dioxide Cells, Report DELET-TR-78-0535-F, U.S. Army Electronics Research and Development Command, Fort Monmouth, New Jersey (1979).

31. Duracell International Inc., Electrochemical cell resistant to cell abuse, Indian Patent No. 290/BOM/80, (June 30, 1984).

32. G. DiMasi and J. Christopoulos, Performance, storage, safety and disposal of Li/SO$_2$ Cells, in *The 1980 Goddard Space Flight Center Battery Workshop,* NASA Conference Publication 2177, pp. 61–71 (1981).

33. P. Bro and S. C. Levy, Lithium sulfur dioxide batteries, in *Lithium Battery Technology* (H. V. Venkatasetty, ed.), John Wiley and Sons, New York (1984).

34. S. C. Levy, Reliability of Li/SO$_2$ cells for long life applications, in *Proceedings of the Symposium on Primary and Secondary Ambient Temperature Lithium Batteries,* Proc. Vol. 88-6, pp. 146–151, The Electrochemical Society, Pennington, New Jersey (1988).

35. S. Abens, Non-Hazardous Lithium Organic Electrolyte Batteries, Report ECOM-73-0242-F, U.S. Army Electronics Command, Fort Monmouth, New Jersey (1974).

36. A. N. Dey and R. W. Holmes, Cell having improved abuse resistance, U.S. Patent 4,450,213 (May 22, 1984).

37. B. E. Jagid, Lithium–sulfur dioxide cell and electrode therefor, European Patent Appl. EP 106,602 (April 25, 1984); U.S. Patent Appl. 466,187 (February 16, 1983).

38. M. G. Rosansky and B. Jagid, Lithium anode comprising copper strip in contact with lithium body and lithium–sulfur dioxide battery using same, U.S. Patent 4,482,615 (November 13, 1984).

39. P. L. Bedder, P. R. Moses, B. Patel, T. F. Reise, and A. H. Taylor, Electrochemical cells, U.S. Patent 4,622,277 (November 11, 1986).

40. L. J. Blagdon and B. Randall, Safety Studies of Lithium Sulfur Dioxide Cells, Report DELET-TR-77-0459-F, Final Report for Period 31 August 1977–31 March 1978, U.S. Army Electronics Research and Development Command, Fort Monmouth, New Jersey (1979).

41. J. F. Zaleski, Lithium cell with internal automatic safety controls, U.S. Patent 3,939,011 (February 17, 1976).

42. A. N. Dey, Lithium–solvent interactions in Li/SOCl$_2$ and Li/SO$_2$ battery systems, in *Proceedings of the Workshop on Lithium Nonaqueous Battery Electrochemistry,* pp. 83–97, The Electrochemical Society, Pennington, New Jersey (1980).

43. A. N. Dey and R. W. Holmes, *J. Electrochem. Soc.* **126**, 1637 (1979).

44. A. N. Dey and R. W. Holmes, Analysis of Pressure Producing Reactions in Lithium–Sulfur Dioxide Cells, Report DELET-TR-77-0472-F, U.S. Army Electronics Research and Development Command, Fort Monmouth, New Jersey (1979).

45. A. N. Dey, *J. Electrochem. Soc.* **127**, 1000 (1980).

46. S. Dallek, S. D. James, and W. P. Kilroy, Exothermic reactions among components of lithium–sulfur dioxide and lithium–thionyl chloride cells, in *Proceedings of the Symposium on Lithium Batteries,* Proc. Vol 81-4, pp. 90–97, The Electrochemical Society, Pennington, New Jersey (1981).

47. W. P. Kilroy and S. Dallek, Thermal Analysis of Some Components of the Li–SO₂ Cell, Report NSWC/WOL TR 78-156, Naval Surface Weapons Center, Silver Spring, Maryland (1978).
48. S. Dallek, S. D. James, and W. P. Kilroy, *J. Electrochem. Soc.* **128**, 508 (1981).
49. D. L. Chua, Validating cell/battery safety, in *The 1982 Goddard Space Flight Center Battery Workshop*, NASA Conference Publication 2263, pp. 85–99 (1983).
50. W. B. Ebner and D. W. Ernst, Safety studies of the Li/SO₂ system using accelerating rate calorimetry, in *Proceedings of the 30th Power Sources Symposium*, pp. 119–124, The Electrochemical Society, Pennington, New Jersey (1983).
51. W. Ebner, Accelerating rate calorimetry: A new technique for safety studies in lithium systems, in *The 1981 Goddard Space Flight Center Battery Workshop*, NASA Conference Publication 2217, pp. 31–43 (1982).
52. J. A. Simmons and W. B. Ebner, Accelerated rate calorimetry studies on partially discharged Li/SO₂ cells, in *Proceedings of the 32nd International Power Sources Symposium*, pp. 239–249, The Electrochemical Society, Pennington, New Jersey (1986).
53. W. B. Ebner, H. V. Venkatasetty, and K. Y. Kim, Lithium–Sulfur Dioxide Battery Safety Hazards Thermal Studies, Contract No. N60921-81-C-0085, Final Report 24 December 1980–5 March 1982, Naval Surface Weapons Center, Silver Spring, Maryland (1982).
54. Y. I. Cho, H. Frank, and G. Halpert, *J. Power Sources* **21**, 183 (1987).
55. H. A. Frank and D. D. Lawson, Safety of Li–SO₂ Batteries under Fleet Conditions, Final Report, Naval Surface Weapons Center, Silver Spring, Maryland (1987).
56. P. Bro and S. C. Levy, *Quality and Reliability Methods for Primary Batteries*, John Wiley and Sons, New York (1990).
57. S. C. Levy, Fault tree analysis: A tool for battery safety and reliability studies, in *Proceedings of the 5th Annual Battery Conference on Applications and Advances*, Paper 90MS-2, California State University, Long Beach (1990).
58. J. Bene, Safety testing of lithium cells, in *The 1980 Goddard Space Flight Center Battery Workshop*, NASA Conference Publication 2177, pp. 111–117 (1981).
59. H. Taylor and B. McDonald, Abuse testing of Li/SO₂ cells and batteries, in *Proceedings of the 24th Power Sources Symposium*, pp. 66–71, PSC Publications Committee, Red Bank, New Jersey (1977).
60. D. D. Lawson and H. A. Frank, Safety of Li/SO₂ Batteries under Pulsed Load Profiles, Contract Review for Naval Surface Weapons Center, White Oak Laboratory, Jet Propulsion Laboratory, Pasadena, California (1984).
61. H. A. Frank, G. Halpert, D. D. Lawson, J. A. Barnes, and R. F. Bis, *J. Power Sources* **18**, 89 (1986).
62. J. Heydecke, *Prog. Batteries Sol. Cells* **6**, 93 (1987).
63. H. K. Street, S. C. Levy, and C. C. Crafts, Safe Design and Assembly of Li/SO₂ Batteries, Report RS 2523/80/1, Sandia National Laboratories, Albuquerque, New Mexico (1980).
64. Handbook, *What You Should Know about Lithium–Sulfur Dioxide Batteries for Army Applications*, U.S. Army Electronics Technology and Devices Laboratory, Power Sources Division, Fort Monmouth, New Jersey (1985).
65. S. C. Levy, *J. Power Sources* **43–44**, 247 (1993).
66. W. J. Moroz, Safety Design Considerations for Lithium Batteries in CF Applications, Technical Note No. 81-11, Defence Research Establishment Ottawa, Department of National Defence, Canada (1981).
67. M. T. Brundage, G. J. DiMasi, L. P. Jarvis, and T. B. Atwater, Significant advantages in the safety and technology of lithium–sulfur dioxide batteries, in *Proceedings of the 32nd International Power Sources Symposium*, pp. 250–258, The Electrochemical Society, Pennington, New Jersey (1986).

68. B. P. Dagarin, J. S. Van Ess, and L. S. Marcoux, Galileo Probe battery systems design, in *Proceedings of the 21st Intersociety Energy Conversion Engineering Conference,* American Chemical Society, Washington, D.C. Vol. 3, pp. 1565–1571 (1986).
69. L. Marcoux, The Galileo Probe Li/SO₂ battery, in *Proceedings of the 18th Intersociety Energy Conversion Engineering Conference,* American Institute of Chemical Engineers, New York, Vol. 4, pp. 1478–1482 (1983).
70. C. Berger, Advanced lithium batteries for command, control and communications, in *Proceedings of the Second Annual Battery Conference on Applications and Advances,* Proc. Vol. 87-16, pp. 220–228, The Electrochemical Society, Pennington, New Jersey (1987).
71. C. E. Schelleng, Battery Box for BA-5590 Battery on KY-57, DELSD-E Report No. 80, U.S. Army Electronics Research and Development Command, Fort Monmouth, New Jersey (1985).
72. S. C. Levy, Transportation container for Li/SO₂ batteries on passenger aircraft, in *Proceedings Lithium '87,* pp. 59–61, Waste Resource Associates, Niagara Falls, New York (1987).

12

Other Soluble Depolarizer Lithium Systems

The lithium/bromine complex (Li/BCX), lithium/sulfuryl chloride, and lithium/chlorine in sulfuryl chloride (Li/CSC) chemistries are liquid depolarizer systems. They are all commercially available but are not in use nearly as much as the liquid depolarizer lithium/thionyl chloride and lithium/sulfur dioxide systems discussed in the previous chapters.

12.1. LITHIUM/BROMINE COMPLEX (Li/BCX)

The lithium/bromine complex system is a lithium/thionyl chloride cell with bromine chloride (BrCl) added to the electrolyte. The concentration of BrCl is usually in the range of 7–20 mol %. Bromine chloride is a corrosive gas having a boiling point of −5°C. Even at comparatively low concentrations, the vapors are highly irritating and painful to the respiratory tract. Liquid BrCl rapidly attacks the skin and other tissues, producing irritation and burns that heal very slowly.[1] The hazardous properties of the other components are given in Chapter 10.

12.1.1. Chemistry

A normal product of the discharge of lithium/thionyl chloride cells is elemental sulfur. As discussed in Chapter 10, reactions involving sulfur contribute to the hazardous behavior of these cells. It has been observed that BrCl reacts with elemental sulfur to form a volatile product or products.[2] Therefore, the addition of BrCl to $Li/SOCl_2$ cells should prevent the accumulation of sulfur, at least in the early stages of discharge. Sulfur formed during the later stages of discharge has been noted, but possessing a modified crystal structure or morphology.

It has also been shown that BrCl reacts with the carbon cathode collector, forming $C-Br$ and $C-Cl$ bonds. This results in a change in the characteristics of the carbon, for example, functional groups, influencing the cell chemistry and possibly improving the system safety.[3]

12.1.2. Safety/Abuse Tests

Short Circuit. Short-circuit tests were carried out on a number of cell sizes. "AA" cells bulged but exhibited no hazardous behavior and a maximum current of 2 A.[2,4] "C" cells also bulged with no hazardous behavior, although some leakage was observed through a cracked glass-to-metal seal. Maximum current was 14 A. "D" cells delivered a maximum current of 20 A with bulging but no venting or hazardous behavior. If insulated, they exploded when the wall temperature exceeded 149°C. "DD" cells having an internal fuse were shorted. The fuses opened, and no hazardous behavior was noted.

Charge. Both fresh and 50% discharged "AA" cells were charged at 100 mA for 10 h with no leakage or venting.[4] Similar results were noted for "C" cells charged at 500 mA for 10 h. However, two of three "DD" cells charged at 2 A for 10 h exploded violently. "D" cells that were either discharged at 1 A or force overdischarged at 1 A and then charged at 1 A showed no signs of hazardous behavior. However, the overdischarged cells exhibited voltage fluctuations when switched to the charging mode.[2,5,6]

Forced Overdischarge. A four-cell pack of "AA" cells was driven into reversal at 125 mA. After 3.2 h, one cell vented through the sidewall. In one study, ~20% of "C" cells overdischarged at 0.22 A vented through the sidewall. Similar behavior was noted for "C" cells that were driven into reversal at 0.02 A for 10 days.[4] However, in other studies, "C" cells were overdischarged at 0.5 A with no safety-related occurrences.[2,5] In addition, seven-cell strings were overdischarged at 3 A without incident. Replacing one cell in the string with a completely discharged cell and discharging at 3 A for 18 h was again without incident.

"D" cells discharged into reversal at 1 A and 5 A for as long as 120 h did not exhibit any hazardous behavior. They did not show any percussion sensitivity when subjected to the shock of a shotgun blast at ~3 m.[2,6] In another study, it was found that "D" cells discharged into reversal at 1 A to their rated capacity were safe, but continued overdischarge beyond rated capacity may lead to venting.[4] As the temperature is raised, more ventings occur. Overdischarge of "DD" cells at 2 A can lead to cell ventings through a cracked glass-to-metal seal.[4]

Heat. Both fresh and 50% discharged "AA" cells were heated in a chamber to 150°C and held for 15 min. The cells bulged but did not leak or vent. Similarly, "C" cells were heated to 149°C and held for 15 min. The discharged cells bulged without leaking, but the fresh cells leaked through the glass-to-metal seal. "DD" cells also leaked at the glass-to-metal seal when heated to 149°C. "D" cells were wrapped in heat tape and heated until failure. Venting occurred at a cell wall temperature of 177°C.[4]

Cells built without a vent were heated over a naphtha fire. They exploded at a temperature of 420°C.[2]

These tests indicate that Li/BCX cells are less hazardous than Li/SOCl₂ cells under certain abusive conditions. However, a limited amount of data is available, and cells should be retested for each specific application.

12.1.3. Design

Lithium/bromine complex cells are normally built having a wound electrode configuration in a stainless steel can. Both electrodes use nickel screen current collectors and are separated by a nonwoven glass separator.[8] A picofuse is located in the positive lead, outside the glass-to-metal seal and beneath a terminal cap located on the end of the cell.

Cells were redesigned for use at elevated temperature (150°C). The cell stack was shortened to increase the void volume. This allowed cells to withstand the increased temperature without leaking through the glass-to-metal seal.[8,9]

12.2. LITHIUM/SULFURYL CHLORIDE

The Li/SO_2Cl_2 chemistry is a high-voltage (open-circuit voltage = 3.90 V), high-energy system. The electrolyte solvent/depolarizer sulfuryl chloride is a colorless liquid. Both the liquid and its vapors are a corrosive irritant to the skin, eyes, and mucous membranes. SO_2Cl_2 boils at 69.1°C and, upon exposure to moisture, forms H_2SO_4 and HCl.[10]

12.2.1. Chemistry

During normal discharge, sulfuryl chloride is reduced to form lithium chloride and sulfur dioxide.[11] No elemental sulfur is formed. Since sulfur has been linked to safety problems in both the lithium/thionyl chloride and lithium/sulfur dioxide systems, the absence of sulfur in the lithium/sulfuryl chloride system points to improved safety.[12,13]

During forced overdischarge, exothermic reactions occur that result in the formation of COS, CO_2, and CS_2.[13]

Sulfuryl chloride thermally decomposes to form sulfur dioxide and chlorine.[12] This reaction is reversible and is catalyzed by the presence of dissolved $AlCl_3$ or $LiAlCl_4$.

12.2.2. Abuse Tests

Forced overdischarge of cathode-limited Li/SO_2Cl_2 cells does not pose a safety hazard. At the point of reversal, there is a temperature and pressure spike, but no explosion, venting, or leakage occurs.[14,15] Charging fresh "D" cells at 1 A is also not hazardous,[15] nor is charging of half-discharged cells.[14] Short circuit of prismatic "D" cells resulted in a maximum current of 28–30 A and case rupture at the weld.[14,15] Shorting of identical $Li/SOCl_2$ cells resulted in thermal runaway and explosion.

Overheating tests of battery packs containing 15 "DD" cells in series were conducted with heat tape.[16] Raising the pack temperature to as high as 390°C resulted in only benign venting.

The limited data available indicate that lithium/sulfuryl chloride cells are indeed safer than lithium/thionyl chloride cells. However, other features of this chemistry, for example, a high corrosion rate and limited shelf life, preclude its widespread use.

12.3. LITHIUM/CHLORINE IN SULFURYL CHLORIDE (Li/CSC)

The Li/CSC chemistry is a high-energy system that employs the addition of ~0.3–0.45M chlorine dissolved in the electrolyte of a lithium/sulfuryl chloride cell. Chlorine is a very irritating, moderately toxic gas. The lowest published lethal concentration for humans is 2530 mg/m^3 for 30 min exposure.[10]

12.3.1. Safety Design Features

Lithium/chlorine in sulfuryl chloride cells are normally built with a frustum reverse buckling vent in the can bottom (see Section 17.1.1). Heat tape tests of these cells at a rate of 3°C/min resulted in all cells venting safely when their external temperature reached 101–109.5°C.[17]

The use of fuses to protect against case rupture during short circuit of Li/Cl_2 in SO_2Cl_2 high-rate "AA" cells has been evaluated.[17,18] Without a fuse, the "AA" cells generate currents of 15–23 A and skin temperatures of 160–200°C. With a 5-A fuse located under the positive terminal cap, four of five fuses opened within 2 s with no temperature rise, and the fifth opened after 17 s with the case temperature rising to 47°C.

12.3.2. Abuse Tests

Forced Overdischarge. Unvented, moderate rate "AA" cells were force overdischarged, five at 50 mA for 60 h and five at 200 mA for 15 h.[17] Two of the cells at 50 mA had heat stains, and one of these cells had a small burn hole. The other eight cells were without incident. Fused "AA" cells have been forced into reversal for >1.5 times capacity at rates from 750 mA (nominal) to 4 A, without rupturing or exploding.[18] Li/CSC "D" cells were force over-discharged to 10 times capacity at 1 A and 150°C with no venting or explosion.[19]

Short Circuit. Li/CSC "D" cells were shorted at several temperatures.[19] At 25°C, a maximum current of 14 A was attained, and currents of 8–12 A were sustained for ~20 min; cells then exploded. At 150–162°C, a maximum current of 18–25 was reached. The cathode-to-pin lead incinerated within 5 min, and no vents or explosions occurred. At 172°C, all cells exploded violently. In another study,[6] "D" cells having an electrode area of 294 cm^2 delivered 30–35 A on short circuit, at which point the internal leads burned. A high-temperature design, having an electrode area of 129 cm^2, yielded 18 A on short circuit at 162°C. The maximum skin temperature attained was 167°C.

Charging. Li/CSC "D" cells have been discharged at 1 A and then charged at 1 A with no hazardous results. However, increasing the charging rate to 10 A resulted in cells venting after ~4 min.[5,6,20,21]

Incineration. No violent reaction has been found to occur upon incineration of Li/CSC cells.[5] Heating of spirally wound cells, without a vent, in a naphtha flame at 800°C resulted only in a mild vent through the glass-to-metal seal.[20,21]

High-Rate Discharge. Li/CSC "D" cells, discharged at 15 A and −2°C, showed an increase in cell wall temperature to ~100°C. As the voltage approached 2 V, the temperature increased rapidly to 120°C and venting occurred through the glass-to-metal seal.[22] Also, at a constant current discharge of >6 A, cells will vent normally with no ruptures, violent vents, or explosions.

Physical Abuse. Penetration of a cell with a bullet results in massive internal shorting. In low-rate designs, a lithium fire may result after a few minutes,[19] while in high-rate designs the cells will explode.[22] Crushing the cell to approximately two-thirds of its initial height will also result in rupture or explosion. Shocking of discharged cells with a 20-gauge shotgun at 2–4 m does not cause the cells to explode.

In general, the Li/CSC system is a relatively safe one when exposed to individual abusive conditions. However, exposure to two or more abuses simultaneously can result in a hazardous situation.[6,22]

REFERENCES

1. Dow Chemical Company, *Bromine Chloride Handbook,* Dow Chemical Company, Inorganic Chemicals Department, Midland, Michigan.
2. C. C. Liang, P. W. Krehl and D. A. Danner, *J. Appl. Electrochem.* **11,** 583 (1981).
3. S. S. Kim, B. J. Carter and F. D. Tsay, *J. Electrochem. Soc.* **132,** 335 (1985).
4. J. B. Trout, Studies of Performance and Abuse Resistance of Lithium–Bromine Complex Cells for Manned Space Use, JSC-20006, NASA Contract No. NAS9-15425, Northrup Services, Inc., Houston (1984).
5. R. Murphy, Performance and safety characteristics of Li/BCX and Li/CSC systems, in *The 1980 Goddard Space Flight Center Battery Workshop,* NASA Conference Publication 2177, pp. 61–71 (1981).
6. R. M. Murphy, P. W. Krehl and C. C. Liang, The effect of halogen addition on the performance characteristics of lithium/sulfur oxychloride battery systems, in *Proceedings of the 16th Intersociety Energy Conversion Engineering Conference,* American Society of Mechanical Engineers, New York, Vol. 1, pp. 97–101 (1981).
7. W. J. Moroz and G. J. Donaldson, Performance and Safety Evaluation Tests on Bromine-Complexed Lithium Cells, Report DREO TN 82-4, Defence Research Establishment Ottawa, Ottawa, Ontario, Canada (1982).
8. Electrochem Industries Inc., Safety Test Report on Li/BCX Cells, Electrochem Industries Inc., Clarence, New York (1980).
9. E. S. Takeuchi, C. F. Holmes and W. D. K. Clark, Thermal properties and effects for Li/ BCX (thionyl chloride) cells, in *Proceedings of the 22nd Intersociety Energy Conversion Engineering Conference,* Vol. 2, pp. 720–724 (1987).
10. N. I. Sax and R. J. Lewis, Sr., *Dangerous Properties of Industrial Materials,* Van Nostrand Reinhold, New York (1989).
11. S. Gilman and W. Wade, Jr., *J. Electrochem. Soc.* **127,** 1427 (1980).
12. K. A. Klinedinst, *J. Electrochem. Soc.* **131,** 492 (1984).
13. K. M. Abraham, M. Alamgir and R. K. Reynolds, *J. Electrochem. Soc.* **135,** 2917 (1988).
14. K. A. Klinedinst and R. A. Gary, *J. Electrochem. Soc.* **134,** 1884 (1987).
15. K. A. Klinedinst and R. A. Gary, Performance and abuse resistance of a high-rate prismatic Li/SO$_2$Cl$_2$ cell, in *Proceedings of the 32nd International Power Sources Symposium,* pp. 463–471, The Electrochemical Society, Pennington, New Jersey (1986).

16. P. B. Hallal and R. F. Bis, Performance and Safety Testing of Lithium Batteries for the Expendable, Mobile, ASW Training Target (EMATT), NUSC Technical Document 6502, Naval Underwater Systems Center, Newport, Rhode Island (1986).
17. S. J. Ebel, D. M. Spillman and W. D. K. Clark, Safety and high power capabilities of the Li/CSC primary battery system, in *Proceedings of the 34th International Power Sources Symposium,* pp. 259–263, IEEE, Piscataway, New Jersey (1990).
18. S. Ebel and W. Clark, *Prog. Batteries Sol. Cells* **8,** 67 (1989).
19. R. M. Murphy and C. C. Liang, *J. Appl. Electrochem.* **13,** 439 (1983).
20. C. C. Liang, M. E. Bolster and R. M. Murphy, The Li/Cl$_2$ in SO$_2$Cl$_2$ inorganic system II: D cell discharge characteristics, in *Proceedings of the Symposium on Lithium Batteries,* Proc. Vol. 81–4, The Electrochemical Society, Pennington, New Jersey (1981).
21. C. C. Liang, M. E. Bolster and R. M. Murphy, *J. Electrochem. Soc.* **128,** 1631 (1981).
22. R. M. Murphy and C. C. Liang, *J. Electrochem. Soc.* **130,** 1231 (1983).

13

Rechargeable Lithium Systems

A number of secondary lithium chemistries are being developed for various applications. Only the lithium/molybdenum disulfide and lithium/manganese dioxide systems have been available commercially. Since most of these cells are in the development stage, little safety information has been published. In addition, a new class of lithium rechargeable batteries, the lithium ion or "rocking chair" systems, are now under intense development. One system, lithium ion/cobalt oxide, can be purchased commercially. In this chapter, the published safety information on the various rechargeable lithium systems will be reviewed.

13.1. LITHIUM/MOLYBDENUM DISULFIDE (Li/MoS₂)

The lithium/molybdenum disulfide system (Molicel) is capable of delivering 100 Wh/kg (240 Wh/liter). It contains an electrolyte of $LiAsF_6$ in a mixture of propylene and ethylene carbonates. Ethylene carbonate (glycol carbonate) is a skin and eye irritant. It is mildly toxic by ingestion, having an oral LD_{50} for rats of 10 g/kg. It is combustible when exposed to heat or flame.[1] The toxicological properties of propylene carbonate and lithium hexafluoroarsenate have been discussed in Chapter 9. Molybdenum disulfide may cause irritation upon contact with the eyes or if inhaled. Based upon animal experiments, molybdenum compounds are highly toxic if ingested and may result in death from heart failure.[2]

13.1.1. General/Design

When Li/MoS_2 cells are cycled at constant current for a fixed time, the voltage limits become progressively wider with cycle number.[3] Upon reversal, cells form permanent shorts. Placing a diode across the cell to limit the voltage reversal to less than -1 V prevents cells from venting.[4] If the cell voltage falls below 0.5 V for more than a few seconds, overheating will result.[5] To protect cells against both overcharge and overdischarge, a safety header has been developed which contains both a positive temperature coefficient (PTC) fuse and diodes.[6]

Large cells have poor thermal conductivity and will overheat on electrical abuse.[7] In the 25–75% state-of-charge condition, constant-current discharge is exothermic while constant-current charge is endothermic.[8]

13.1.2. Safety/Abuse Tests

Several series of abuse tests have been carried out on the Li/MoS_2 system, ranging from the 600-mAh "AA" size to 45-Ah "BC" cells (approximately the size of a beer can). The results are summarized in the following paragraphs.

Short Circuit. Shorting of "AA" cells after 0, 100, and 300 charge/discharge cycles was accomplished at both 21 and 55°C.[3,9] The maximum current delivered was 16 A at 21°C and 27 A at 55°C. The maximum temperature reached was 125–135°C. At 21°C, new cells and those with 300 cycles showed a <5% probability of venting at 21°C. Cells with 100 cycles have a 20% probability of venting at 21°C and 30% probability at 55°C. In all cases, the venting was a mild pressure relief. "C" cells shorted at 22°C did not vent, and a wall temperature of 140°C was reached.[4]

Short circuit of fresh 45-Ah "BC" cells in a 21°C heat sink block yielded a maximum current of 248 A and had a maximum temperature of 129°C. After ~19 min, mild venting occurred. On cells experiencing 40 cycles, a similar test resulted in a maximum current of 275 A and a maximum temperature of 300°C. Venting with fire occurred.[7]

Forced Overdischarge. Both fresh and cycled (100 cycles) "AA" cells were force discharged 100% at the $C/10$, $C/5$, and C rates. Maximum skin temperature increase was 32°C above ambient (either 21 or 55°C). The occurrence of venting is not dependent on cycle life or ambient temperature and is rated at ≤10%.[3,9] "C" cells that were cycled 10 times were force discharged to 400% of capacity and vented with flame at all rates. After cycling to a 50% loss in capacity, the cells vented with flame after being force discharged to 250% of capacity.[4] "BC" cells were force discharged at 10 A during both the 1st and 40th cycles. Vigorous venting occurred after 211% capacity on cycle 1 and after 156% capacity on cycle 40.[7]

Overcharging. "AA" cells were 100% overcharged at the $C/10$ rate. The maximum temperature increase was ~6°C. No cells vented at 21°C, but

~10% of the cells vented at 55°C.[3] Cells that have been cycled 100 times do not behave any differently than fresh cells.[9] In all cases, ventings were a mild pressure release accompanied by a small amount of liquid electrolyte. Overcharge of "C" cells resulted in venting at $>C/5$ rate.[4] When "C" cells were charged at >280 mA, they vented with flame at 80% overcharge, unless heat-sinked.[6] "BC" cells were overcharged on cycles 1 and 40 at the $C/10$ rate.[5] Venting with flame occurred after 80% capacity on cycle 1 and after 119% capacity on cycle 40.[7]

Heat. Heat tests were conducted on fresh "AA" cells and on cells that had been cycled 100 and 300 times.[3,9] No safety problems were noted at $<140°C$. At temperatures between 140 and 180°C, most cells vented mildly, releasing pressure and a small amount of liquid electrolyte. Above 180°C, cells vented with flame but no explosion. Incineration in a propane flame caused the safety vent to open in ~10 s. The cells then ignited and burned vigorously, with some sparks produced. "C" cells were heated on a hot plate at 5°C/min until venting with flame occurred.[6] Cells vented between 115 and 180°C, the cells with the higher cycle numbers venting at the lower end of the temperature range.

Crush. "AA" cells were crushed radially in a hydraulic press to a force of 10,000 pounds, in both the fresh state and after 100 cycles.[3,9] No fire or explosion occurred, only mild venting and electrolyte leakage. Crushing of "C" cells causes the glass-to-metal seal to crack. Internal shorts form and result in venting with fire.[6]

Shock and Vibration. Tests were conducted on "AA" cells, both in the fresh state and after 100 cycles. Shocked cells were subjected to an acceleration of $>75g$ for 3 ms, with a peak of $125-175g$, in three axes. No physical or electrical changes were noted. Dropping 6 ft onto concrete twice, once on the cell top and once on the cell bottom, did not cause any safety problems either. Similar cells were subjected to sinusoidal vibration of 0.03-in. amplitude with a frequency varying from 10 Hz to 55 Hz to 10 Hz at 1 Hz/min in three mutually perpendicular axes. Again, no physical or electrical changes were noted.[3,9]

13.2. LITHIUM/TITANIUM DISULFIDE

The Li/TiS_2 chemistry is a high-energy system capable of delivering 125 Wh/kg and 300 Wh/liter. It is still in the developmental stage, and a number of electrolytes have been used by various laboratories. Electrolytes studied to date include dioxolane with either LiSCN or LiClO$_4$; LiPF$_6$ in 4-methyl-1,3-dioxolane/dimethoxyethane; and LiAsF$_6$ in either tetrahydrofuran (THF), THF/2-methyl-THF, THF/2-methyl-THF/dioxolane, or propylene carbonate/ethylene carbonate/triglyme. Most titanium compounds are considered physiologically inert.[1] However, TiS$_2$ will release SO$_2$ on combustion in air and will hydrolyze slowly in water, forming H$_2$S. LiSCN has a low acute toxicity, but prolonged absorption may produce various skin eruptions, running nose, and occasionally dizziness, cramps, nausea, vomiting, and disturbance of the nervous system.[1] LiPF$_6$ is an irritant to the skin, eyes, and mucous membranes and is corrosive by ingestion and inhalation. Its threshold limit value is 2.5 mg/m^3 as fluorine.[10] THF is a flammable liquid with a boiling point of 65.4°C and a flash point of -17°C. It is moderately toxic by ingestion and inhalation and is an irritant to the eyes and mucous membranes. It can be narcotic in high concentrations. THF is a very dangerous fire hazard if exposed to heat, flame, or oxidizers. Unstabilized THF forms thermally explosive peroxides on exposure to air. The lowest published toxic concentration for inhalation by humans is 25,000 ppm.[1] 2-Methyl-THF is a flammable liquid with a boiling point of 80°C and a flash point of -11°C. It is a very dangerous fire hazard when exposed to heat, flame, or oxidizers. It is mildly toxic by ingestion, inhalation, and skin contact and is an eye irritant. The LC$_{50}$ for inhalation by rats is 6000 ppm for 4 h.[1] Triglyme (2,5,8,11-tetraoxadodecane) can produce dangerous human reproductive effects, as can all glycol ethers. It boils at 216°C and its flash point is 111°C. On heating to decomposition, triglyme emits acrid smoke and irritating fumes.[1] 4-Methyl-1,3-dioxolane is probably similar in toxic properties to 2-methyl-1,3-dioxolane, which is mildly toxic by ingestion and inhalation and emits acrid smoke and irritating fumes when heated to decomposition.[1] Properties of the other electrolyte components used with Li/TiS$_2$ cells have been discussed previously.

13.2.1. Safety-Related Chemistry

The first electrolyte used with the Li/TiS$_2$ system contained LiClO$_4$ as the conductive salt. Safety problems were noted, with respect to the explosive nature of LiClO$_4$ in the presence of organics, leading to the investigation of other suitable electrolytes.[11]

Care must be taken if cells using the LiSCN/1,3-dioxolane electrolyte are involved in a fire or in any situation that may result in cell overheating. The thermal decomposition of this electrolyte results in the formation of HCN, which is a highly toxic gas.[12]

The thermal stability of the electrolyte LiPF$_6$ in 4-methyl-1,3-dioxolane/dimethoxyethane can be increased by the addition of 18 vol % triethylenephosphoramide.[13]

13.2.2. Design

Li/TiS$_2$ cells have been designed using a fusible separator. The separator (trademark SafeTsep) consists of two or more microporous layers with at least one of the layers capable of transforming to a nonporous state at a temperature between 100 and 150°C.[14] The cells also contain a 225-psi rupture disk. However, cells were found to vent prematurely owing to electrolyte expansion. A decrease in electrolyte fill volume of 0.1 ml in the "AA" cells increased the venting temperature by $\sim$30°C.

One of the major failure modes of Li/TiS$_2$ cells is the formation of lithium dendrites and subsequent shorting.[15] Use of Li–Al alloys containing small amounts of Bi or Nb as the anode in Li/TiS$_2$ cells can eliminate the formation of Li dendrites on cycling.[16] However, the use of δ-Li–Al alloys, containing from 60 to 85% lithium, was found to have no distinct safety advantage over pure lithium anodes.[17]

13.2.3. Overcharge Protection

Overcharge protection has been achieved in Li/TiS$_2$ cells by the addition of n-butylferrocene to the electrolyte.[18] During overcharge, the cell voltage is maintained at 3.25 V, whereas without the additive, cell voltage reaches >5 V.

The addition of a large excess of LiI (0.5–1.0M) to cells using LiAsF$_6$ in THF as the electrolyte will also afford overcharge protection to Li/TiS$_2$ cells.[19] Since there is excess LiI, not all of it is oxidized to I$_2$ during overcharge. The remaining material combines to form LiI$_3$, which reacts further with the lithium to restore the cell to its original configuration.

13.2.4. Safety/Abuse Tests

A limited number of abuse tests have been reported on Li/TiS$_2$ cells. They are summarized in the following paragraphs.

Short Circuit. Shorting of "AA" cells after five cycles yielded a maximum current of 13.5 A and a maximum temperature of 106°C. No safety hazards were noted.[17] Shorting of 2-Ah prismatic cells after 24 cycles gave currents of 7–8 A and a skin temperature of $\sim$50°C. After 55 cycles, shorting at 85°C gave 6 A and a temperature of $\sim$110°C. No venting occurred in any cell.[20,21]

Forced Overdischarge. Prismatic cells (2 Ah) were force overdischarged overnight at 400 mA, into reversal. No can deformation or venting was noted.[20,21]

Overcharge. Prismatic cells (2 Ah) were overcharged at 200 mA at room temperature. A plateau was reached at 3.8–4.1 V, followed by a rise to the power supply voltage of 40 V. After some time, the voltage dropped and the cells acted as resistors. A mildly exothermic reaction occurred, but there was no can deformation or venting.[21] Tests were also conducted on a 12-cell string, charging at 200 mA to 40 V. No venting occurred.[20]

Heat. Heating of "AA" cells resulted in an exotherm beginning at ~150°C and reaching a maximum temperature of 300°C. The cells exploded, rupturing the weld between the cover and can.[17] Heating a fully charged prismatic cell resulted in venting at 130–165°C. The internal pressure rose to >800 psi in <0.7 s, and the cell wall temperature was >650°C during the pressure peak. Heating of a partially discharged prismatic cell to 290°C did not result in a venting. An exothermic reaction and internal shorting occurred at 125–130°C.[20,21]

Puncture. No significant reaction was noted when cells were penetrated by a nail. Although the cells were shorted by the nail, there was only a 20°C temperature rise and no venting.[21]

13.3. LITHIUM/MANGANESE DIOXIDE

The rechargeable Li/MnO_2 battery is a high-energy system capable of delivering >125 Wh/kg and >300 Wh/liter. It is still in development and may be found with various electrolytes, such as $LiPF_6$ in propylene carbonate/ 2-methyl-THF. The toxicological properties of these materials, as well as those of MnO_2, have been discussed previously.

13.3.1. Design

It has been found that in cells having a wound or coiled electrode assembly, the use of an anode which is wider than the cathode will prevent the growth of lithium dendrites on the anode.[22]

To prevent gas evolution during overdischarge in cells containing a lithium-insertable alloy anode, the amount of lithium in the anode in the charged state must be >1% by weight of the lithium-insertable alloy and ≤20% by weight of the MnO_2 in the cathode.[23]

13.3.2. Chemistry

To increase the safety of Li/MnO_2 batteries in case of internal heat generation, the electrolyte comprises $LiPF_6$ salt and the cathode contains 0.01–

0.5% by weight H_2O. Upon overheating, the $LiPF_6$ reacts with the H_2O and decomposes, increasing the internal resistance of the battery. Short-circuit tests of cells containing MnO_2 cathodes, treated to control the moisture content to 0.1%, and $1M$ $LiPF_6$ in propylene carbonate/2-methyl-THF electrolyte, showed no significant deformation of the case. Similar tests with cells containing 10 ppm H_2O in the electrolyte had 88% of the cells showing case deformation.[24]

13.3.3. Abuse Tests

Li/MnO_2 cells were crushed until cell thickness was one-fourth of the original diameter.[25] Cells that had not been cycled extensively did not experience any serious safety problems. However, cells that had been cycled many times, especially large-diameter cells, caught fire or exploded. This behavior is thought to be due to reaction of high-surface-area lithium, accumulated on the anode surface during cycling, initiated by heat generated by high currents through internal short circuits caused by the crushing.

13.4. LITHIUM/SULFUR DIOXIDE

The major difference between primary and secondary Li/SO_2 cells is the absence of an organic cosolvent (acetonitrile) in the rechargeable version. Liquid sulfur dioxide is used as both the active cathode and electrolyte solvent. The conductive salt is $LiAlCl_4$, at a concentration to yield either $LiAlCl_4 \cdot 3SO_2$ or $LiAlCl_4 \cdot 6SO_2$. A high-surface-area carbon is used as the cathode collector.

These cells are pressurized, even when fully discharged. A "C" cell, with a $LiAlCl_4 \cdot 6SO_2$ electrolyte, has an internal pressure of ~ 55 psi at 0 V.[26] The safety of rechargeable Li/SO_2 cells is sensitive to their design and the materials used for stack insulation and separators, as well as to cathode density and stack compression.[27] Dendrite shorting, which may lead to explosions, is strongly affected by the pore size and pore size distribution in the separator.[28]

Use of a copper foil current collector at the anode has been linked to cell rupture in cycled cells.[29,30] The hazardous behavior was noted if cells were deep discharged (to <1 V) after numerous cycles or received extended overcharge. The anode of a cycled cell will react violently if exposed to air. In one study, a cycled cell (32 cycles) was frozen in liquid nitrogen and cut open for analysis. The electrode package detonated ~ 10 s after being exposed to the atmosphere. Two possible causes for this behavior have been postulated: either dendritic lithium reacted explosively with moisture condensed on the frozen stack, or some combination of unstable reaction products was present on the surface of the lithium electrode.[27]

13.4.1. Chemistry

The rechargeable Li/SO_2 system has a built-in mechanism to protect it against overcharging[29,31]:

At the positive:

$$AlCl_4^- \rightarrow AlCl_3 + \tfrac{1}{2}Cl_2 + e^- \qquad (13.1)$$

At the negative:

$$Li^+ + e^- \rightarrow Li^0 \qquad (13.2)$$

Then the following recombinations occur, forming the starting material:

$$\tfrac{1}{2}Cl_2 + Li \rightarrow LiCl \qquad (13.3)$$

$$LiCl + AlCl_3 \rightarrow LiAlCl_4 \qquad (13.4)$$

The addition of a small amount of Br_2 (2–5%) to the electrolyte has been shown to improve the safety characteristics of Li/SO_2 rechargeable cells.[28] The bromine lowers the charging potential, since Br_2 is formed rather than Cl_2. This results in increased separator stability, which in turn leads to a safer cell, since several hazards have been traced to internal shorts.[32]

The addition of $>0.2M$ lithium iodide to the electrolyte reduces the explosion hazard on forced discharge.[33] The iodide is spontaneously oxidized to I_2 by the SO_2. Under conditions of forced discharge, iodide is oxidized at the anode and I_2 is reduced at the cathode. Also, the discharge product is more uniformly deposited in the cathode, reducing the plugging of the pores. This results in a safer cell under high-rate forced discharge, even when the cell is not lithium limited.

The safety of Li/SO_2 cells can be improved by substituting Li–Al alloy for pure Li as the anode.[34] The high-melting alloy (600–700°C) does not form dendrites on charging and is less reactive to gases and moisture than lithium metal.

13.4.2. Safety/Abuse Tests

A limited number of abuse tests have been carried out on "C" size Li/SO_2 secondary cells. The results of these tests are summarized in the following paragraphs.

Short Circuit. Fresh cells will vent upon shorting whereas cycled cells will experience thermal runaway.[26,31] Storing fresh cells for three months and then shorting will also result in venting,[29] with a maximum current of 29 A and a maximum temperature of 74°C.[30]

Overcharge. Cells have been overcharged at 0.1 A for 24–240 h. Some cells experienced a mild venting, whereas others did not vent at all.[29,30] In another test, overcharging of fresh cells at moderate rates caused no safety problems.[26]

Heat. Fresh cells placed on a hot plate will vent, whereas cycled cells will rupture.[29,30] External heating of cycled cells results in thermal runaway.[26]

13.5. LITHIUM/COBALT OXIDE

The lithium/cobalt oxide chemistry is a high-energy system having an energy density of >150 Wh/kg and >300 Wh/liter. The actual cathode material is lithiated cobalt oxide, Li_xCoO_2, with $x \leq 1$. Various investigators have used different electrolytes with this system, including $LiPF_6$ in propylene carbonate/dimethoxyethane, $LiClO_4$ in 3-acetyl-THF, $LiClO_4$ in 3-cyano-THF, and $LiAsF_6/LiBF_4$ in methyl formate/methyl acetate. The specific toxicity of lithiated cobalt oxide is unknown. However, cobalt compounds are suspected carcinogens of the connective tissues and lungs.[35] No threshold limit data are available. Methyl acetate is a flammable liquid with a boiling point of 57°C. It is a dangerous fire hazard when exposed to heat or flame. Methyl acetate is an eye, mucous membrane, and skin irritant. It is also a central nervous system depressant. Absorption may lead to metabolism to methyl alcohol, producing methyl alcohol poisoning. Exposure to 10,000 ppm is immediately dangerous to life or health.[36] Toxicological properties of the other common electrolytes have been given previously.

13.5.1. Design Features

Safety improvements have been made to Li/Li_xCoO_2 cells through the use of multiple layers of microporous polyethylene separators which inhibit the growth of dendrites that cause internal shorts[37] or by using melting-type shutdown separators.[38] Another design has a porous polymer film formed on the anode and/or cathode, on the side facing the other electrode. This film prevents overheating caused by external shorts or by separator damage due to dendrite growth.[39]

Cells that experienced >50 cycles at full depth of discharge were analyzed after failure due to dendrite shorts.[37] It was found that the loss of lithium was less than anticipated. Therefore, cells were built with one-half to one-fourth of the excess lithium normally used, without seriously affecting cycle life. These cells should be less of a safety hazard than those with the normal excess of lithium.

13.5.2. Gas Evolution

The Li/Li_xCoO_2 system will generate enough gas under certain conditions to cause rupturing of the cell case. By using an electrolyte of $1M$ $LiClO_4$ in 3-acetyl-THF[40] or $1M$ $LiClO_4$ in 3-cyano-THF,[41] cells could be charged to 4.5 V with no explosions due to gas evolution.

13.5.3. Abuse Tests

A very limited amount of abuse testing has been performed on the Li/Li_xCoO_2 system. The results below were obtained on prototype "D" cells with an electrolyte of $LiAsF_6/LiBF_4$ in methyl formate/methyl acetate.[37]

Overcharge. Abusive overcharging (>300% excess charge capacity at rates up to 2 A) resulted in no safety hazards.

Overdischarge. Overdischarge to below the normal 1.5-V cutoff did not result in any adverse effects.

Internal Shorts. Cycling of cells for several weeks after internal shorting occurred did not result in any cell ventings, overheating, or other safety hazards.

13.6. LITHIUM/NIOBIUM TRISELENIDE

The $Li/NbSe_3$ chemistry is a high-energy system, yielding an energy density of >120 Wh/kg and >300 Wh/liter. An electrolyte of $LiAsF_6$ in the ternary mixture ethylene carbonate/propylene carbonate/triglyme is used in these cells. Since the $NbSe_3$ was experimentally prepared for these development cells, there are no published toxicological data. However, niobium compounds can cause kidney damage, and selenium compounds are poisonous and experimental carcinogens. Long-term exposure to selenium compounds may be a cause of amyotrophic lateral sclerosis (Lou Gehrig's disease) in humans.

Exposure to inorganic selenium compounds can cause dermatitis.[1] Properties of the electrolyte materials have been discussed previously.

13.6.1. Safety/Abuse Tests

A limited number of abuse tests have been conducted on "AA" $Li/NbSe_3$ cells. In general, it was found that fresh, uncycled cells are unaffected by the various abuse conditions. Therefore, all the reported data are on cells that have been cycled 20 times.[42,43] Additional work is planned on more heavily cycled cells.

Short Circuit. Cells vent benignly when shorted. The short-circuit current is 10–15 A, and the cell temperature increases by $\sim 60°C$ for a short time.

Overcharge. Cells were overcharged a minimum of 500% at currents up to 800 mA. A slight temperature increase was noted, and cells vented benignly.

Overdischarge. Cells were driven into reversal below 0 V, to at least 100% of capacity, at currents up to 1.6 A with no venting.

Heat. Cells subjected to either slow (5°C/min) or rapid (drop into 300°C chamber) heating will vent with flame after the melting point of lithium has been reached. Thermal cycling between −40 and +70°C (100 cycles) does not result in any safety problems.

Crush. Crushing the "AA" $Li/NbSe_3$ cells results in a temperature rise of $\sim 20°C$ and benign venting.

Drop. Cells were dropped ten times from 6 ft onto concrete and did not pose a safety hazard.

13.7. LITHIUM/VANADIUM OXIDE

Lithium/vanadium oxide batteries can comprise either of two forms of cathode material, V_2O_5 or V_6O_{13}. The electrolyte used with this chemistry is typically $LiAsF_6$ in a mixture of propylene carbonate and dimethoxyethane. The vanadium oxide cathode is also used in polymer electrolyte systems. Contact of V_6O_{13} with the skin may cause irritation and dermatitis. Inhalation can cause cough, irritation, bronchitis, chest pain, metallic taste in mouth, and greenish discoloration of the tongue. Ingestion may cause vomiting, diarrhea, and convulsions. The oral LD_{50} in rats is 549 mg/kg.[44] The toxicological properties of V_2O_5 have been given in Chapter 9.

Very few safety data have been published on this system. One patent describes a V_2O_5 cell which utilizes a Li–Al alloy as the anode.[45] The alloy contains ≤8% by weight lithium, compared to V_2O_5, and ≥4% by weight lithium versus Al. The alloy also contains 2% by weight Mn. This alloy anode limits dendrite formation on charge.

Safety tests were performed on "C" and "AA" cells having a V_6O_{13} cathode. The cells gave good safety performance on long-term shorting, abusive charge, and discharge.[46]

13.8. LITHIUM/CHROMIUM OXIDE

Cells having chromium oxide (CrO_x) cathodes and a lithium salt-containing electrolyte use a Li-insertable alloy as the anode. In the charged state, the amount of lithium in the alloy is ≥1% by weight of the alloy and ≤7% by weight of the chromium oxide in the cathode. This configuration prevents gas evolution on overdischarging the cell.[47]

13.9. LITHIUM/NIOBIUM PENTOXIDE

Cells have been built containing a Nb_2O_5 cathode, a lithium salt-containing electrolyte, and a Li-insertable alloy anode. The amount of lithium in the anode in the charged state is ≥1% by weight of the alloy and ≤8% by weight of the Nb_2O_5 in the cathode. This composition prevents gas evolution during overdischarge of the cells.[48]

13.10. LITHIUM ION SYSTEMS

Lithium ion systems are a new class of rechargeable battery. They can use either a form of lithium-intercalating carbon or a lithium-insertable transition metal oxide as the anode. In either case, the safety of the system is improved considerably over that of cells using metallic lithium as the anode. A lithium-insertable cathode is typically used in these systems, for example, transition metal oxides or sulfides, and either a liquid or polymeric lithium-ion-conducting electrolyte. A number of investigators are also looking at organic cathode materials. The cells operate by moving lithium ions from the anode to the cathode on discharge and from the cathode to the anode on charge, hence the name "rocking chair battery." A number of lithium ion chemistries are under development, but only one, the $Li_xC/Li_{1-x}CoO_2$ system, is commercially available.

13.10.1. Lithium Ion/Cobalt Oxide

The $Li_xC/Li_{1-x}CoO_2$ cell is a high-energy system, having an energy density of ~ 115 Wh/kg and 250 Wh/liter. Several electrolytes may be used with this chemistry, including $LiAsF_6$ in propylene carbonate/dimethoxyethane and $LiPF_6$ or $LiClO_4$ in diethyl carbonate/propylene carbonate. Diethyl carbonate is a flammable liquid with a boiling point of 126°C and a flash point of 25°C. It is a dangerous fire hazard when exposed to heat or flame. Exposure to diethyl carbonate can result in redness of the skin, irritation of the eyes, and sore throat, coughing, shortness of breath, and headaches, if it is inhaled.[49] Properties of the other cell components have been described previously.

Since these cells are so new, few safety data have been published. However, if the cell voltage is allowed to exceed 4.5 V during charging, a rapid rise in internal pressure occurs and can lead to case rupture.[50] An antiovercharging safety device has been developed to shut down the charging current if the voltage rises above 4.5 V. It consists of an aluminum burst disk which bulges, breaking electrical contact within the cell.[50] When a small amount of Li_2CO_3 is added to the cathode, it decomposes during overcharge to form CO_2, causing an increase in internal pressure which activates the burst disk before the temperature rises to dangerous levels.[51]

It has been found that subjecting cells to 1–5 charge/discharge cycles prior to sealing decreases the gas evolution within the cell after sealing.[52] Cells were cycled 10 times and stored at 60°C for 15 h. Precycled cells appeared normal, but cells that had not been conditioned vented.

A limited number of safety tests have been reported for small cylindrical cells.[51] Crushing resulted in a maximum temperature of 78–98°C with some electrolyte leakage. Nail penetration resulted in electrolyte leakage and gas evolution, with a peak temperature of 77–105°C. Cells appeared to be safer with increasing cycle life. Overcharging was terminated by activation of the protective current shutoff device with the temperature below 70°C. On short circuit, the cell temperature reached 150°C, but no fire or explosion occurred.

13.10.2. Other Systems

The use of MoO_2 and WO_2 as intercalation anodes has been studied using $LiCoO_2$ as the cathode.[53] $LiPF_6$ in propylene carbonate was the electrolyte. Operating voltages of the systems are 2.5 V for the MoO_2 anode and 3.2 V for the WO_2 negative. In another study, it was found that coating the transition metal oxide anode with a thin film of a lithium ion-conducting solid will prevent explosions and/or electrolyte leakage on extended cycling.[54] Cells having a WO_3 anode sputtered with a 2-μm layer of $LiTaO_3$ were fabricated. After 200 cycles, cells behaved normally, but control cells without the coating leaked electrolyte after 50 cycles.

REFERENCES

1. N. I. Sax and R. J. Lewis, Sr., *Dangerous Properties of Industrial Materials,* Van Nostrand Reinhold, New York (1989).
2. Material Safety Data Sheet, "Molybdenum Disulfide," Johnson Matthey Catalog Chemicals, Ward Hill, Massachusetts (1990).
3. D. T. Fouchard, Safety studies of the Molicel AA cell, in *Proceedings of the 32nd International Power Sources Symposium,* pp. 218–225, The Electrochemical Society, Pennington, New Jersey (1986).
4. J. A. Stiles, Performance and safety characteristics of lithium–molybdenum disulfide cells, in *The 1983 Goddard Space Flight Center Battery Workshop,* NASA Conference Publication 2331, pp. 65–82 (1983).
5. Moli Energy Ltd., Electrical Characteristics of Molicel A-Type Cells, Moli Energy Ltd., Burnaby, British Columbia, Canada (1987).
6. K. Brant, D. Fouchard and J. A. R. Stiles, Safety aspects of a rechargeable lithium C cell, in *Proceedings of the 31st Power Sources Symposium,* pp. 104–113, The Electrochemical Society, Pennington, New Jersey (1985).
7. J. B. Taylor, D. Fouchard, D. Wainwright and L. Ruggier, in Evaluation of the performance of large rechargeable lithium cells, in *Proceedings of the 2nd Annual Battery Conference on Applications and Advances,* Proc. Vol. 87-16, pp. 94–110, The Electrochemical Society, Pennington, New Jersey (1987).
8. C. J. Johnson, Importance of heat transfer in Li/MoS_2 batteries for aerospace applications, in *Proceedings of the Symposium on Primary and Secondary Ambient Temperature Lithium Batteries,* Proc. Vol. 88-6, pp. 459–463, The Electrochemical Society, Pennington, New Jersey (1988).
9. Moli Energy Ltd., Report on Results of Safety and Abuse Testing of 'AA' Cells, Moli Energy Ltd., Burnaby, British Columbia, Canada (1986).
10. Material Safety Data Sheet, "Lithium Hexafluorophosphate," Johnson Matthey Catalog Chemicals, Ward Hill, Massachusetts (1981).
11. G. H. Newman, Electrolyte structure and lithium rechargeability, in *Proceedings of the Workshop on Lithium Nonaqueous Battery Electrochem.,* Proc. Vol. 80-7, pp. 143–157, The Electrochemical Society, Pennington, New Jersey (1980).
12. B. M. L. Rao and G. E. Milliman, Safety-related investigation of a lithium battery electrolyte— thermal decomposition of LiSCN in 1,3-dioxolane, *Proceedings of the Symposium on Lithium Batteries,* Proc. Vol. 81-4, pp. 216–222, The Electrochemical Society, Pennington, New Jersey (1981).
13. K. Yoshimitsu, K. Kajita and T. Manabe, Organic-electrolyte batteries, Jpn. Kokai Tokkyo Koho JP 61,208,758 (September 17, 1986).
14. D. Zukerbrod, R. T. Giovannoni and K. R. Grossman, Life, performance and safety of Grace rechargeable lithium–titanium disulfide cells, in *Proc. of the 34th International Power Sources Symposium,* pp. 172–175, IEEE, Piscataway, New Jersey (1990).
15. R. Somoano, Characterization of prototype secondary lithium battery, in *The 1979 Goddard Space Flight Center Battery Workshop,* Publication N80-20821, pp. 147–154 (1980).

16. A. Okayama, Y. Yanai, T. Yasuda and T. Kuroda, Aluminum–lithium anode for secondary nonaqueous batteries, Jpn. Kokai Tokkyo Koho JP 63,136,467 (June 8, 1988).

17. D. M. Pasquariello, E. B. Willstaedt, R. J. Hurd and K. M. Abraham, Secondary δ-LiAl/TiS$_2$ cells, in *Proceedings of the 35th International Power Sources Symposium*, pp. 192–196, IEEE, Piscataway, New Jersey (1992).

18. K. M. Abraham, D. M. Pasquariello and E. B. Willstaedt, *J. Electrochem. Soc.* **137**, 1856 (1990).

19. W. K. Behl and D. T. Chun, Electrolyte additive for secondary lithium battery with organic electrolyte, U.S. Patent Appl., U.S. 153,611 (July 15, 1988).

20. D. Schwartz, P. Hill, R. Hurd and P. Rebe, Design and performance of a 24 volt, two amperehour lithium rechargeable battery, in *Proceedings of the 32nd International Power Sources Symposium*, pp. 239–249, The Electrochemical Society, Pennington, New Jersey (1986).

21. D. A. Schwartz, P. B. Hill, P. E. Rebe and G. Wilson, Rechargeable 24-Volt, 2.0 Ampere-Hour Lithium–Titanium Disulfide Battery, Report SLCET-TR-84-0414-F, U.S. Army Laboratory Command, Fort Monmouth, New Jersey (1988).

22. M. Yokogawa, Secondary nonaqueous battery, Jpn. Kokai Tokkyo Koho JP 01,128,371 (May 22, 1989).

23. N. Koshiba, K. Momose and T. Sawai, Secondary lithium batteries, Jpn. Kokai Tokkyo Koho JP 01,157,067 (June 20, 1989).

24. K. Inada, K. Ikeda, H. Nose, K. Tsuchia and Y. Mochizuki, Secondary nonaqueous batteries, Jpn. Kokai Tokkyo Koho JP 01,286,266 (November 17, 1989).

25. T. Nagaura, *Prog. Batteries Battery Mater.* **10**, 209 (1991).

26. A. N. Dey, H. C. Kuo, D. Foster, C. Schlaikjer and M. Kallianidis, Inorganic electrolyte rechargeable Li/SO$_2$ system, in *Proceedings of the 32nd International Power Sources Symposium*, pp. 176–184, The Electrochemical Society, Pennington, New Jersey (1986).

27. R. C. McDonald, P. Harris, F. Goebel, S. Hossain, R. Vierra, M. Guentert and C. Todino, Advanced Rechargeable Lithium Sulfur Dioxide Cell, Report SLCET-TR-88-0849-F, U.S. Army Laboratory Command, Fort Monmouth, New Jersey (1991).

28. H. C. Kuo, M. Kallianidis, P. Piliero and A. N. Dey, High Capacity Rechargeable Lithium Sulfur Dioxide Cell, Report SLCET-TR-86-0003-F, U.S. Army Laboratory Command, Fort Monmouth, New Jersey (1989).

29. A. N. Dey, H. C. Kuo, P. Keister and M. Kallianidis, Inorganic electrolyte Li/SO$_2$ rechargeable system: Development of a prototype hermetic C cell and evaluation of the performance and safety characteristics, in *Proceedings of the Symposium on Primary and Secondary Ambient Temperature Lithium Batteries*, Proc. Vol. 88-6, pp. 343–362, The Electrochemical Society, Pennington, New Jersey (1988).

30. A. N. Dey, H. C. Kuo, P. Piliero and M. Kallianidis, *J. Electrochem. Soc.* **135**, 2115 (1988).

31. A. N. Dey, H. C. Kuo, D. Foster, C. Schlaikjer and M. Kallianidis, *Prog. Batteries Sol. Cells* **6**, 73 (1987).

32. R. J. Mammone and M. Binder, Rechargeable lithium–sulfur dioxide cell, in *Proceedings of the 21st Intersociety Energy Conversion Engineering Conference*, pp. 1073–1076, The American Chemical Society, Washington, D.C. (1986).

33. J. F. Connolly and R. J. Thrash, Lithium–sulfur dioxide electrochemical cell with an iodine catalyzed cathode, U.S. Patent 4,889,779 (December 26, 1989).

34. C. De la Franier, M. Brule and A. C. Harkness, *Prog. Batteries Sol. Cells* **7**, 262 (1988).

35. Material Safety Data Sheet, "Lithium Cobalt (III) Oxide," Johnson Matthey Aesar Group, Seabrook, New Hampshire (1989).

36. Material Safety Data Sheet, "Methyl Acetate," Occupational Health Services, Inc., New York, New York (1990).

37. K. W. Beard, Hardware development and application specifications for Li/CoO$_2$ electrochemical cells, in *Proceedings of the 34th International Power Sources Symposium*, pp. 160–163, IEEE, Piscataway, New Jersey (1990).

38. K. W. Beard, W. A. DePalma and J. P. Buckley, Advancements in Li/LiCoO$_2$ hardware cell performance, safety and reliability, in *Proceedings of the 35th International Power Sources Symposium*, pp. 201–205, IEEE, Piscataway, New Jersey (1992).

39. A. Yoshino and K. Inoue, Batteries with increased safety, Jpn. Kokai Tokkyo Koho JP 03,263,758 (November 25, 1991).

40. T. Matsui, Nonaqueous secondary batteries, Jpn. Kokai Tokkyo Koho JP 03,108,275 (May 8, 1991).

41. T. Matsui, Nonaqueous secondary batteries, Jpn. Kokai Tokkyo Koho JP 03,108,276 (May 8, 1991).

42. F. A. Trumbore, *J. Power Sources* **26**, 65 (1989).

43. B. Vyas, High energy density lithium niobium triselenide cells, in *Proceedings of the 33rd International Power Sources Symposium*, pp. 101–106, The Electrochemical Society, Pennington, New Jersey (1988).

44. Material Safety Data Sheet, "Vanadium Oxide, V$_6$O$_{13}$," CERAC, Inc., Milwaukee, Wisconsin (1988).

45. N. Koshiba, T. Ikehata and K. Takata, Lithium secondary battery, U.S. Patent 4,874,680 (October 17, 1989).

46. X. Feng, D. Wang and D. Liu, The cathode reactions of vanadium oxide (V$_6$O$_{13}$(NS)) in rechargeable lithium battery, in *Proceedings of the Symposium on Primary and Secondary Ambient Temperature Lithium Batteries*, Proc. Vol. 88-6, pp. 511–519, The Electrochemical Society, Pennington, New Jersey (1988).

47. N. Koshiba, K. Momose and T. Sawai, Secondary lithium batteries, Jpn. Kokai Tokkyo Koho JP 01,157,066 (June 20, 1989).

48. N. Koshiba, K. Takada and K. Takai, Secondary lithium batteries, Jpn. Kokai Tokkyo Koho JP 01,157,068 (June 20, 1989).

49. Material Safety Data Sheet, "Diethyl Carbonate," Occupational Health Services, Inc., New York (1990).

50. T. Nagaura, *Prog. Batteries Battery Mater.* **10**, 218 (1991).

51. K. Ozawa and M. Yokokawa, Cycle performance of lithium ion rechargeable battery, in *Proceedings of the Tenth International Seminar on Primary and Secondary Battery Technology and Application*, Florida Educational Seminars, Boca Raton, Florida, 1993.

52. T. Yamahira and M. Anzai, Secondary nonaqueous batteries, Jpn. Kokai Tokkyo Koho JP 01,294,372 (November 28, 1989).

53. J. J. Auborn and Y. L. Barberio, *J. Electrochem. Soc.* **134**, 638 (1987).

54. Y. Mizuno and S. Kondo, Secondary nonaqueous batteries, Jpn. Kokai Tokkyo Koho JP 61,263,069 (November 21, 1986).

14

High-Temperature Systems

Several battery systems have been developed which operate at elevated temperatures and contain one or more components in the molten state. These chemistries have been developed for specific applications, for example, electric vehicle propulsion and military ordnance. A few of the more common systems will be discussed in this chapter.

14.1. SODIUM/SULFUR

The sodium/sulfur system is a rechargeable battery which operates at 320–350°C. This technology has been developed for nonmilitary, large-scale energy storage applications. In the charged state, its active components are molten sodium as the negative electrode and molten sulfur as the positive, separated by a ceramic β''-alumina electrolyte. The electrochemical reactions during discharge include the oxidation of sodium and the reduction of sulfur to form sodium polysulfides of the general formula Na_2S_x ($x = 2.7$–5). The final value of x depends on the end-of-discharge voltage, 1.9 V corresponding to $x = 4$ (i.e., Na_2S_4) and 1.78 V corresponding to $x = 3$ (Na_2S_3). Many cells are designed with a 1.9-V discharge limit, since Na_2S_4 is less corrosive than Na_2S_3.[1] Because of their high-temperature operation, sodium/sulfur batteries are relatively large (>20 kWh) and are composed of many monopolar cells within a single thermal enclosure.

14.1.1. Chemistry

During normal use, the condition that can lead to the most hazardous condition in a Na/S battery is the direct reaction between molten sodium and molten sulfur.[2] This can only occur at the cell level if the β''-alumina electrolyte fractures, allowing the active materials to come in contact. Another reaction that can lead to leakage of molten materials is the corrosion of the cell case or its seals by molten sodium polysulfides. To ensure a safe battery, these two reactions must be prevented from occurring, or at least minimized.

Other potential hazardous chemical reactions that may occur with Na/S batteries include reaction between sodium and water, which is a violent, exothermic reaction, and reaction between sodium and oxygen, which, at the temperatures involved in a thermal runaway, may yield toxic, gaseous Na_2O. Reactions involving molten sulfur include the formation of SO_2 and H_2S from reaction with oxygen and water and the formation of chromic sulfide from reaction with the chromium normally used in a corrosion-resistant layer on the cell container.[3]

14.1.2. Cell Design

Three primary design principles related to safety must be addressed to ensure that thermal runaway or loss of corrosive products does not occur[2]:

1. Minimize the amount of sodium and/or sulfur available for reaction if the β''-alumina fails.
2. Keep the bulk of the reactants separated, and restrict their flow to the reaction sites.
3. Protect the outer cell case and seals against corrosion by sodium polysulfides.

The first two principles can be accomplished by minimizing the cell size (quantity of reactants) and maintaining the molten sodium in a protected safety can that has a restricted opening. A gap is present between the can and electrolyte into which the sodium flows, making only a fraction of the sodium available for direct reaction.[4] The amount of sodium available for immediate reaction is minimized by designing the gap between the displacement and electrolyte tubes to be as small as possible.[2] The gap may be filled with an inert material to further reduce the volume of sodium. Care must be taken to avoid using container or filler materials that are readily corroded by sodium polysulfides in case of electrolyte failure, since a significant amount of sodium may then become available for reaction.[2]

Restricting the movement of molten sulfur is a secondary consideration because the preferred use of low-density carbon felts in the sulfur electrode inhibits free movement of sulfur. The use of sulfur electrodes with felt-free channels is therefore not recommended.[2]

The third safety principle, protection of the cell case from corrosion, is the key to preventing long-term toxic reactant leakage and the associated potential for electrical shorting. The performance requirements include:

1. Long-term chemical compatibility with sodium polysulfide at operating temperature ($\sim 350°C$)
2. Thermal-shock resistance
3. Short-term protection against sodium polysulfide corrosion at temperatures above 400°C
4. Electrical conductivity

A single-component container has proven impractical because the corrosion-resistant, electrically conductive metals (e.g., molybdenum and chromium) are generally too expensive and/or difficult to fabricate. Additionally, Mo sometimes produces irreversible cell polarization problems. Aluminum has excellent durability in the corrosive sodium polysulfides but forms a nonconducting sulfide product layer. Although a few nickel-based superalloys and stainless steels have fair resistance to molten sodium polysulfides, their corrosion rate is too high for the very long life desired for these cells (more than five years).

The dilemma of not being able to identify acceptable single-component containers has forced the selection and use of composite materials: usually an inexpensive substrate (e.g., aluminum, carbon steel, stainless steel) that has been coated, plated, or sheathed with at least one corrosion-resistant material. The key to success of these composites is to ensure that the corrosion-resistant layer is defect free, thus preventing undermining, substrate attack, and spalling from occurring. Screening of a wide range of candidate materials has led to the selection of a chromium-containing layer as the preferred corrosion barrier.

14.1.3. Multicell Battery Design

The safety hazard in multicell batteries is greater than in single cells, since failure of one cell can propagate to other cells in the battery. Therefore, safe designs are aimed at mitigating failure propagation. The primary cause of failure propagation is thermal runaway. A poorly designed cell can reach very high temperatures following electrolyte failure (>600°C). If unchecked, temperatures in adjacent cells can then become high enough to fail seals (often containing aluminum), leading to cell breaching. Two additional mechanisms are possible for failure propagation if cell breaching occurs[2]:

1. Rupture of adjacent cell cases due to corrosion by sodium polysulfides ejected from the failed cell

2. Failure of adjacent cells due to thermal shock of the electrolyte caused by ejection of hot sodium polysulfides

Cell failure propagation can be minimized by utilizing a modular construction in an insulated outer box. Applying an external protective coating to each cell and/or separating cells and modules with a material that is impermeable to hot sodium polysulfides will hinder failure propagation within each module. Care must be taken that coatings do not interfere with thermal management of the battery. Of course, use of a safe cell, that is, one that incorporates the three safety principles discussed above, is the primary technique to ensure safe performance at the battery level.

14.1.4. Safety/Abuse Tests

As mentioned previously, failure of the β''-alumina electrolyte results in the most hazardous consequence for Na/S batteries under normal operating conditions. Three safety test methods have been used to evaluate the response of cells to catastrophic electrolyte failure[2]:

- *Electrical breakdown*: The cell is discharged and then recharged at a high rate. At the top of charge, high voltage is applied until dielectric breakdown of the electrolyte occurs (gross fracturing).
- *Mechanical fracture*: An external stimulus is used to rupture the electrolyte. Either a sharp or a blunt stake is placed on the side of the cell and impacted with a force sufficient to indent the cell wall and fracture the electrolyte, while the cell is on test.
- *Precrack*: β''-Alumina tubes are fractured prior to cell manufacture. Assembled cells are then heated until reaction occurs.

The electrical breakdown method is the simplest and most commonly used technique. It has several advantages over the other methods:

1. No modifications to the cell are required.
2. No remote actuation is required.
3. Cells can first be tested to ensure that they are performing satisfactorily.
4. This failure mode represents the "worst case" of what actually can occur in an operational battery.

Some of the reported results for a particular 10-Ah electric vehicle cell are summarized below.

Electrolyte Failure. The electrical breakdown method was used, and cells were cycled after failure to ensure that all the sodium reacted. The maximum temperature attained was 440°C, and analysis of the data predicted that the probability of a cell exceeding 450°C on failure is less than 1 in 10,000.[4] In tests of several hundred cells containing safety features, no escape of materials from the cell case has been observed.

Puncture. Fully charged and partially discharged cells were punctured through the side with a metal rod, fracturing the electrolyte. The worst case occurred with the fully charged cells, but in no case was any leakage of materials noted. To evaluate the effectiveness of the safety features, four cells were built without added safety features and punctured. Two of the cells leaked molten reactants, and the other two had high-temperature exotherms.[4]

Thermal Shock. Previously cycled cells were heated from ambient to ~350°C within 5 min with no adverse effects (i.e., thermal excursions or cell breaching). Cells were then quenched in a water bath, the case temperature dropping to 70°C in 2 s. Again, no adverse safety effects were noted. Repeating the thermal cycling tests caused no noticeable changes in performance or safety characteristics.[4]

A number of safety/abuse tests have also been reported for full-size (~22 kWh), multicell batteries. These tests are summarized below.

Short Circuit. Short circuit of Na/S batteries resulted in very high currents, on the order of 1200–1500 A. However, internal connectors melted before a hazardous situation arose.[5]

Crush. Batteries were deformed with a pneumatic press to ~25% of their diameter. Many cells breached, and an internal fire occurred, but the battery case remained intact and no hazardous conditions resulted.[5]

Shock. Batteries were accelerated to a speed of 50 km/h and then stopped within ~50 cm to simulate a vehicle crash. The mean value of deceleration was ~18 g for 80 ms. No adverse safety or performance effects were observed.[5]

Electrolyte Failure. A large test battery comprised of 48 central sulfur cells was fabricated. Forty-two of the cells incorporated all the safety features related to the three safety principles discussed in Section 14.1.2. Six cells did not contain these features. The six unprotected cells were deliberately failed. Five of the six ruptured, but none of the remaining 42 cells did.[6]

The following conclusions were drawn from this test:

(a) Cell failure did not propagate through the battery, even though one in eight cells failed within a short time.
(b) Cells adjacent to the ruptured cells failed because of thermal shock but did not present any safety problems.
(c) No toxic fumes were emitted from the battery.
(d) Heat generated by the failed cells was rapidly dispersed throughout the battery, resulting in significantly lower temperatures than expected from single-cell tests.

Fire. A full-size sodium/sulfur battery was on test in an automobile when a fire broke out at the facility. After the fire was extinguished, the car was totally destroyed, except for the battery. The only damage to the battery was a small hole, caused by a fireman's tool when removing it from the wrecked automobile. Electrical tests indicated the battery was still able to operate satisfactorily after this incident.[7]

14.1.5. Safety Status

A major part of the development of the sodium/sulfur technology has been devoted to ensuring that its safety performance will be adequate. The results to date have demonstrated that, with proper design and engineering, the chemical and thermal risks associated with the presence of molten sodium, sulfur, and sodium polysulfides can be satisfactorily prevented. Thus, large sodium/sulfur batteries operating at 350°C can be designed to operate safely, even under limited abusive conditions. The response to gross mishandling or operation under extreme conditions remains unresolved.

14.2. LITHIUM/IRON SULFIDE

The lithium/iron sulfide system has been developed as a rechargeable battery for electric vehicle propulsion. It consists of a Li–Al alloy anode, a LiCl–KCl eutectic electrolyte (mp, 352°C), and an FeS cathode. A boron nitride felt is used as the separator. Other electrolytes have been used in this system, for example, LiF–LiCl–LiBr, as well as other separators (addition of MgO to the electrolyte pellet to maintain rigidity while molten). The lithium monosulfide battery, as it is sometimes referred to, is intrinsically safe. The electrodes are both solid and are completely saturated with a nonflammable molten salt, which protects them from exposure to air even if the cell case is breached. No gas is generated during the cell reaction, and the cells are sealed within a strong metal container, which in turn is sealed in a double-walled insulating container.

14.2.1. Design

Due to the intrinsic safety of the lithium/iron sulfide system, electrically abusive conditions such as short circuit, overcharge, and overdischarge will not cause a hazardous situation to occur. One hazardous condition was observed, however, when cells were cooled below the melting point of the electrolyte and then reheated to operating temperature. The pressure increase due to melting of the electrolyte caused leakage of the molten electrolyte through the cell feedthrough.[8] This problem can be eliminated by designing cells with an adequate void space at the top of the cell.

14.2.2. Safety Tests

A limited number of safety tests have been reported on Li–Al/FeS cells. A brief summary is given below.

Short Circuit. Cells that had developed a short circuit were continued on cycle. After ~50 cycles, heat generation peaked at 90–100 W on discharge. After another 15–30 cycles, the heat generation rate declined. No hazardous condition occurred.[9] In another life test of Li–Al/FeS cells, cells that failed as a result of short circuits formed across the separator did not pose any safety problems.[8]

Crush. Static crush tests were performed on both hot and cold cells. The hot cells ruptured at ~8000 lbf. The salt dispersal was minimal, and solidification occurred within a few seconds. The cold cells were very stiff, with little deflection up to ~100,000 lbf.[10]

Shock. Heated cells were placed on the bottom baseplate of a crash tower. A 76-lb weight was dropped on the cells at a velocity of 23.5 mph. The cell cases were breached and salt dispersed. The odor of SO_2 was detected immediately after the test.[10] Analysis of the data from these tests suggested that an electric vehicle battery could be designed to withstand the force of a 30 mph impact and still contain the molten electrolyte.

A 60-cell battery module was under cycle test when failure occurred. Initial failure was a short circuit in one cell that resulted in electrolyte leakage. Propagation of the failure was due to corrosion by the electrolyte, resulting in conductive paths between cells and between the cells and the metallic cell tray. However, failure of the large module did not result in an external breach of the outer case. There was no release of a noxious gas, and no thermal or electrical effects were noted on an adjacent module.[11]

14.3. LITHIUM/IRON DISULFIDE "THERMAL BATTERIES"

The lithium/iron disulfide system, commonly known as a thermal battery, is a primary reserve battery used for the most part in military ordinance applications. The anode can be either lithium–silicon, lithium–aluminum alloy, or pure lithium immobilized with iron powder (LAN). The electrolyte is most commonly LiCl–KCl eutectic containing MgO as a binding material, allowing the electrolyte/binder layer to also act as the separator. The cathode is FeS_2, usually mixed with electrolyte or electrolyte/binder. Heat is supplied to the battery by "heat pellets," a mixture of iron and $KClO_4$. The heat pellets are located between the bipolar cells and contain excess iron so they can act as intercell connectors after ignition. The batteries are activated by firing a squib or primer (percussion cap), located in the header. Hot sparks from the squib or primer can ignite the heat pellets directly, when fired into a hole in the center of the cell stack, or they can ignite a heat paper strip which runs across the top of the cell stack and down the sides. The heat paper consists of a $Zr/BaCrO_4$ mixture which is very ignition sensitive.

14.3.1. Design

The $Li–Si/FeS_2$ thermal battery is intrinsically safe, as discussed previously for the $Li–Al/FeS$ secondary system. However, in some designs an epoxy potting is bonded to the header to secure the squib and leads. If excessive temperatures are generated inside the battery, this material may ignite or pyrolyze, resulting in the generation of gases. The increased internal pressure can cause the case to bulge and the glass-to-metal seals in the header to fail. This may result in acrid smoke leaking from the battery.[12] Since the electrode materials are solid and the electrolyte is immobilized with MgO, no other components will leak from the battery.

Under certain conditions, such as excessive heat input caused by over-weight heat pellets or direct chemical reaction of the anode and cathode, a thermal runaway may result. When this happens, the internal temperature will build rapidly, resulting in melting of the stainless steel case and exposure of the battery cell stack to outside air. A shower of sparks and debris will be ejected from the battery, posing a serious hazard to nearby personnel or equipment, if the battery is not confined. The hazards are greatly increased if the battery is potted in foam, owing to the flammability of this material.

14.3.2. Safety/Abuse Tests

The following are typical results obtained when thermal batteries are subjected to abusive conditions.

Short Circuit. Li–Si/FeS$_2$ thermal batteries are capable of very high short-circuit currents, resulting in the melting of internal leads. This acts as an internal fuse to prevent thermal runaway.[12,13]

Charge. Charging activated thermal batteries does not pose a safety hazard. Batteries were charged with currents as high as 4.5 A at 35 V with no adverse reactions noted.[12,13]

Heat. Thermal batteries were externally heated to >500°C. Units with epoxy potting in the header swelled and the glass-to-metal seals fractured. Smoke poured from the ruptured seals.[12] Units with no expoxy potting showed no adverse effects and discharged normally when activated after cooling.[13]

Construction Errors. A series of Li–Si/FeS$_2$ thermal batteries were built with various errors in construction to determine their effects on performance. Errors included excess internal heat (110-cal/g vs. 92-cal/g heat pellets), omitting the collectors between the anode and heat pellet, omitting the separator in every other cell, reversing the anode and cathode, and assembling with excessive stack force and with no stack force. Performance was severely affected, but no safety hazards were noted.[13]

REFERENCES

1. J. W. Braithwaite, Sodium beta batteries, in *Handbook of Batteries,* 2nd ed. (D. Linden, ed.), McGraw-Hill, New York (in press).
2. A. R. Tilley, Safety, in *The Sodium Sulfur Battery,* (J. L. Sudworth, and A. R. Tilley, eds.), Chapman and Hall, London (1985).
3. F. Stodolsky, Safety Considerations for Sodium–Sulfur Batteries for Electric Vehicles, Report SP-793—Recent Advances in Electric Vehicle Technology, Society of Automotive Engineers, Warrendale, Pennsylvania (1989).
4. F. M. Stackpool, Safety qualification of sodium sulfur cells, in *Proceedings: DOE/EPRI Beta (Sodium/Sulfur) Battery Workshop VIII,* U.S. Department of Energy, Washington, D.C., and Electric Power Research Institute, Palo Alto, California (1990).
5. W. Dorrscheidt, T. Shiota, and M. Ishikawa, Safety of Beta Batteries, in *Proceedings: DOE/ EPRI Beta (Sodium/Sulfur) Battery Workshop VII,* Electric Power Research Institute, Palo Alto, California (1988).
6. M. D. Hames, D. G. Hartley, and N. M. Hudson, Some aspects of sodium–sulfur batteries, in *Power Sources 7, Proceedings of the 11th International Symposium on Research and De-*

velopment in Non-Mechanical Electrical Power Sources, (J. Thompson, ed.), Academic Press, London (1979).

7. Anon., ABB Na–S battery survives fire at BMW, *Batteries International* **1993** (January), 11.

8. F. J. Martino, E. C. Gay, and W. E. Moore, *J. Electrochem. Soc.* **129,** 2701 (1982).

9. D. L. Barney, R. K. Steunenberg, A. A. Chilenskas, E. C. Gay, and J. E. Battles, Lithium/ Iron Sulfide Batteries for Electric-Vehicle Propulsion and Other Applications, Progress Report for Oct. 1981–Sept. 1982, Report ANL-83-62, Argonne National Laboratory, Argonne, Illinois (1983).

10. A. W. Biester, Design Study Project for the Application of the Lithium/Metal Sulfide Battery to Electric Road Vehicles, Phase I Battery/Vehicle Interface, Report No. 0434, The Budd Co., Fort Washington, Pennsylvania (1977).

11. V. M. Kolba, J. E. Battles, J. D. Geller, and K. Gentry, Failure Analysis of Mark IA Lithium/ Iron Sulfide Battery, Report ANL-80-44, Argonne National Laboratory, Argonne, Illinois (1980).

12. J. A. Gilbert, Characteristics and Development Report for the MC3816 Thermal Battery, Report SAND88-1418, Sandia National Laboratories, Albuquerque, New Mexico (1989).

13. P. G. Neiswander and A. R. Baldwin, Characteristics and Development of the MC3815 Thermal Battery, Report SAND87-0143, Sandia National Laboratories, Albuquerque, New Mexico (1988).

V

ACCIDENT PREVENTION

Each and every battery accident has a cause and can, in principle, be avoided. The battery accident reports reviewed in Chapter 2 indicated that most of the accidents were due to mishandling, abuse, or operating errors on the part of users and were not due to battery defects. The responsibility for avoiding accidents falls, therefore, primarily on battery users. The reports also indicated that most of the accidents, especially the more serious accidents, involved automotive lead–acid batteries. The more serious injuries were caused by battery explosions attributed to the ignition of hydrogen formed during battery charging. Less serious injuries were caused by sulfuric acid spills attributed to careless handling and to the existence of gas pressures inside the batteries. Some of the reported accidents involved primary alkaline cells, but the associated injuries were less severe than in the case of accidents involving lead–acid batteries. This is due, in part, to the smaller size and lower hazard potential of alkaline cells. Interestingly, many of the accidents in this group were attributed to faulty circuitry in battery-operated toys. None of the reported accidents involved lithium batteries even though lithium battery accidents have occurred. To date, the number of accidents involving lithium batteries has been relatively insignificant,[1,2] but some of these accidents have been serious. As the use of lithium batteries increases during coming years, the number of accidents involving these batteries may be expected to increase, particularly as the use of the larger size lithium batteries increases. They have a greater hazard potential than aqueous electrolyte batteries of comparable size and rate capability.

Although it is unlikely that accidents and injuries due to abuse and mishandling of batteries will ever cease to occur, it is desirable that every reasonable effort be made to reduce the occurrence of such accidents and injuries, whatever their cause. The burden of achieving such a reduction falls primarily on battery users, but battery manufacturers and device manufacturers also have a role to play in reducing the incidence of injuries. Battery manufacturers need to develop and manufacture abuse-resistant batteries, based on the types of abuse their products may be exposed to, and device manufacturers need to employ fault-free circuitry and tamper-resistant structures in their devices, especially in battery-operated toys for children. Users have a responsibility to handle batteries safely in accordance with the manufacturers' recommendations and by use of common sense. An example where the latter seems to have been absent is the following case[3]: "Victim hit his car battery with a hammer in an attempt to make it discharge. Battery exploded and electrolyte splashed into victim's eye." It is not reasonable to require of battery manufacturers that they anticipate each and every kind of abuse that creative users invent. It is reasonable, however, to require that they render their batteries resistant to normal abuse, to the extent that the measures that must be taken to do so are consistent with the requirements of battery performance and cost considerations.

Designing for safety begins with a consideration of the contributions of all the components of a battery-operated system and of their interactions to the operational safety of the system as a whole. In particular, it requires an analysis of the contributions of the electrical circuitry of the system, the device structure, the electrochemical system, the battery package, any battery control circuitry, the battery/device interface, and all of their interactions. Although battery chargers are not battery-operated devices, they should be included in the analysis since they may contribute significantly to battery hazards if designed or operated improperly. The total system design is particularly important in applications that involve large batteries because of their greater hazard potential compared with that of small batteries.

An essential aspect of battery safety is the quality and the effective control of battery manufacturing operations. Just as safe designs can be negated by flawed manufacturing and by flawed materials, unsafe designs cannot be rendered safe by flawless manufacturing and by flawless materials. We do not discuss the many important and interesting aspects of battery manufacturing and their impact on battery safety. Experience has shown that most battery manufacturers do produce high-quality batteries. Manufacturing and material flaws have made only minor contributions to battery accidents. This is particularly true for well-established manufacturing operations. It is not unusual, however, for some problems to occur during the developmental phases of new battery products, but such problems are generally eliminated by the time the new products enter into regular production.

REFERENCES

1. N. Marincic, Hazardous behaviour of lithium batteries, Case histories, in *Proceedings of the 1982 Goddard Space Flight Center Battery Workshop,* NASA Conference Publication 2263, Greenbelt, Maryland (1983).
2. P. K. Raj, Safety Issues Related to the Storage of Chemicals in Advanced Missile Bases for Power Generation and Life Support Systems, Technical Report, Volume 2, Technology and Management Systems, Inc., Burlington, Massachusetts (1989).
3. U.S. Consumer Products Safety Commission, Summarization of 82 In-Depth Investigations of Product-Related Injuries, U.S. Consumer Products Safety Commission, Bureau of Epidemiology, Wet Cell Batteries (0849), Washington, D.C. (1976).

15

Safety Evaluation

Before using a battery in an application or starting a test program, one should be certain that the battery will perform safely. This is especially true for lithium batteries, which are more energetic than the aqueous systems. A series of safety/abuse tests should be conducted to evaluate battery safety. The purpose of these tests is twofold: to define the level of abuse that the battery can tolerate safely and to determine how the battery will behave under the most severe abuse conditions. The first piece of information can be used to define a "safety envelope," that is, conditions under which the battery will operate safely. The second type of information should be used to determine what safeguards are necessary to protect personnel and equipment in the event the battery is inadvertently placed outside its "safety envelope."

15.1. TYPES OF SAFETY/ABUSE TESTS

Two general categories of abuse test are normally performed on cells and batteries: electrical and environmental. These tests should be carried out whenever a new chemistry is being used, a modification to an existing system has been made, a new design or size is used, a new manufacturer is making the cells, or, in the case of lithium batteries, there has been a period of greater than one year since the last lot of cells has been built.

15.1.1. Electrical Tests

Four basic tests should be conducted to define the safety hazards associated with electrical abuse: short circuit, high-rate discharge, charge, and voltage reversal.[1] The individual cell's current, voltage, and skin temperature should be monitored continuously, for at least one hour beyond removal of the electrical load. A video record of each test should be made, whenever feasible. A minimum of three replications should be carried out for each test.

Short Circuit. This test consists of applying a direct short ($<0.05\ \Omega$) across the positive and negative leads of the cell. It should be performed at the maximum temperature the cell or battery will see during its qualifying tests or in actual use. Upon completion of the test, the cells should be left untouched for several hours.

High-Rate Discharge. This test is used to determine the maximum safe continuous discharge current (I_{mc}) for each cell. This is the largest continuous current the cell can deliver which does not result in any cell venting, rupture, or fire. If the cell does not vent under the short-circuit test described above, then the short-circuit current is I_{mc}. If the cells contain a manufacturer-supplied internal fuse, then I_{mc} shall be 80% of the fusing current, assuming no venting occurs at that current. If the cells vent at 80% of the fuse value, then I_{mc} shall be the maximum current at which venting does not occur. This test should also be carried out at the maximum temperature the cell or battery will see during electrical testing and use.

Charge. Charging of a cell can occur if the battery is in some way connected to a power supply or another larger battery, or if the individual battery contains cells or strings of cells connected in parallel. Three types of charging test should be performed (but if the worst case is known for a particular chemistry, only that test need be performed):

(a) *Electrochemical hazard charging.* This is a low-rate charge ($\sim C/24$) for an extended period of time. It is designed to determine the hazards associated with any unstable chemical species that may form during charging. The test, performed on a fresh cell, should run for 24 h, corresponding to 100% of the cell's normal discharge capacity at that rate.

(b) *Polarization (heat) hazard charging.* This is a high-rate charge to determine if any hazards are associated with internal heat generated during this process. A fresh cell should be charged at I_{mc} for 100% of its rated capacity.

(c) *Partially discharged hazard charging.* This test is designed to determine the hazards associated with any unstable chemical species that may form during charging a partially discharged cell at moderate rates for long periods of time. The cell is first discharged at the $C/24$ rate for 21.5 h (90% discharge) and then charged at the $C/24$ rate for 24 h.

Voltage Reversal. A cell can experience voltage reversal if it is connected to a power supply or other external source of current or if it is located in a series string of cells. In the latter case, if one cell is depleted in capacity before the rest of the cells in the string, the working cells will drive the depleted cell into reversal. Current passes in the normal direction, but the polarity of the electrodes is reversed. Two voltage reversal tests should be performed on each cell (but if the worst case is known, only that test need be performed):

(a) *Electrochemical hazard reverse voltage.* This is a moderate-rate, long-term test to determine the hazards associated with unstable chemical species that may form during voltage reversal. The test consists of discharging a cell at the $C/24$ rate for 60 h using an external power supply, which will drive the cell into reversal for 150% of its normal discharge capacity. This corresponds to the worst case possible in a series string, one cell having no capacity and the other cells at 100% capacity driving that cell into reversal for 100% of its normal capacity, plus a 50% margin beyond that level.

(b) *Polarization (heat) hazard reverse voltage.* This test is designed to determine the hazards associated with internal heat generated as a result of high reverse voltage currents for short times. A cell shall be discharged at the $C/24$ rate using a resistive load until the cell reaches 0 V (~ 24 h). Then the cell is connected to a power supply and a constant current equal to I_{mc} is applied for a duration equal to 150% of the rated capacity at the I_{mc} rate.

15.1.2. Environmental Tests

If a battery is to be used in an application in which it will be subjected to severe environments, tests need to be conducted to ensure that it will perform safely. Environmental tests should be performed on multicell battery packs, rather than individual cells, since additional safety considerations may arise from packaging the cells. There are a number of environmental tests that need to be considered: high temperature, shock, vibration, puncturing, crushing, etc. Even if the application consists of a benign environment, tests should be considered to account for rough handling, transportation, and possible accident scenarios.

High Temperature. To determine the effect of high temperature on battery safety, units can be tested in actual fires, in furnaces, on hot plates, or wrapped in heat tape. Fires are difficult to control, and many laboratories do not have the facilities for conducting fire tests. Furnaces are more readily controlled, but it is difficult to observe the batteries during the test. Hot plates are useful for small cells or batteries but are not suitable for large units. Heating occurs from one side only, which may or may not be suitable for all circumstances. Therefore, wrapping batteries in heat tape is the preferred method of high-temperature testing. The rate of temperature increase can be controlled, thermocouples can be attached to various sections of the battery, and the test can be recorded on video for further study. It should be noted that even the most benign lithium batteries will react violently during a heat test. The purpose of this type of test is to compare different systems on a relative basis. How violent is the energy release? At what temperature does safety become an issue? What is the effect of the rate of temperature increase? One must remember that even a sealed can of water will rupture violently when heated!

Shock. Although many applications do not have a severe shock environment, batteries can be dropped when being handled or loaded onto vehicles for shipment. Therefore, shock tests are important to guarantee the safety of people handling the batteries. There are a number of commercial testers available that can subject batteries to the most severe shock that one can estimate the battery may be subjected to during its lifetime. The hazards associated with severe shock are cracking of cans or seals allowing toxic or corrosive electrolyte to leak, breaking of internal or external connections causing a short circuit, or deformation of internal cell components resulting in a short circuit.

Vibration. Other than in certain military and aerospace applications, vibration of batteries is usually associated with their transportation. Even though the actual vibration environment may be mild, in some instances the construction of the batteries might result in an amplification of the vibration in the interior of the package. Commercial testers are available that can subject batteries to the appropriate vibration environment. Hazards associated with vibration are related to short circuits that may occur if welded or soldered joints come apart, electrode materials flake off and migrate to the opposite electrode, or separators tear.

Puncture. Normally, one would not expect a battery to be punctured in a typical application. However, there are instances where one could conceive of this happening. A battery could be dropped on a sharp metallic object during handling, a forklift could puncture a crate of batteries during loading onto a vehicle, or a bullet could pierce a radio used by a soldier. Usual puncture tests involve piercing the cell with a nail or shooting a bullet into the battery. The main concern is the formation of internal shorts. Leakage of toxic or corrosive materials may also be a problem.

Crush. In the rare instances when a battery will get crushed, the battery is probably not a major concern. However, violent reactions due to shorts and leakage of electrolyte may add to an already hazardous situation. Pneumatic presses and/or vises may be used to simulate this severe environment.

15.2. TEST PROTOCOLS FOR LITHIUM BATTERIES

Various organizations have established test protocols that all lithium batteries must undergo in order to be used in any application. Several are summarized in the following sections.

15.2.1. U.S. Navy

Before any lithium battery can be used in a U.S. Navy application, it must undergo the following safety/abuse tests[2]:

1. *Constant-current discharge and reversal, bypassing any safety devices installed in the battery.* Discharge current is the fuse value, and the power supply voltage is the open-circuit voltage of the battery. The battery is driven into reversal to 1.5 times its ampere-hour capacity. The voltage, current, temperature, and pressure are continuously monitored.

2. *Short circuit at ambient temperature.* Resistance is to be <0.02 Ω for ≥24 h. The voltage, current, temperature, and pressure are continuously monitored.

3. *High temperature.* Batteries are to be wrapped in heat tape and heated at a rate of 20°C/min to 500°C. The voltage, pressure, and temperature are continuously monitored.

4. *Charge.* Batteries are first discharged to 50% of capacity at the fuse current. After standing for ≥72 h, they are charged at the fuse current to 100% capacity. The voltage, current, pressure, and temperature are continuously monitored.

5. *Safety device test.* All batteries used by the U.S. Navy must contain an overcurrent protection device in the ground leg of each series string and in each tap that fails open. Parallel strings must be protected with blocking diodes. This final test is conducted with all safety devices in place. The batteries are discharged at 80–90% of the fuse current into reversal, to 1.5 times their capacity. The voltage, current, pressure, and temperature are monitored continuously. No venting of any kind is allowed in this test.

15.2.2. Underwriters Laboratory (UL)

The Underwriters Laboratory will give their UL stamp of approval to lithium batteries containing ≤0.5 g of lithium per cell and ≤1.2 g of lithium per battery, and which do not contain any pressurized vapor or liquid, if they pass a series of tests.[3,4] In general, passing a test means that no fire or explosion occurs as a result of the abuse.

Prior to testing, the batteries are subjected to various preconditioning steps to simulate aging. This is done to determine if storage or use of a battery will make it unstable.

Groups of batteries are each subjected to one of the following preconditioning tests:

1. Aging in an oven at 71°C for 90 days
2. Temperature cycling between 71°C and −54°C
3. 50% depth of discharge at room temperature
4. 50% depth of discharge at 71°C
5. Complete discharge at room temperature
6. Complete discharge at 71°C

Batteries intended for use in technician-replaceable applications are then subjected to the following tests:

1. Batteries shall be short-circuited at room temperature and 65°C. Cells shall not reach a temperature of 150°C nor shall they vent, unless it can be demonstrated that the gas released from the cell is nontoxic.
2. Batteries shall be heated to 165°C at a rate of 5°C/min, then held at temperature for 10 min.
3. Batteries shall be crushed between a flat surface and a cylindrical surface to a thickness less than one-fourth their original thickness.
4. Batteries shall be exposed to a relative humidity of 95% at 65°C for 6 h. The temperature of the chamber is then to be reduced to 30°C in 16 h while maintaining a relative humidity of at least 85%. This procedure is to be repeated 10 times.
5. Batteries shall be vibrated at an amplitude of 0.8 mm, at a frequency ranging between 10 and 55 Hz, varying at the rate of 1 Hz/min and returning in 90–100 min. Batteries are to be tested in three mutually perpendicular axes.
6. The batteries from the vibration tests (best 5) shall be dropped 10 times from a height of 1.9 m onto a concrete floor in randomly oriented positions.
7. Batteries shall be force discharged using a 12-V DC power supply to 2.5 times their capacity at a current recommended by the manufacturer. If the battery is to be used in a series connection, a completely discharged battery should be connected in series with fresh batteries of the same kind. The number of fresh batteries to be used shall be one less than the number of cells to be connected in series. Tests shall be conducted at room temperature and at 71°C. Cells that were driven into reversal shall be removed and short-circuited at room temperature.
8. Batteries shall be charged by a 12-V DC power supply to 2.5 times their capacity at a current recommended by the manufacturer.

Batteries intended for user-replaceable applications are subjected to the same tests, with some differences in the requirements. In addition, they are put through fire exposure tests. There are two reasons for the fire testing. First, since the users are not familiar with the hazards of lithium batteries, the batteries are more likely to be abused and could be disposed of in a fire. The fire exposure tests are required to ensure that such disposal will not result in flames or spraying sparks or explosion with sufficient force to produce shrapnel and cause injury. Second, since these tests are very abusive, any tendency for a particular battery design to react violently will be discovered.

The fire exposure tests are performed as follows:

1. Cells are exposed to an open flame from a propane burner.
2. Cells are placed in an explosion chamber and heated.
3. Cells are exposed to an open flame with a piece of cheesecloth positioned 3 ft away.
4. Cells are subjected to a direct flame while inside an aluminum screen cage.

The first fire exposure test is conducted to determine if further fire exposure testing is necessary. The explosion test is conducted in a 6-in. diameter chamber with a 0.75-in. vent on the side and a 30-lb weighted cover. Batteries must not explode with sufficient force to lift the weighted lid.

If a battery is found to emit sparks or flame in the fire test, it must undergo test 3. Batteries must not emit sparks or flames that ignite the cheesecloth. The final test is conducted on batteries that explode and emit flying particles. The battery is placed in a 2 ft by 2 ft by 2 ft cage made of aluminum window screen. Upon heating, no part of the case is allowed to penetrate the wire screen.

15.3. TEST PROTOCOLS FOR NONLITHIUM BATTERIES

The Underwriters Laboratory is considering a test protocol to be used with household and commercial batteries (nonlithium). Both primary and rechargeable batteries, ranging from coin, pin, and AAA- to D-size cells will be included.[5] Fresh cells or batteries and cells or batteries that have been preconditioned by oven exposure and temperature cycling (see Section 15.2.2) will be used.

Batteries for user-replaceable applications will be subjected to the following tests:

1. Short circuit at room temperature and at $60 \pm 2°C$. The test is to continue until a fire or explosion is obtained or until the battery is completely discharged and the case temperature has returned to near ambient.
2. Heat at a rate of $5 \pm 2°C$ per minute to $150 \pm 2°C$ and hold at this temperature for 10 min.
3. Crush between two flat surfaces using a hydraulic ram with a 1.25-in.-diameter piston until a force of 3000 lb is applied.
4. Impact with a 20-lb weight by placing the battery on a flat surface with a $\frac{5}{16}$-in.-diameter bar across its center. The weight is to be dropped 2 ft onto the sample.
5. Charge at a voltage between 12 and 25 V at 200 W power until the cell or battery explodes or vents or the case temperature returns to ambient.
6. Force discharge by connecting in series with a 12-V DC power supply to 2.5 times the capacity if the battery voltage is lower than 12 V or by connecting a completely discharged cell in series with fresh cells of the same kind if the battery voltage is higher than 12 V. The number of fresh cells shall be equal to one less than the number of cells to be used in series in the application. The test shall be performed both at room temperature and at $71 \pm 2°C$. Cells that are driven into reversal during this test shall be removed from the battery and short-circuited at room temperature. Cells not intended for use in series are exempted from this test.
7. Charge at three times the manufacturer's recommended charging rate using a 12-V DC power supply. For batteries containing a protective device, the charging current should be just below that at which the safety device will operate.
8. Batteries will be exposed to a fire test similar to that described in Section 15.2.2. for lithium batteries.

Batteries passing all of the above tests will be certified acceptable for all applications. Those cells or batteries passing all but the fire test will be certified but restricted from use in those applications that were shown to cause fire or explosion.

At the time of this writing, the proposed UL protocol was still under review and subject to change.

15.4. SAFETY TESTING DURING DESIGN

The design process of a new cell prototype should include safety/abuse testing as part of the program. At one battery manufacturer, a comprehensive safety test program is practiced as part of the design process.[6] During the R & D stage, cells undergo short-circuit, high-rate discharge, and forced over-discharge testing. If any hazardous behavior is noted at this stage, the cells are redesigned and the safety tests repeated. Upon passing the safety tests, the cell design is then transferred to engineering.

Engineering prototypes are fabricated with production equipment. The prototype cells are then put through a more severe set of safety/abuse tests. These tests include high-current/high-temperature discharge, short circuit, forced overdischarge, charge, crush, puncture, drop, and incineration. The cells must also pass the U.S. Department of Transportation (DOT) required tests for shipping.[7] If the cells pass these tests, a drawing package is prepared and transferred to production.

If any changes are made to the design, the safety tests are repeated. In addition, safety tests are periodically performed on cells that have been stored under various conditions.

15.4.1. Accelerating Rate Calorimetry

A useful tool for evaluating the safety of batteries during the design phase is accelerating rate calorimetry. This technique consists of heating a sample (either a single component, several components, or a complete cell) in a bomb under adiabatic conditions. As the sample undergoes thermal decomposition due to self-heating, it is held under adiabatic conditions and the temperature–pressure–time relationship for the runaway process is recorded.[8]

The accelerating rate calorimeter operates by increasing the temperature incrementally (usually in 10°C steps) until there is a measurable rate of self-heating as evidenced by a slow, steady temperature increase. The sample is then left adiabatic until its thermal excursion is complete. From the recorded data, the following parameters can be calculated:

(a) Time/temperature runaway curve
(b) Adiabatic temperature rise
(c) Temperature of maximum rate
(d) Time to maximum rate
(e) Self-heat rate at any temperature
(f) Pressure at any temperature
(g) Pressure rate at any temperature

From these data, it is possible to make predictions about the cell in various states. These can then be used to make an accurate hazard prediction.

15.4.2. Cell Pressure Measurements

During the design phase of a cell, it is important to determine the pressure characteristics of the system and, if necessary, to design a cell case with a venting mechanism that operates at the appropriate pressure.

A technique for monitoring the pressure in Li/SO_2 "D" cells, utilizing a pressure transducer, has been reported.[9] It was found that use of a vent operating at <200 psi may prevent cell explosions upon shorting of Li/SO_2 cells.

15.4.3. Other Considerations

Other areas to consider during the design of a cell include corrosion, seal integrity, and aging characteristics. Corrosion of the case may lead to leakage of toxic materials. A number of techniques may be utilized to determine if any corrosion reactions will occur when a metal comes in contact with a conductive solution. The use of the electrochemical techniques: potentiodynamic polarization, potentiostatic polarization, and complex impedance analysis has been described.[10] The galvanic corrosion current may also be measured by using a potentiostat as a zero-resistance ammeter.[11] Microcalorimetry is a very sensitive technique that has been used to detect the presence of corrosion in various battery systems by measuring the heat flux, which is due to chemical reactions (i.e., corrosion) occurring in the system.

Integrity of the seal over the life of the battery is also crucial to preventing the leakage of toxic materials from the system, particularly in the case of crimp seals. Tests should be run to ensure that the grommet material is not degraded by the electrolyte or attacked by other components at the cell potential. Good quality control procedures must be followed to obtain a good, leak-free crimp seal.[10]

The aging characteristics of every component must be considered when designing a safe cell. Areas of special concern include parasitic side reactions or decompositions that may lead to the formation of a gas, which could lead to cell rupture; formation of reactive species that might corrode the cell container; changes to the grommet material that may compromise the seal integrity; or the generation of reactive intermediates that may trigger violent chemical reactions.

15.5. SAFETY PROCEDURES

This section addresses the issue of how to safely handle a cell or battery that has experienced some form of abuse, or was manufactured erroneously, and shows signs of hazardous behavior.[12]

15.5.1. Hot Cells

Immediately upon identification of a hot cell or battery pack, the area should be evacuated. Using a remote infrared temperature probe, the cell should be periodically monitored until either it starts to cool, it vents, or it explodes. The subsequent course of action to be taken in each of these eventualities is given in the following paragraphs.

Cell Cools. If the cell starts to cool, wait until it returns to ambient temperature, and then remove it from the area. Dispose of the cell, following all hazardous waste regulations.

Cell Vents. If the cell leaks or vents, ventilate the area until the cell is removed and no odor is detectable. Allow the cell to cool to ambient temperature and then, wearing the appropriate safety gear (lab coat, rubber gloves, safety glasses, respirator), remove the cell to a well-ventilated area. Seal each cell in double plastic bags. Place the double-bagged cell in a third bag containing an absorbent material (e.g., vermiculite) and seal. Neutralize any spilled or vented electrolyte with an appropriate material, and then wash the area with copious amounts of water. Dispose of the vented cell and contaminated cleaning materials following all hazardous waste disposal regulations.

Cell Explodes. If a cell explosion occurs, initiate ventilation and continue until the cell is removed and no odor is detectable. Allow the cell to cool to room temperature before attempting to handle. Wearing the appropriate safety gear (lab coat, rubber gloves, safety glasses, respirator), clean area using the appropriate neutralizing material for the type of cell. Place all residues and contaminated cleaning materials in plastic bags and seal. Seal the plastic bags in glass jars, and dispose of the jars in accordance with all hazardous waste disposal regulations. Wash the area with copious amounts of water.

15.5.2. Fires

Batteries, particularly lithium batteries, involved in a fire can lead to an extremely hazardous situation. The first action should be to completely evacuate all personnel from the area and sound the fire alarm. Ventilation should be initiated and continued until no odor remains. Trained personnel should enter the area and completely bury the burning cell with a class D fire-extinguishing material (e.g., Lith-X). Using the appropriate extinguishing agent, they should then attend to any secondary fires. After the residue has been allowed to cool, it should be carefully turned over to ensure that reignition does not occur. If it does, the Lith-X should be reapplied. The area should be cleaned as noted above.

REFERENCES

1. Sandia National Laboratories, Standard Electrical Abuse Tests, Internal memo (1986).
2. U.S. Navy, Technical Manual for Batteries, Navy Lithium Safety Program Responsibilities and Procedures, S9310-AQ-SAF-010, Naval Sea Systems Command, Washington, D.C. (1988).
3. Underwriters Laboratories, Standard for Safety—Lithium Batteries, UL 1642 (1985).
4. J. P. Allen, Underwriters Laboratories test program for lithium batteries, in *Proceedings Lithium '87*, Waste Resource Associates, Niagara Falls, New York (1987).
5. Underwriters Laboratories, Investigation for Household and Commercial Batteries, Subject 2054, Draft (1993).
6. R. C. Stinebring and P. Krehl, Improvements in safety testing of lithium cells, in *The 1984 Goddard Space Flight Center Battery Workshop*, NASA Conference Publication 2382, pp. 115–121 (1985).
7. A. I. Roberts, Exemption DOT-E 7052, U.S. Department of Transportation.
8. Columbia Scientific Industries Corp., ARC™ Information, Columbia Scientific Industries Corp., Austin, Texas.
9. A. N. Dey, Sealed Primary Lithium–Inorganic Electrolyte Cell, Report DELET-TR-74-0109-F, U.S. Army Electronics Research and Development Command, Fort Monmouth, New Jersey (1978).
10. P. Bro and S. C. Levy, *Quality and Reliability Methods for Primary Batteries*, John Wiley and Sons, New York (1990).
11. S. C. Levy, Modified Li/SO₂ cells for long life applications, in *Proceedings of the 29th Power Sources Conference*, pp. 96–98, The Electrochemical Society, Princeton, New Jersey (1980).
12. Wilson Greatbatch Ltd., Safe Handling of Lithium Batteries under Inordinate Conditions, Product Safety Procedure PRS-001, Wilson Greatbatch Ltd., Clarence, New York (1992).

16

Aqueous Electrolyte Batteries

We begin this chapter with a brief review of design improvements aimed at making batteries more abuse resistant and safe for users, and we then discuss some of the safety precautions that users need to take to prevent accidents. With few exceptions, most of the design improvements we mention have already been incorporated into batteries where necessary. This means that most of today's batteries are robust and quite resistant to abuse. It is useful to keep in mind, however, that packaged energy in any form has an intrinsic hazard potential. This is as true of a box of matches, a tank of gasoline, and cans of paint solvents as it is of a battery. All require that they be handled with due regard for their intrinsic hazard potential if accidents are to be avoided.

16.1. DESIGNING FOR SAFETY

We limit our discussion in this chapter to the more important design features that are needed to increase the abuse resistance of batteries.

Mechanical Aspects. An understanding of the processes discussed in Chapter 2 and an appreciation of their consequences for battery safety are prerequisites for the effective design of safe and abuse-resistant batteries. Safe designs are based on the premise that the more effectively the battery contents can be contained by a battery structure and the smaller the driving forces tending to expel the contents from a battery, the safer is the battery. The requirement of the mechanical integrity of battery packages under all reasonable conditions of use and abuse was discussed in Chapters 3 and 4 and needs no further elaboration at this point. Since batteries are not subject to the

American Society of Mechanical Engineers (ASME) Pressure Vessel Code, battery manufacturers are free to design battery housings and containing structures to satisfy the requirements of various applications as they think best.

Closely related to the mechanical integrity of the battery container and seals is the mechanical integrity of the interior battery components. The electrodes, separators, internal connectors, and insulators must be designed and constructed to resist damage due to shock and vibration. Failure to maintain their integrity may lead to internal short circuits, possibly resulting in thermal runaways and battery ruptures. The ruggedness that needs to be designed into the interior components depends to some extent on the battery applications. All batteries must resist damage due to the forces a battery may experience during normal handling and shipping, including shocks due to the possible dropping of shipping containers. Automotive batteries need to be more rugged since they are exposed to much more sustained shock and vibration than are smaller batteries used, for example, in portable communication devices.

Interior ruggedness is ensured by the use of inert mechanical matrices or suitable current collectors to contain or hold the electroactive materials in place, electrode bonding agents, strong separators, and heavy-duty electrode contacts. In applications where severe shocks may be encountered, the electrode structures are reinforced and prevented from shifting within the cells by auxiliary restraining structures. If motion of the electrodes were to occur, it would increase wear of the separators, and short circuits might also occur. For similar reasons, insulators are interposed between the electrode stack and the cell case and between the electrodes and the top and the bottom of the cell to prevent the electrodes from shorting via the cell case. Considerable attention is devoted to the positioning and immobilization of the interior electrode contacts to prevent the possibility that they may short the electrodes in any way as a result of shocks and vibration. The mechanical integrity of seal structures was discussed in Chapter 3.

Vented lead–acid batteries have a special requirement in that they must permit the escape of low-pressure gases generated on charge and permit access to the individual cell compartments to replenish water lost by electrolysis. Specially designed, removable vent caps are used for this purpose. They have built-in liquid traps that permit gas to escape and electrolyte to be entrained and returned to the cells.

Electrical Controls. Excessive heat may be generated in batteries if they are discharged continuously at rates much higher than their rated current capabilities or if they are short-circuited for extended periods of time, provided they have a high current capability (but see the discussion on control of rate capability below in the section on Electrochemical Designs). Since high temperatures may cause ruptures of ambient temperature batteries, excessive heating by any means must be avoided. The simplest way to prevent excessive heat generation by high-rate discharges and short circuits is to incorporate fuses in the electrical circuits as an integral part of the battery structure itself or as part of the battery-operated circuitry. Slow-blow fuses may be used to prevent battery incapacitation by accidental high-current discharges of short duration. The fuses are selected to activate at currents slightly greater than the service ratings of the batteries. The safety record of primary alkaline batteries indicates that fusing of such batteries is unnecessary for most applications.

Internal short circuits present a different problem. In their case, no external means can be provided to limit the discharge current. Instead, an internal means in the form of fusible separators may be used. They have been found to be effective current limiters. In their normal operating states, the separators are thin sheets of highly porous, inert polymers or woven fabrics that are interposed between the electrodes. If their temperature is increased either locally or uniformly as a result of heat generation inside the cells, the polymeric separators fuse and congeal to the point where they become dense, impenetrable sheets that reduce or block the passage of current. The degree of current blocking depends upon the temperature and the fusion characteristics of the polymer, which can be varied over an appreciable range by control of its composition and molecular weight distribution. When the current is limited by the fusion of such separators, the rate of heat generation decreases and may be reduced to safe levels. Regardless of the method used to limit the cell current and the associated rate of heat generation, whether an external or an internal method, once such fuzing or fusing devices have been actuated, the cells may no longer serve any useful purpose. Fusible separators are employed in many lithium batteries but are rarely used with aqueous electrolyte batteries.

Cell reversal (see Section 3.1.2) is a process that may occur in high-voltage stacks of series-connected cells, and, if maintained for any length of time, it may create hazardous situations. The severity of reversals can be much reduced by employing diodes to let the current bypass a defective cell or a cell that has exhausted its capacity before the other cells in the stack have done so. The protective action of the diode is shown schematically in Fig. 16.1. During normal operation (A in the figure), most of the current passes through the cell while it is still functional, and a small leakage current passes through the diode. When the cell goes into reversal, its voltage goes negative and the diode switches to a conductive mode (B in the figure). Now, most of the battery current flows through the diode, and only a small fraction of the current flows through the exhausted cell. Thus, reversal is not entirely eliminated, but its severity is much reduced. The extent of protection afforded by a diode depends upon the characteristics of both the diode and the cell. It

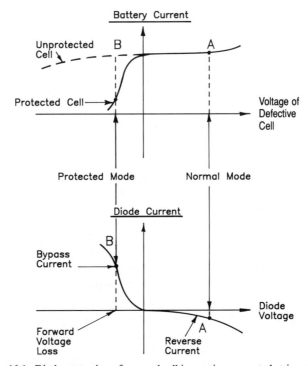

Figure 16.1. Diode protection of reversed cell in a series-connected string of cells.

may be seen that some current is likely to flow through the diode on open circuit and slowly discharge the protected cell. The rate of this parasitic discharge depends upon the characteristics of both the diode and the cell. In most applications, this self-discharge is relatively unimportant, but it may promote reversal in long-term applications.

In parallel circuits of batteries, the impedances of the parallel branches provide an intrinsic mechanism for balancing the branch currents, but all the branches operate at the same load voltage. If for any reason the voltage of one of the parallel branches falls below the load voltage, this branch will be exposed to a charging voltage. The charging can be prevented by connecting a diode in series with each branch of the parallel circuit (Fig. 16.2). Since diodes have significant forward voltage drops, about 0.4 V for germanium diodes and about 0.8 V for silicon diodes, some energy is lost by the use of such protective circuitry. The energy loss may be an appreciable fraction of the total battery energy for low-voltage batteries but is less significant for high-voltage batteries.

Current

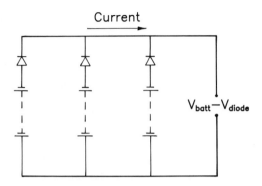

$V_{batt} - V_{diode}$

Figure 16.2. Use of diodes to block charging currents in a parallel battery stack.

Electric protection is not generally incorporated in rechargeable batteries, except for special applications such as spacecraft power systems. The most important protection of rechargeable batteries is afforded by the proper design and operation of their chargers. In the case of multicell, high-voltage rechargeable batteries for high-reliability applications, circuitry may be provided for charging each cell individually rather than the common practice of charging the high-voltage stack as a single unit. Most of the rechargeable battery accidents that occur are associated with lead–acid batteries and can be attributed in some way to the charging process. It is essential that the correct charging devices be employed for the various types of rechargeable batteries and that the charging devices provide the appropriate charging regimes to match the requirements of the batteries for safe charging. It is the responsibility of the device engineers to design chargers with charging characteristics that are appropriate and safe for the various types of batteries. This can be done most effectively in cooperation with battery engineers to ensure that the charging circuits perform in the safest possible manner, keeping in mind not only the battery characteristics, but also the needs of the battery owners who use the chargers. The availability of low-cost microprocessors makes it possible to design smart chargers that minimize potentially adverse effects due to improper charging regimes. Chargers should never be used in attempts to restore the capacity of primary batteries. This will almost certainly lead to accidents and possibly severe injuries.

Electrochemical Designs. There are many aspects to the prevention of battery accidents and injuries. Given that different electrochemical systems have different hazard potentials, the choice of what battery system to use in a particular application is also a choice of the battery hazards that will be associated with that application. Thus, designers of battery-operated devices who specify the batteries to be used with their devices play an important role in the prevention of battery accidents, whether or not they are cognizant of that role. The safety of battery-operated systems begins with the design of circuits and device structures with the requisite safety features for the applications in question. The device designers have a choice between particular electrochemical systems and between batteries of various sizes and shapes, voltages and capacities, and rate capabilities. Given the desired voltage and capacity, the next important question they face is the question of rate capability, that is, the power rating of the battery. The greater the power rating of a battery, the greater is its hazard potential. Therefore, designers should select batteries with no more than the required power rating for their applications.

Battery engineers have several options for the control of the power ratings of their batteries, and, hence, of their relative safety. The larger the area of the electrodes in a battery in a given volume, the greater is their rate capability, but also, for a fixed total current, the greater the area, the lower is the heat generation. This means that high-rate batteries generate less heat at a given current than do low-rate batteries at the same current. However, the high-rate batteries have a greater hazard potential unless electrically protected from short circuits or inadvertent high discharge currents. Another way of controlling the rate capability of cells is by the use of separators of various thicknesses and porosities: the thicker the separators and the lower their porosities, the greater their impedance and the lower the rate capability of the cells. A similar approach involves the use of low-conductivity electrolytes.

Overcharge creates potential hazards. In the case of aqueous electrolyte cells, the hazards due to electrochemically generated gases can be reduced by the use of catalytic gas recombination elements inside the cells. This has the added advantage of permitting the cells to be sealed with no need to replenish water lost from the cells in the form of electrolysis products, hydrogen and oxygen. As mentioned earlier, it also prevents the accumulation of explosive gas mixtures. Another benefit of sealed cells is that the likelihood of electrolyte spills and associated injuries is much reduced.

Chemical Protection. Various chemical means are available to reduce the potential hazards of batteries. We have already mentioned the use of gas recombination elements in aqueous electrolyte rechargeable batteries. A similar approach was discussed in Chapter 3 that involves the chemical reduction of oxygen by the lead electrodes and the chemical oxidation of hydrogen by the lead dioxide electrodes in lead–acid batteries. Similar gas scavenging reactions can be exploited in all aqueous electrolyte rechargeable batteries. The scavenging reactions protect the cells from excessive pressures whatever the source of the hydrogen and oxygen, whether generated by overcharge or reversal. In order for this means of pressure reduction to be effective, the cells must be designed to facilitate the transport of the gases from the sites of their generation to the electrodes where they are consumed. This is accomplished most readily by the use of electrode/electrolyte fill ratios that leave some part of the gas-scavenging electrodes exposed to the gas phase, that is, by the use of incompletely flooded electrodes. In addition, the relative amounts of electroactive materials are adjusted to compensate for the need to allocate some of their capacity to gas consumption. By this means, the battery designers have some control over which gases are formed on overcharge and on reversal. This may be illustrated by the following regularities:

	State of charge of:		Gas formed at:	
Process	Positive	Negative	Positive	Negative
Reversal	Capacity consumed	Residual capacity	Hydrogen	
	Residual capacity	Capacity consumed		Oxygen
	Capacity consumed	Capacity consumed	Hydrogen	Oxygen
Overcharge	Fully charged	Not fully charged	Oxygen	
	Not fully charged	Fully charged		Hydrogen
	Fully charged	Fully charged	Oxygen	Hydrogen

In Ni/Cd batteries, for example, cadmium hydroxide may be added to the positive electrode to serve as a scavenger for oxygen when the hydroxide has been fully converted to cadmium metal by reduction during normal discharge.

Another protective feature of a chemical nature that is reasonably effective in reducing accidents due to electrolyte spills is the use of gelled electrolytes in both primary and secondary cells. By maintaining the electrolyte in an essentially immobilized and/or highly viscous state, the rate of electrolyte leakage would be much reduced if any leakage path should be present.

Separators are important battery components, and their chemical stability in the presence of strongly oxidizing positive electrode materials may impact battery safety. An illustration is provided by silver/zinc batteries of both the primary and secondary types. Silver oxide is a strong oxidizing agent in alkaline solutions, where it is partly soluble. If cellulosic separators are used in these batteries, the silver oxide will react slowly with the cellulose and form metallic silver inside the separator structure as well as other decomposition products. Given sufficient time, and especially at elevated temperatures where the reaction is more rapid, the metallic silver may short-circuit the cells. Generally, this leads to no more than an increased rate of capacity loss by self-discharge, but, in extreme cases, it may also lead to an excessive rate of heat generation in the cells due to high self-discharge rates. The positive electrodes in Ni/Cd cells are also strongly oxidizing, and for long-term applications it may be necessary to employ chemically resistant separators. The principal effect of the chemical deterioration of separators is a deterioration of battery performance. Only rarely does it create hazards.

Thermal Controls. The control of battery temperatures is of little concern for the majority of battery applications, except, of course, that ambient temperature batteries should never be exposed to high temperatures, flames, or fires. Applications that involve high temperatures require the use of batteries designed for that specific purpose. No features are incorporated into the design of general-purpose batteries to control their heat dissipation.

16.2. BATTERY OPERATING PROCEDURES

Since the majority of accidents occur while users are working with batteries or actively using batteries, a lowering of the battery accident rate has to be sought by improving the manner in which users employ or work with batteries. This is particularly important for lead–acid batteries since they are involved in most of the serious accidents. The importance of following the safety rules for working with lead–acid batteries that are provided by their manufacturers, not only in their general product literature, but also printed directly on the batteries, cannot be overemphasized.

Many of the more serious injuries can be avoided if users follow a few simple rules when jump-starting their cars. Frequently, this type of operation is carried out during the winter months, when inclement weather conditions may induce some carelessness. Protective gear such as eye and face shields should be worn whenever there is a possibility that electrolyte may escape from lead–acid batteries, and rubber gloves should be worn when acid-contaminated parts such as fill caps are handled. Jump-starting generally takes place by starting a stalled car that has a dead or low battery by means of battery power from another car with a running engine. Since the engine of the assisting car is running, all necessary steps should be taken to prevent injury from contact with the moving parts of the running engine and to avoid entanglement of the jump-start cables with the running engine. The jump-start cables should have physically separated positive and negative cables. If the cables are webbed together, accidental shorting may occur owing to uncontrolled motions of dangling cable clamps. The following is a summary adapted from the recommendations of the Independent Battery Manufacturers' Association (IBMA) and the Battery Council International (BCI).*

---

* Readers interested in obtaining more information may contact IBMA, 100 Larchwood Drive, Largo, Florida 33540, or BCI, 401 North Michigan Avenue, Chicago, Illinois 60611. (Their *Battery Service Manual* on lead–acid batteries is highly recommended.)

In all the steps described below, and on other occasions when one is working with lead–acid batteries, it is essential that connections be made to lead–acid batteries in such a way that if any sparks occur, they must not occur near the battery, where explosive gases may be present, or near the carburetor or any of the fuel lines, where gasoline vapor may be present. Similarly, other sources of ignition such as lighted cigarettes or lighters must not be used in the vicinity of the battery or the fuel lines. The following is the recommended sequence of steps to be adhered to when jump-starting a car:

1. Make sure that the voltages of the two batteries are the same. Never try to charge a low-voltage battery with a high-voltage battery.
2. The vent caps of vented lead–acid batteries should be properly tightened and a moistened cloth placed over the vent caps of both batteries before charging begins.
3. The cars should be positioned to prevent physical contact between them. There should be no possibility of any electrical contact between the two cars other than that provided by the jump-start cables.
4. Properly and fully insulated jump-start cables should be used, with no worn insulation or exposed bare parts and with proper spring-loaded clamps.
5. The positive (+) cable clamp should be connected to the positive (+) terminal of the dead battery. No metallic parts of the two clamps of the positive cable should touch any metal of either car other than the positive (+) terminals.
6. The other end of the positive (+) cable clamp should be connected to the positive (+) terminal of the live battery. No metallic parts of the two clamps of the positive cable should touch any metal of either car other than the positive (+) terminals.
7. Then, one end of the negative (−) cable should be connected to the negative (−) terminal of the booster battery.
8. Next, the other end of the negative (−) cable should be connected to the engine block of the stalled engine. *Do not connect it to the negative terminal of the dead battery.*
9. Stand away from the two batteries when starting the stalled car.
10. When the engine of the stalled car is running, remove the jump-start cables in the reverse order to that indicated above, i.e., go from 8 to 7 to 6 to 5, and make sure the cable clamps do not contact any metal parts during the removal operations.

These are simple rules, and, if observed, will eliminate many serious accidents and injuries.

Another cause of accidents may be attributed to the use of improper procedures in charging of lead–acid batteries with chargers powered by household current or some other power source such as a motor generator. In the case of vented batteries, if they are not mounted upright in the car during charging, they should be placed in a level and stable position before charging begins. The fill caps should be removed carefully to slowly release any internal pressure and without spilling any sulfuric acid that may adhere to the fill caps. It is recommended that a moist cloth be placed over the fill ports to prevent escape of sulfuric acid mists and droplets that may be formed during charging. It is also advisable that operators wear eye protection and rubber gloves when handling the batteries and battery parts. The use of eye protection, in particular, is very important. Before the charger is connected to its power supply and before it is connected to the battery, the voltage selector of the charger should be set to the correct charging voltage. Then, the charger output cables should be connected to the battery terminals, with all the precautions given above for jump-starting being followed. It is imperative that the charger cables be connected to the battery with the correct polarity. The polarity should be double-checked before charging power is applied to the battery. Only then is the charger connected to the power supply. These procedures are reversed at the completion of the charging process. It is particularly important that sparking be avoided before, during, and after the charging and that no sources of ignition be present in the vicinity of the battery since explosive gas mixtures are formed during charging. Also, the space in which charging is carried out must be well ventilated to prevent the accumulation of explosive gas mixtures.

If it is necessary to add water to the batteries to replenish the electrolyte, it should be done slowly using nonmetallic water dispensers to prevent accidental short circuits. As a matter of practicality, and unrelated to battery safety, distilled water should be used as the impurities present in nondistilled water might affect the battery performance adversely. Overfilling should be avoided. Since electrolyte mists may escape from the fill ports during and after water replacement, protective eye wear should be worn when the electrolyte level is checked visually and when the electrolyte density is checked with a suction-operated pycnometer. At the completion of charging and/or water replacement, the fill ports should be closed again with the fill caps, and the caps should be securely tightened. Any excess electrolyte and water present on the top of the battery or elsewhere should be removed with a moist cloth and the battery wiped clean. The operator should wear rubber gloves when cleaning the battery to prevent acid burns.

The same general rules apply to the charging of any vented rechargeable battery. The only difference is that exposure to electrolyte may now involve alkaline burns rather than acid burns if the operators are careless. In the case of sealed rechargeable batteries, the same precautions should be observed except those that relate to the opening of the battery and the replenishing of the electrolyte, which are not necessary in this case.

The precautions that must be taken to avoid accidents and injuries with primary batteries have been discussed in many of the preceding chapters, and we summarize the essential points briefly. Except for the small button and coin cells that have no space for warning labels, most primary batteries carry labels that warn users against the principal abuses that are likely to cause injuries. The rules apply to all cells and batteries whether large or small, whether marked with warning labels or not. Most of these precautions apply to rechargeable batteries as well, but with the obvious exception related to the warning not to recharge the batteries. The essential precautions are:

1. Do not dispose of batteries in fires.
2. Do not recharge the batteries.
3. Do not install batteries with the wrong polarity.
4. Do not short-circuit the batteries.
5. Do not mix used batteries with fresh batteries.
6. Do not mix different types of batteries.

Failure to observe these precautions may cause battery leakage, rupture, and/ or explosions, all of which are likely to cause injuries.

There are other precautions as well whose observance contributes to the safe usage of batteries. Among them, we mention precautions concerning the exposure of batteries to high temperatures from any source, the use of excessive mechanical force in handling batteries, and attempts to attach electrical leads to batteries by welding or soldering. Users who assemble high-voltage batteries from single cells should be aware that batteries should never have intermediate voltage taps. The use of intermediate voltage taps will give rise to an imbalance between the different segments of a battery, a primary cause of cell reversal and the associated hazards.

All the precautions we have given and others that could be given may be condensed into the simple statement that common sense and thoughtfulness are required to avoid accidents and injuries with batteries.

16.3. REGULATORY ASPECTS

Although most accidents occur during active use and use-related activities, accidents may also happen during other phases of battery life because of improper handling and care. These other phases include transportation, storage, and disposal of batteries, and we address some of the precautions that need to be observed during the inactive parts of the battery life cycle.

The first considerations involve the labeling of batteries and of shipping containers and the construction of shipping containers. Since batteries contain toxic and/or corrosive substances, battery labeling, packaging, and transportation are all regulated by law. In the United States, the relevant rules and regulations have been promulgated by agencies of the U.S. Government in cooperation with industrial representatives. Similar regulations may be expected to apply in other countries, especially since batteries are an item of international trade, but regulations in other countries may differ in some of their details.

The regulations that apply within the United States, regardless of the origin of the batteries, are based on the Federal Hazardous Substances Act (FHSA) and the Poison Prevention Packaging Act (PPPA). These are comprehensive documents, and we discuss only some of their regulations with no pretension of completeness. Interested readers should consult the relevant documents for specific details. The Consumer Product Safety Commission of the Federal Trade Commission has jurisdiction over the poison and explosive labeling requirements for the batteries and the cartons in which they are shipped. The Department of Transportation has jurisdiction over the shipping carton requirements and the labeling requirements for the shipping cartons. The applicable regulations are concerned with all types of batteries, and the most detailed labeling requirements pertain to lead–acid batteries because of their predominance in the accident statistics relative to other types of batteries. The relevant documents are*:

1. Federal Hazardous Substances Act (FHSA)
2. Federal Hazardous Substances Act Regulations
3. Poison Prevention Packaging Act (PPPA)
4. Federal Trade Commission, Fair Packaging and Labeling Act
5. U.S. Department of Commerce, National Bureau of Standards, Model State Packaging and Labeling Regulations

Insofar as batteries are concerned, the FHSA applies to batteries that contain toxic, corrosive, irritational, flammable or combustible, and pressure-generating substances. The definitions of these terms are stated in the FHSA, and it is clear that all batteries are covered by the regulations without exception. The Act also applies to toys and other articles intended for use by children if

---

* These documents may be obtained from Superintendent of Documents, U.S. Government Printing Office, Washington, D.C. 20402.

they present electrical, mechanical, or thermal hazards. This means that battery-operated toys are covered by the Act.

The labeling regulations require that the batteries and the shipping containers identify the manufacturers and their addresses and that the batteries carry adequate warning labels that identify the nature of their hazards, whether they are due to flammable materials, poisonous materials, materials that may cause burns, or other adverse reactions that may be caused by the release of materials from the batteries or if the batteries or the containers should be mishandled in some way. Also, the battery labels should inform users of actions to be avoided (see preceding sections) or actions to be taken to avoid accidents. Not only do the regulations identify the information that must be given on the labels, they also require that the labels conform to an acceptable format in regard to the size of the printed matter and the contrast of the label markings.

The labeling requirements for the shipping containers differ for different products. For example, shipping containers for dry-charged lead–acid batteries that do not contain any sulfuric acid should not carry the warning "Poison— Causes Severe Burns." This warning is required on shipping cartons used for batteries that do contain sulfuric acid. The word "Poison" must appear in 18-pt. size type. The shipping containers must be resistant to the materials they contain. Quite detailed test procedures are spelled out in the regulations (PPPA) to determine whether or not the containers satisfy the requirements of the Act.

16.4. BATTERY DISPOSAL

The disposal of aqueous electrolyte batteries presents few problems for the general user unless he deliberately mishandles or otherwise maltreats the batteries. Very few accidents have been reported that are related directly to the handling of discarded batteries. There are good reasons for this that are related to the types of batteries involved and the manner of their disposal. With few exceptions, the larger size batteries, principally lead–acid batteries, are returned to the vendors for recycling, and the vendors have either adopted or developed their own safe handling procedures to avoid acid spills, acid burns, and exposures to toxic lead and lead compounds. The same holds for the larger sizes of Ni/Cd batteries. In the case of the smaller batteries that constitute by far the major portion of batteries discarded by the general user, once the batteries have been used, their hazard potentials are low, unless the batteries are abused. The hazards are associated primarily with the toxic and corrosive materials they contain and the possible presence of pressurized hydrogen in the cells. As long as the small batteries remain structurally sound, their hazard potential is low.

The worst mistake that can be made in the disposal of batteries, regardless of their state of charge, is disposal by casual incineration. This will almost certainly cause battery ruptures or explosions with a consequent scattering of hot electrolyte beyond the incinerator that may inflict severe chemical and thermal burns on nearby personnel. The force of the explosions is due to the high pressures generated by the steam formed by the heating of the electrolyte and due to secondary explosions of any hydrogen that may be present in the cells. An additional consideration is that the force of explosions may scatter burning material beyond the incinerator and start fires if any flammable material is within reach.

A less severe, but still serious, mishandling is the mechanical destruction of batteries by crushing, puncturing, or other means, whether deliberate or inadvertent. This may occur during casual disposal of batteries with other wastes in household waste bins or in the larger waste collection units employed by community waste collectors. Since most alkaline cells are pressurized owing to the presence of hydrogen inside the cells, the rupture of the cells is likely to generate a spray of electrolyte that may injure nearby personnel.

Clearly, accident prevention during battery disposal is a matter of following the simple rules printed on the batteries and summarized in Section 16.2.

A discussion of the environmental impact and recycling of discarded batteries and the associated hazards and precautionary steps are subjects beyond the purview of this text. These aspects of battery disposal were discussed briefly in Chapter 2 and more fully in the references cited at the end of that chapter. Interested readers are encouraged to study the available literature on battery disposal and recycling.

17

Ambient Temperature Lithium Systems

The majority of nonaqueous battery systems contain metallic lithium as the anode material. All lithium batteries are high-energy systems and under certain conditions may present a safety hazard. These batteries may be classified as either liquid cathode, solid cathode, or solid state. In general, the liquid cathode systems contain the highest energy and are capable of the highest rates. They, therefore, present the greatest risk for hazardous behavior. The solid-state systems are more benign, primarily because of their higher internal impedance. However, under certain abusive conditions, even the "safest" lithium battery is capable of releasing a large amount of energy in an unsafe manner.

When evaluating the safety hazards associated with a particular lithium battery, several factors need to be considered. These include: (1) the total lithium content, (2) the toxicity, flammability, and stability of component materials and reaction products, (3) the cell balance and design, (4) thermal management properties, (5) the system's tolerance to abuse, and (6) the use of safety features, for example, vents, thermal and electrical fuses, diodes, and seals.[1]

Hazardous behavior in a lithium battery may either be due to manufacturing defects or be user induced.[2] Manufacturing defects can usually be attributed to poor quality control. They include poor welds, defective separators, impure materials, equipment out of specification, and insufficient testing. User-induced hazardous behavior can be caused by various means: use in equipment or under conditions for which the battery was not designed, exposure to abusive conditions, for example, high temperature, short circuit, charging, or overdischarging, or improper design of multicell batteries.

17.1. PRIMARY CELLS

17.1.1. Designing Cells for Safety

Cell designs can be grouped into two major categories, low rate and high rate. The low-rate cells are characterized by thick electrodes and low surface area. They can be built in button, prismatic, or cylindrical configurations. High-rate cells have much thinner electrodes and a large surface area and are usually found in cylindrical or prismatic configurations. The low-rate designs are inherently safer and should be used whenever possible.

Numerous design features have been developed to improve the safety of lithium cells. Hermetically sealed cells will normally contain a safety vent, designed to open and relieve the pressure, as well as to eject active material, when the internal pressure reaches a value corresponding to a temperature somewhat below the melting point of lithium (181°C). This allows the cell to benignly relieve its pressure and eliminates active material before the lithium melts. Molten lithium is extremely reactive, since its passive film does not remain intact on the surface. Therefore, if lithium melts in a cell having all its active components still within the case, runaway reactions and catastrophic failure may result. Various vent designs have been developed. Some are scored in the cell case either as a line or in various designs, others involve the glass- or ceramic-to-metal seal in the header, and some are specially designed vents that are welded into the cell case. An example of the latter kind is the frustum reverse buckling vent.[3] It is compression loaded so that the pressure causes reverse buckling. Prescored lines on the disk cause it to open without fragmenting. The pressure at which the vent opens can be varied. Care must be taken when welding these vents into the case, since excess heat may result in a change in the pressure at which the vent opens.

Another safety feature used in primary lithium cells is the fusible separator.[4–6] The fusible separator is a two-layer material, one layer with a high melting point and the second with a low melting point (below the melting point of lithium). If the internal temperature of the cell increases, for example, as a result of a high current drain due to an external short, the low-melting component melts and forms a high-resistance barrier. This limits the current and prevents the cell from exploding. Polypropylene has been used as the high-melting material, which acts as a support. A low-melting polyethylene has been used as the fusible component.

Some cells capable of delivering very high currents have been built with an internal fuse.[7,8] The fuse comprises a narrowed portion of the tab connecting one of the electrodes to its terminal. If too high a current is drawn from the cell, the thin portion of the tab melts, opening the circuit and preventing temperature and pressure buildup. Another design that can be used in cylindrical cells is a circular disk having a groove concentric to the center of the disk. The central portion is connected to the cell terminal, while the outer portion is connected to one of the electrodes. A potentially hazardous situation may arise with the use of fusible tabs in nonaqueous cells. If the cell is positioned such that the electrolyte is in contact with the fusible link, the high local temperature prior to the fuse opening may result in an unsafe condition. Boiling of the solvent can cause high internal pressure, or decomposition of the electrolyte may occur with the formation of unstable or reactive compounds.

Another type of internal fuse, or safety switch, has been developed for use with solid cathode systems.[9] It utilizes a current collector fabricated from a Ni–Ti shape memory alloy. The base portion has extending legs that make electrical contact with the terminal in the cover, while the base is in contact with the electrode. When the internal temperature of the cell rises, the legs retract to their original position in the plane of the base plate, breaking the electrical circuit in the cell.

Some lithium cells have been built with external fuses, located between the header and a false top. Devices made from conductive polymers having a positive temperature coefficient of resistance (PTCR) have also been used to protect lithium cells and batteries.[10] PTCR devices have a resistivity of ~ 1 Ω cm at 20°C, ~ 10 Ω cm at 100°C, and $>10^6$ Ω cm at 125°C. They can protect a cell or battery in either the current limit or thermal limit mode. If the device is closely coupled to the cells, it will be thermally activated. If it is remotely located, it will be activated by current. A combination of the two will give more reliable protection. When connected in series, the PTCR device will pass normal currents. If a fault occurs that results in excessive current or overheating, the PTCR device will switch to its high-resistance mode and limit the current until the temperature returns to normal.

A number of other design features have been developed to improve the safety of lithium cells. One involves the use of a conductive composite material located between the cathode and its collector.[11] The electronically conductive composite consists of a metal powder mixed with a fluorocarbon powder. As the temperature increases, the fluorocarbon expands more rapidly than the metal. At a predetermined temperature, the interparticle contact between the metal particles is lost and an open circuit occurs. The process is reversible as the cell cools, and, as the temperature drops below its critical level, the cell will again deliver current.

In spirally wound cells having lithium on the outside and the tab on the outside end of the lithium, the outside wrap acts as a reservoir of lithium if the cell is driven into reversal.[12] This lithium can then plate on the cathode in the form of dendrites and lead to shorts and thermal runaway. If the lithium is attached away from the end, the outer wrap becomes electrically disconnected at the end of life, and lithium is not available to plate on the cathode during reversal. Also, placing the cathode on the outside (known as cathode outer wrap design, or COW) imparts further safety during reversal.

Other safety innovations include applying polyethylene microcapsules containing LiCl to the separator in spirally wound cells to prevent inflammation upon short circuit or overheating,[13] adding polyethylene microcapsules containing mineral oil to poison the cell when overheated,[14] coating the current collectors with a counterion-containing polymer which loses conductivity when the active material is exhausted and prevents polarity reversal,[15] connecting a heavy metal having a dissolution voltage higher than the lithium and lower than the nickel collector to the anode, to prevent polarity reversal during overdischarge,[16] and the use of various electrolyte solutions and additives to improve the general safety during abusive conditions.[17-20]

Physical features that have been developed for improved safety include an expandable case fabricated from two overlapping cylindrical casings that are press-fit together. At a predetermined pressure, the cylinders glide-expand to accommodate the increase in pressure with no case rupture.[21] Also, the use of a hollow terminal pin containing a liquid (e.g., H_2O) that is sealed under reduced pressure or vacuum improves the heat dissipation efficiency of the cell.[22]

Other safety features, designed specifically for particular cell chemistries, have been discussed in the sections relating to those chemistries.

17.1.2. Safe Handling of Lithium Cells and Batteries

Certain precautions need to be taken when handling high-energy cells and batteries to prevent injury to personnel or damage to equipment.[23,24] Everyone who will be handling these units, whether to assemble cells into multicell batteries or to install them in equipment, should be fully trained in the safety procedures to be followed. These procedures should be reviewed at regular intervals. All work must be done on nonconductive surfaces, metal benches should be covered with an insulating material, and the work area should be clean and free of sharp objects. Workers should remove jewelry, such as rings, wristwatches, pendants, etc., that could come in contact with the battery terminals. When cells are not in their original packages, they should be neatly arranged to prevent shorting and should not be stacked. All tools used with the cells should be nonconductive. Cells should be transported in plastic trays on pushcarts to prevent dropping.

Operational precautions include (a) cut only one lead at a time, (b) make all connections by welding whenever possible, (c) check welding efficiency several times a day—use a shear rather than a peel test, (d) if you must solder, use cells with high-temperature jackets, place shields on solder pots, and do not let soldering irons come in contact with the cell case, (e) make leads as short as possible, (f) tape leads when they are not connected to their appropriate contact, (g) use plastic calipers, and (h) avoid stacking finished components with exposed cells.

Cells should be stored in their original containers in a well-ventilated, dry area that is isolated from combustible materials. A class D fire extinguisher and respirators should be available. Never stack heavy objects on top of boxes containing batteries, and do not store excessive quantities in any storage area.

17.1.3. Multicell Batteries

Safe cells do not ensure a safe battery! Safety must be designed into battery packs.[25,26] If the battery pack consists of cells that are designed to vent, for example, liquid cathode chemistries, care must be taken not to inhibit the vents from opening when necessary. A volume must be included for the cells to vent into, and a path must be available for the gas to escape. Thermal management must be considered in the design of multicell battery packs, especially those containing many cells. The location of the battery during its intended use must also be taken into account when considering thermal management. In many instances, conditions approaching adiabatic may be seen by a battery or by cells in the interior of a battery.

When designing a battery, cells with the lowest power output necessary to meet requirements should be chosen. Different cell chemistries should never be mixed in a battery pack, and cells used in series or parallel connections should always be of the same size. Cells from the same lot and history should be used, and partially discharged cells should never be mixed with fresh cells. A center tap should not be incorporated in a string of series-connected cells, since this may result in cells becoming unequally discharged.

Depending on the battery design and application, one should consider the use of fuses, current-limiting resistors, or thermal switches. When using thermal switches, one must ensure good thermal contact with the cells. If parallel strings of cells are used, blocking diodes should be incorporated into the design to prevent one string from charging another. Factors to consider when choosing a diode are (a) the forward current, which is the maximum current the diode can pass, must be greater than the current the battery is to deliver, (b) the reverse current, which is the amount of current the diode will pass in the reverse direction, should be less than 0.1 mA to alleviate the effects of charging currents, (c) the breakdown voltage, which is the level of reverse voltage that the diode can withstand before breakdown occurs, should be greater than the open-circuit voltage of any one string, and (d) the forward voltage drop, which is the voltage lost due to the diode in the circuit, must be considered in terms of the output voltage of the battery. Under some conditions, reverse-bias diodes may be necessary on individual cells in a series string to prevent them from being driven into reversal. Diodes with a low internal resistance should be used to prevent cells from premature discharge through the diode.

Care must also be taken in the welding or soldering of all intercell connections and leads. Strong joints are necessary to prevent shorting in case of environmental stress during transportation, handling, or use.

A number of safety features have been developed for use with multicell lithium batteries. Thin sensors have been used to continually monitor the pressure, temperature, and output voltage of individual cells in a battery.[27] The sensors are connected to an alarm/control network that monitors the condition of the circuit and allows corrective action to be taken before an explosion or venting occurs. An exit gas drying and flame safety unit has been designed for use with lithium batteries.[28] Gases from venting cells must pass through the unit, located on the side of the case, prior to leaving the battery. The first part of the receptacle removes moisture and acid particles. The flame safety unit is formed from fused porous polymer, sponge metal, or metal fleece. Any flashback of a flame into the battery, caused by an outer ignition, is prevented by fast combustion of gases in a flame extinguisher chamber. A leakproof battery case has been developed for liquid cathode batteries. The case contains a shock-absorbing material, for example, polyurethane foam, which is impregnated with $NaHCO_3$ to neutralize any materials escaping from the cells within the case.[29]

NASA has devised a number of safety features that are used with batteries on all manned space vehicles. These consist of both internal designs and external devices to prevent hazardous conditions from occurring.[30] To help prevent short circuits within the batteries, the inner surfaces of metal battery cases are coated with an insulating paint that is resistant to the electrolyte. This prevents battery grounds to the case in the event of electrolyte leakage. The cell terminals are protected from contacting other surfaces by potting or by use of a nonconducting barrier. The portions of the terminals inside, or passing through, the case are insulated. The parts of the terminals outside the case are protected from accidental bridging by use of female connectors. All wires inside the battery case are insulated and physically constrained from moving. In addition, various current interrupters, for example, fuses, circuit breakers, and thermal switches, are located close to the battery, in the ground leg, and are rated well below the short-circuit current of the battery.

Overheating of the batteries that might lead to electrolyte leakage is prevented by protection against short circuits, as described above, and by the use of heat sinks or cooling loops in conjunction with thermal switches. The batteries are also insulated from external heat. The risk of circulating currents occurring inside the batteries is minimized by preventing electrolyte leakage.

Also, a few large cells are used rather than many smaller cells connected in parallel. A small, conservatively current-rated Schottky barrier rectifier is placed in each parallel leg as a blocking diode.

Overdischarge protection is provided if more than one cell is connected in series. All cells have shunt diodes across them to remove the load whenever a single cell nears 0 V. Charge prevention is accomplished by not paralleling batteries, strings of cells, or batteries with external power supplies. If paralleling must be done, blocking diodes are used.

All batteries used on manned space flights must be two-failure tolerant; that is, failure of two hazard controls shall not result in any hazard to the crew or to other equipment.

17.1.4. Fire Safety

Several studies have been conducted to determine the most effective means of extinguishing a fire in which quantities of lithium batteries are present.[31,32] The major contributor to a lithium battery fire is the flammable electrolyte. Graphite powder fire extinguishers (Lith-X) are effective only during the early stages of a lithium battery fire. If graphite is used, the extinguished residue should be washed with excess water to react with any remaining lithium. Carbon dioxide extinguishers are not effective, and Halon extinguishers should never be used on a lithium fire. Applying copious amounts of a fine water spray has been found to be the most effective way to combat a lithium battery fire, for all types of lithium batteries. The spray helps extinguish the fire, destroys any unreacted lithium, absorbs some toxic gases, and helps keep adjacent batteries cool, reducing the probability of these batteries venting or exploding.

17.2. RECHARGEABLE CELLS

Lithium rechargeable cells are relatively new, and not much has been reported concerning their safety. Many of the safety considerations are similar to those of primary lithium cells, since the same or similar materials (e.g., lithium metal and organic electrolyte solvents) are used in primary and rechargeable lithium cells. However, there are important differences that one needs to be aware of. In particular, the hazardous behavior of rechargeable lithium cells may increase with increasing cycle life. This is due to the fact that lithium forms dendrites on recharge, and many of these needlelike particles will disengage from the bulk of the anode. Thus, most rechargeable lithium cells contain a three- to sixfold excess of lithium to improve their cycle life. Near the end of useful life, the anode contains a reactive sponge of lithium, having a high surface area and interspersed with solution reaction products. When dry, this residue may be shock sensitive.[33] Safety tests, therefore, should not only be performed on fresh and discharged cells but also early and late in the cell's cycle life.

A number of design features, both chemical and physical, have been developed to improve the safety of rechargeable lithium cells. These are discussed in the following sections.

17.2.1. Chemical Considerations

Rechargeable lithium cells utilizing an electrolyte, with at least one component comprising an ether, can produce peroxides and radicals on overcharge, which may lead to an explosion. The addition of either alkali halides, alkaline earth halides, tetraalkylammonium halides, or haloalkyls to the cathode or electrolyte, in amounts corresponding to 2–200mM concentration in the electrolyte, will eliminate the peroxides and radicals and prevent explosions on overcharge.[34]

The presence of water in batteries having a chalcogenide cathode material may lead to hazardous behavior. The use of an electrolyte salt of $LiPF_6$, which reacts with water at elevated temperatures, has been found to reduce the danger of explosion.[35]

Use of nontypical cathode materials has also been reported to result in safe rechargeable lithium cells.[36] Onium salts derived from nitriles, either liquid or polymeric, and an iodine cation in an organic electrolyte have provided safe, high-voltage cells with short-circuit currents as high as 17.5 mA/cm^2.

The addition of phosphate esters to the organic electrolyte has been shown to reduce the fire hazard of rechargeable lithium batteries in the event cells vent under abusive conditions.[37]

17.2.2. Overcharge Protection

If sufficiently large overcharge currents are pushed through rechargeable lithium cells, they can react violently.[33] Redox agents, for example, metallocenes, have been used to provide overcharge protection. In addition to preventing violent behavior, they also help to maintain capacity balance among individual cells in a series string. The metallocenes, dissolved in the electrolyte, are used as a redox shuttle.[38] They remain unreactive until the cell is fully charged. Then, at a potential slightly above the normal charge cutoff voltage of the cell, the redox shuttle is electrochemically converted (oxidized) to products which diffuse to the negative electrode, where they are reduced or react with lithium to form the starting materials. The cell potential is fixed at the oxidation potential of the redox shuttle during the overcharge period. The particular material that is best suited for use as a redox shuttle depends on the system voltage; for example, cobaltocene is best suited for cells that are fully charged by 1.70 V, ferrocene for cells fully charged at 2.20 V, and nickelocene for cells fully charged by 2.65 V.

Another approach that has been taken to protect lithium cells against overcharge is to choose a solvent in which the lithium–solvent reaction is reversible. Then, the reaction products can be recharged back to lithium and the solvent.[39] This reversibility allows the cell to be overcharged without any loss of active materials from the system.

The addition of LiBr to the electrolyte has been shown to provide overcharge protection to systems that are fully charged by 3.50 V.[40] In these cells, the LiBr is oxidized at potentials positive to 3.5 V, protecting electrolyte solvents which are oxidized at higher potentials.

17.2.3. Short-Circuit Prevention

Several design features have been devised to prevent short circuit of cells after repeated cycling. One design utilizes anodes having a larger surface area than the cathodes, so, upon stacking, the edges of the anode are not superimposed with the cathode.[41] A design for use in spirally wound cells consists of an anode comprised of a sheet of a lithium-insertable, low-melting alloy, covered by a narrower sheet of lithium foil. Narrow strips of the alloy on both the upper and lower edges are left exposed, and the separator is wider than the electrodes.[42] Another design used in wound cells consists of a microporous polyolefin separator that is wider than the electrodes. The portions of the separator protruding from the electrodes are bent to the center of the spiral by a hot air flow to cover the electrodes.[43]

17.2.4. Lithium Dendrite Suppression

Several design features to suppress lithium dendrite growth in rechargeable lithium cells have evolved. One method involves cathodically protecting the case by electrically connecting it to the anode. The case must then be insulated, or the cell stack encased in an inert liner, to prevent the case from becoming a center for lithium deposition.[44] Alternatively, the case may be made from a material that does not corrode at the normal operating potential of the cell, such as Ni, Mo, Cr, or stainless steel. The anode tab must be coated with an insulating material, however, to prevent lithium deposition on it.[44]

A cell has been designed to prevent dendrite growth on the anode as well as deformation of the anode. It consists of a cathode that is wider than the anode and a coating on the edges of the anode with an insulator that is the same thickness as the anode. A separator is placed between the electrodes.[45]

Another design found to suppress dendrite formation utilizes an anode comprised of a lithium layer on an aluminum alloy sheet, with the aluminum layer facing the cathode, and a separator between the electrodes.[46]

Lithium dendrite growth has also been found to be suppressed in cells having a wound design by using an anode having lithium foil pressed on both sides of a metal sheet that does not react with lithium, for example, a sheet of Ni, Cu, or stainless steel.[47]

17.2.5. Multicell Batteries

In order for rechargeable lithium batteries to safely deliver an acceptable capacity, all the cells in a series string need to remain in a narrow range of state of charge over the entire useful life of the battery.[48] This requires the use of electronic protection devices against inadvertent overcharge or overdischarge. One battery has been designed with diodes, either light-emitting diodes (LED) or Zener diodes, connected in parallel across each cell to prevent overcharging.[49]

REFERENCES

1. R. F. Bis, J. A. Barnes, W. V. Zajac, P. B. Davis and R. M. Murphy, Safety Characteristics of Lithium Primary and Secondary Battery Systems, Report NSWC TR 86-296, Naval Surface Weapons Center, Silver Spring, Maryland (1986).
2. G. Halpert, S. Subbarao and J. J. Rowlette, The NASA Aerospace Battery Safety Handbook, *Jet Propulsion Laboratory Publication* 86-14, Jet Propulsion Laboratory, Pasadena, California (1986).
3. B. C. Navel, An innovative rupture disk vent for lithium batteries, in *Proceedings of the Symposium on Primary and Secondary Ambient Temperature Lithium Batteries,* Proc. Vol. 88-6, pp. 169–174, The Electrochemical Society, Pennington, New Jersey (1988).
4. K. Yamamoto, T. Murata and M. Ishikura, Lithium batteries, Jpn. Kokai Tokkyo Koho JP 61,232,560 (October 16, 1986).
5. D. J. Steklenski, Fusible separators for lithium batteries, European Patent Appl. EP 352,617 (January 31, 1990).
6. A. Yoshino, K. Nakanishi and A. Ono, Batteries with improved safety and resistance to short circuiting, Jpn. Kokai Tokkyo Koho JP 03,203,159 (September 4, 1991).
7. D. Biegger, Battery terminal fuse, U.S. Patent 4,879,187 (November 7, 1989).
8. J. P. Descroix, J. L. Firmin and J. P. Planchat, Battery with a lithium anode and a liquid cathode, French Patent Appl. 86/17,426 (December 12, 1986); European Patent Appl. EP 271,086 (June 15, 1988).
9. P. Georgopoulos and J. Langkau, Electrochemical cell with internal circuit interrupter, U.S. Patent 4,855,195 (August 8, 1989).

10. A. Tomlinson, New means of Li/SO$_2$ battery protection using positive temperature coefficient of resistance (PTC) polymeric materials and devices, in *Proceedings of the 30th Power Sources Symposium,* pp. 130–133, The Electrochemical Society, Pennington, New Jersey (1983).

11. R. C. McDonald, J. Pickett and F. Goebel, *Proc. IECEC* **26**(6), 74 (1991).

12. J. B. McVeigh, T. F. Reise and A. H. Taylor, Spirally wound electrochemical cells, U.S. Patent 4,707,421 (November 17, 1987).

13. K. Yamamoto, T. Murata, Y. Harada, T. Mizuno and H. Nagura, Nonaqueous batteries with safety devices, Jpn. Kokai Tokkyo Koho JP 63,86,355 (April 16, 1988).

14. D. H. Fritts, Over-temperature battery deactivation system, U.S. Patent 4,075,400 (February 21, 1978).

15. Y. Hino, Y. Harada, M. Nakanishi, M. Kitakata and H. Nagura, Lithium batteries with coated current collectors, Jpn. Kokai Tokkyo Koho JP 02,144,854 (June 4, 1990).

16. S. Ubukawa and T. Amezutsumi, Nonaqueous batteries, Jpn. Kokai Tokkyo Koho JP 02,189,866 (July 25, 1990).

17. K. Shinoda, K. Yamamoto, Y. Harada and M. Kitakata, Nonaqueous batteries with mixed electrolyte solvents, Jpn. Kokai Tokkyo Koho JP 01,200,562 (August 11, 1989).

18. K. Shinoda, K. Yamamoto, Y. Harada and M. Kitakata, Nonaqueous batteries, Jpn. Kokai Tokkyo Koho JP 01,200,563 (August 11, 1989).

19. U. Alpen, Nonaqueous organic electrolyte for lithium batteries, Ger. Offen. DE 3,611,123 (October 8, 1987).

20. V. Z. Leger and J. W. Marple, Nonaqueous electrolyte, U.S. Patent 4,952,330 (August 28, 1990).

21. H. Sauer, Sealed lithium battery, Ger. Offen. DE 3,443,454 (May 28, 1986).

22. Matsushita Electric Industrial Co. Ltd, Battery with a nonaqueous or alkaline electrolyte, Jpn. Kokai Tokkyo Koho JP 58,164,150 (83,164,150) (March 25, 1982).

23. H. K. Street, S. C. Levy and C. C. Crafts, Safe Design and Assembly of Li/SO$_2$ Batteries, Report RS 2523/80/1, Sandia National Laboratories, Albuquerque, New Mexico (1980).

24. M. J. Brookman, *Electronic Products* **1984** (November 15), 79.

25. S. C. Levy, *J. Power Sources* **43–44**, 247 (1993).

26. J. J. Ciesla, Safety considerations for fabricating lithium battery packs, *J. Power Sources* 18(2/ 3), 101 (1986); *The 1985 Goddard Space Flight Center Battery Workshop,* NASA Conference Publication 2434, pp. 75–91 (1986).

27. U.S. Department of the Army, Monitoring electrochemical cells, U.S. Patent Appl., U.S. 831,027 (June 6, 1986).

28. E. Buder, H. U. Metz and D. Uebermeier, Multiple-cell battery with a combined exit-gas drying and flame-safety unit, Ger. Offen. DE 3,503,014 (July 31, 1986).

29. L. F. Urry, Nonaqueous battery construction, U.S. Patent 4,184,007 (January 15, 1980).

30. National Aeronautics and Space Administration, Manned Space Vehicle Battery Safety Handbook, JSC-20793, Lyndon B. Johnson Space Center, Houston (1985).

31. A. Attewell, The behaviour of lithium batteries in a fire. *J. Power Sources* **26**, 195 (1989).

32. W. N. C. Garrard, An Introduction to Lithium Batteries, MRL-GD-0018, Department of Defence, Materials Research Laboratory, Melbourne, Australia (1988).

33. K. M. Abraham and S. B. Brummer, Secondary lithium cells, in *Lithium Batteries* (J. P. Gabano, ed.), Academic Press, London (1983).

34. F. Kita and K. Kajita, Secondary nonaqueous batteries containing halide compounds, Jpn. Kokai Tokkyo Koho JP 02,114,464 (April 26, 1990).

35. Y. Mochizuki, K. Tsuchia and K. Inada, Secondary nonaqueous batteries, Jpn. Kokai Tokkyo Koho JP 02 65,071 (March 5, 1990).

36. T. Shono and S. Kashiwamura, Nonaqueous electrolyte secondary battery, Jpn. Kokai Tokkyo Koho JP 60,182,661 (September 18, 1985).

37. M. Ue, Fire-resistant electrolytes for lithium batteries, Jpn. Kokai Tokkyo Koho JP 04,184,870 (November 19, 1990).
38. K. M. Abraham and D. M. Pasquariello, Overcharge protection of secondary, non-aqueous batteries, U.S. Patent 4,857,423 (August 15, 1989).
39. A. N. Dey, W. L. Bowden, H. Kuo, M. L. Gopikanth, C. Schlaikjer and D. Foster, All inorganic ambient temperature rechargeable lithium battery—II, in *Proceedings of the 31st Power Sources Symposium,* pp. 89–97, The Electrochemical Society, Pennington, New Jersey (1984).
40. W. K. Behl, Rechargeable lithium–organic electrolyte battery having overcharge protection, U.S. Patent Appl., U.S. 744,344 (September 27, 1985).
41. J. Yamaki, Y. Sakurai, S. Yamada and T. Oosaki, Secondary lithium nonaqueous battery, Jpn. Kokai Tokkyo Koho JP 01,128,370 (May 22, 1989).
42. Y. Nishikawa, T. Morita, Z. Ito, T. Fujii, J. Yamaura and S. Hamada, Secondary nonaqueous lithium battery, Jpn. Kokai Tokkyo Koho JP 01,143,140 (June 5, 1989).
43. T. Fujii, S. Hamada, K. Komatsu, Z. Ito, Y. Nishikawa and J. Yamaura, Cylindrical secondary lithium batteries, Jpn. Kokai Tokkyo Koho JP 01,307,176 (December 12, 1989).
44. O. K. Chang, J. C. Hall, J. Phillips and L. F. Sylvester, Cell design for spirally wound rechargeable alkaline metal cell, U.S. Patent 4,863,815 (September 5, 1989).
45. K. Yamamoto, Y. Hino, Y. Harada and H. Nagura, Secondary nonaqueous lithium batteries, Jpn. Kokai Tokkyo Koho JP 03,129,678 (June 3, 1991).
46. A. Watanabe and H. Muramatsu, Secondary lithium batteries, Jpn. Kokai Tokkyo Koho JP 01,134,875 (May 26, 1989).
47. K. Ikegami and T. Morita, Secondary lithium batteries with coiled electrode stack, Jpn. Kokai Tokkyo Koho JP 02,144,861 (June 4, 1990).
48. I. Faul and J. Knight, *Chem. Ind. (London)* **24,** 820 (1989).
49. N. Koshiba, Y. Okuzaki and K. Momose, Series circuit for secondary lithium batteries, Jpn. Kokai Tokkyo Koho JP 61,206,179 (September 12, 1986).

18

Storage, Transportation, and Disposal of Lithium Cells

Special care must be taken in the storage, transportation, and disposal of lithium batteries because of their high energy content and the reactivity of lithium. In the United States the transportation and disposal of lithium batteries are regulated by the Department of Transportation (DOT) and the Environmental Protection Agency (EPA). Similar regulations are in effect in other countries.

18.1 STORAGE

Lithium batteries are classified as "flammable solid," and care must be taken to prevent them from causing, or being exposed to, a fire environment. In addition, many lithium systems contain toxic or other hazardous materials that may cause injury or death if released in large quantities. Therefore, the following guidelines are presented to reduce the risk associated with storing quantities of lithium batteries[1-3]:

(a) If large quantities of lithium batteries are to be stored, storage should be in a remote area. No other hazardous or combustible materials should be stored in the vicinity.
(b) The storage area should be well ventilated, cool (temperatures should never exceed 54°C), and protected with a sprinkler system. If a sprinkler system is not available, a noncombustible storage building should be used.

(c) Fresh batteries should be stored in their original shipping containers. Spent batteries should be kept separate from new ones.
(d) The quantity of stored batteries should be kept to a minimum.
(e) No smoking or open flames should be allowed in the area.

18.2. TRANSPORTATION

Lithium cells, lithium batteries, and items containing lithium cells and/or batteries are considered as hazardous materials for the purpose of transportation. Shipment in the United States is regulated by the DOT in accordance with the Code of Federal Regulations (49 CFR). Shipment outside the United States by air is regulated by either the International Civil Aviation Organization or the International Air Transport Association.

Lithium cells and batteries are classified for transportation according to the amount of metallic lithium contained in them.

18.2.1. Cells Containing Less than 0.5 g of Lithium

Lithium cells containing up to 0.5 g of lithium and batteries containing up to 1 g of lithium are exempt from the regulations if the following requirements are met[4]:

1. Each cell is hermetically sealed.
2. Cells are separated so as to prevent short circuits.
3. Batteries are packaged in strong outside packaging, except when installed in electronic devices.
4. If a multicell battery contains ≥ 0.5 g of lithium or lithium alloy, it may not contain a liquid or gas that is a hazardous material, unless the liquid or gas would be completely absorbed or neutralized by other materials in the battery if released.

18.2.2. Cells Containing More than 0.5 g of Lithium

Cells containing more than 0.5 g of lithium are authorized for commercial transportation by motor vehicle, rail freight, cargo vessel, and cargo-only aircraft under Exemption DOT-E 7052. This exemption is for primary cells containing the following cathode materials: vanadium pentoxide (V_2O_5), manganese dioxide (MnO_2), sulfur dioxide (SO_2), poly(carbonmonofluoride) $[(CF_x)_n \ (x \approx 1)]$, thionyl chloride ($SOCl_2$), and sulfuryl chloride (SO_2Cl_2). Secondary lithium cells containing titanium disulfide (TiS_2) and molybdenum disulfide (MoS_2) are also included in this exemption.

Shipment of lithium batteries under DOT-E 7052 is allowed if the following conditions are met:

1. The maximum amount of lithium per cell is 12 g (46 Ah), and the maximum amount of lithium per container is 500 g.
2. Batteries containing cells, or series strings of cells, connected in parallel must have appropriately positioned diodes.
3. Each cell or battery must either be of a design that will preclude violent rupture under any conditions incident to transportation or incorporate a safety venting device.
4. Packaging requirements include:

   (a) All inner containers and all inner surfaces must be separated by at least one inch of vermiculite or equivalent material for shipment in drums.
   (b) An effective means to prevent external short circuits for each cell and battery must be provided.
   (c) For shipment by cargo-only aircraft, the outside container must be a removable head drum of DOT specification 17H or 17C or equivalent, equipped with a gastight gasket.
   (d) For shipment by motor vehicle, rail freight, or cargo vessel, the outer container must be either a strong wooden box, DOT specification 12B fiberboard box, DOT specification 21C fiber drum, or DOT specification 17H or 17C metal drum.
   (e) Packages must be labeled FLAMMABLE SOLID.

5. The following tests must be performed on each type of cell or battery:

   (a) Short-circuit test of three representative cells from each week's production.
   (b) Exposure of a minimum of ten cells and one battery of each week's production to 75°C for 48 h with no evidence of leakage, distortion, or internal heating.

6. The following cells are exempt from the requirement for weekly testing and use of DOT specification 17H or 17C drums: hermetically sealed cells having up to 12 g of lithium and containing the following cathode materials: V_2O_5, $(CF_x)_n$, MnO_2, $SOCl_2$, SO_2, $SOCl_2/BrCl$, SO_2Cl_2/Cl_2, TiS_2, or MoS_2. However, prior to the first shipment, ten cells or four batteries must pass the following series of tests:

(a) The batteries or cells shall be stored for 6 h at a pressure corresponding to an altitude of 50,000 ft (1.7 psi) at $24 \pm 4°C$.
(b) The same cells shall then be stored at $75°C$ for 48 h.
(c) The cells or batteries shall be subjected to the vibration test regime given in paragraph 8e iii of DOT-E 7052.
(d) The cells or batteries shall be subjected to the shock test regime given in paragraph 8e iv of DOT-E 7052.

To pass the above tests, cells or batteries must not exhibit any evidence of leakage, outgassing, weight loss, or distortion.

18.2.3. Special Exemptions

Special exemptions have been issued by the DOT to specific companies. Included among these are:

1. DOT-E 8141 and DOT-E 8979 for the transport of lithium/thionyl chloride cells containing >12 g of lithium per cell.
2. DOT-E 8564 and DOT-E 9181 for the transport of lithium/thionyl chloride reserve batteries containing >12 g of lithium per battery.
3. DOT-E 8457, DOT-E 9348, and DOT-E 9355 for the transportation of small lithium cells aboard passenger-carrying aircraft.
4. DOT-E 10190 for the transportation of lithium/sulfur dioxide batteries on passenger-carrying aircraft, provided they are packaged in a specially designed container containing an SO_2 getter.[5]

18.2.4. Discharged Cells

Primary lithium cells that have been discharged to the extent that the open-circuit voltage is <2.0 V, or batteries containing such cells, are not authorized for shipment under DOT-E 7052. Such cells and batteries are regulated by 49 CFR 173.1015. This regulation allows lithium batteries to be transported to permitted storage facilities or disposal sites by motor vehicle only, provided the following conditions are met:

1. When new, the battery contained no more than 12 g of lithium per cell.
2. The battery is equipped with an effective means for preventing external short circuits. This can be accomplished by placing each cell in a plastic bag, by taping the terminals or lead wires, or by reusing the original packing materials.

3. The battery is classified as ORM-C (Other Regulated Material; C signifies that the waste lithium possesses inherent characteristics that make it unsuitable for shipment unless properly identified and prepared for transportation).
4. The batteries are overpacked in a strong fiberboard box or metal or fiber drum in compliance with 49 CFR 173.24.

These conditions do not apply to lithium batteries that were exempt from regulation under the Half Gram Rule, CFR 49 173.206f, when new.

18.2.5. Shipment outside the United States

Transportation of lithium cells and batteries internationally by cargo-only aircraft is regulated by either the United Nations affiliated International Civil Aviation Organization (ICAO) or the International Air Transport Association (IATA). Both organizations adhere to the same requirements for packaging, testing, and maximum lithium weight limits for transport by cargo-only aircraft as given in DOT-E 7052. IATA does allow lithium cells and batteries of every solid cathode material to be shipped under its regulations. IATA also uses DANGEROUS WHEN WET labels rather than the FLAMMABLE SOLID labels required by DOT-E 7052.

18.3. DISPOSAL

Used lithium batteries are considered hazardous waste by the U.S. Environmental Protection Agency (EPA) and, as such, are regulated under the Resource Conservation and Recovery Act (RCRA) of 1976, as well as by the individual states. RCRA defines hazardous waste in two general categories: listed in 40 CFR 261 or by characteristics. The characteristics of hazardous waste are identified as (1) ignitability, (2) corrosivity, (3) reactivity, and (4) extraction procedure toxicity. Lithium batteries are not listed in 40 CFR 261 but fall under the characteristic of reactivity. However, waste generators have the ultimate responsibility for determining whether or not their batteries are hazardous waste (40 CFR 262.11). To dispose of a hazardous waste, it must first be treated to render it nonhazardous. In addition, the Department of Transportation regulates the movement of all hazardous materials, of which hazardous waste forms a subsection.

18.3.1. Treatments

The following are some treatments that have been used to render waste lithium batteries nonhazardous.[6]

Incineration normally results in the expulsion of components or the release of toxic or reactive vapors. The residue requires analysis for the presence of toxic materials prior to disposal. The main problems associated with this treatment are explosions and the release of dense smoke. Venturi scrubbers need to be used to remove particulates, and packed tower scrubbers are used for gases.

Hydrolysis or neutralization can be hazardous. Cells need to be shredded or otherwise opened and then treated with copious amounts of water. The aqueous sludge can be sent to a commercial wastewater treatment facility or needs to be treated before being discharged to the sewer system. Concerns with hydrolysis of lithium batteries are related to the generation of hydrogen gas, dense smoke (similar to incineration), exploding cells, burning lithium, and damage to equipment. Emissions need to be controlled, as with incineration. Hydrogen gas can be controlled by adjusting the feed rate and exhaust ventilation.

Encapsulation can be accomplished if the cells do not contain any EPA listed hazardous waste and are hazardous only because they possess an EPA characteristic. These wastes can be mixed with nonhazardous waste, the mixture then becoming the waste. This mixture should not exhibit any of the characteristics of hazardous waste. Concrete is acceptable as a material with which the cells can be mixed. However, for large volumes of cells, the long-term effectiveness is uncertain.

Ocean dumping is permitted only in case of emergency and is limited to waste batteries generated at sea. Disposal of waste batteries in the ocean that have been generated on land requires an EPA permit.

Detonation is now prohibited. It had been used by certain military organizations as a last resort to dispose of waste lithium batteries that could not be transported to a regulated disposal site.

Recycling of lithium batteries is not presently practiced. Although much effort is presently being expended worldwide on the recycling of other battery types,[7] no financially sound process is available for lithium systems.

18.3.2. Technologies

A technology that has been developed to deactivate and dispose of lithium batteries has been approved by the EPA and local authorities.[8,9] This technology consists of safely opening the batteries and then reacting the contents hydrolytically. The batteries are placed on a conveyor belt which passes through an opening in the wall of a specially constructed room, designed to resist explosions. The batteries are dropped into a hopper and pass into a hammermill. While the hammermill opens the batteries, they are deluged with a 12% NaOH solution. As the particulate size is reduced to <6 mm, they fall through a grid in the bottom of the mill into a reaction tank filled with 12% NaOH solution. After the reaction is complete, the solids are separated from the liquid. The liquid is sent to a commercial wastewater treatment facility, and the solids are drummed and sent to a secure landfill. All gases emitted by this process are scrubbed. A venturi is used to remove particulates, and an alkaline packed tower is used to remove the acidic gases.

The U.S. Army has devised a method for allowing used Li/SO_2 batteries to be disposed of in landfills. A resistor is placed inside the battery case with a switch on the outside that connects the resistor across the terminals. Before a battery is to be disposed of, the resistor is switched on and the battery is allowed to discharge completely to 0 V. Since the Army uses Li/SO_2 batteries having a balanced design, that is, containing approximately equal quantities of lithium and sulfur dioxide, this procedure essentially reacts all of the active materials. The EPA has ruled "a Li/SO_2 battery that has been fully discharged to zero volts would contain substantially reduced quantities of reactive materials such that the battery is not likely to exhibit any of the properties of the reactivity characteristic."

A special facility has been built to deactivate and dispose of very large (10,000 Ah) $Li/SOCl_2$ batteries.[10,11] To ensure that a minimum amount of H_2 gas is generated during treatment, the facility only accepts batteries with a lithium-limited design. The facility consists of three pump houses, three deactivation tanks, and an evaporation pond. A one-ton monorail traverses the entire facility. When batteries arrive, they are checked for open-circuit voltage. If the open-circuit voltage is greater than zero, they are discharged completely in a special area. Batteries exhibiting zero volts (either upon receival or after discharge) are placed in the deactivation facility. The fill ports are exposed and carefully opened. Deactivation hoses are attached to the fill ports, and the unit is lowered into a deactivation tank that is partially filled with water. Each cell is then pressurized with N_2 until it bursts. Water is then pumped through each cell for 21 days, at which time Na_2CO_3 is added to the water, and circulation is continued for another three days. The pH is checked, and circulation is continued until the pH is above 6. At that point, the battery is raised and the type of burst determined. If the unit burst at the top, it is returned to the tank for another 21 days of circulation. If it burst along the bottom (normal), it is drained, flushed, and removed to a landfill. The water remaining in the tanks is then pumped to the evaporation pond, which is lined with high-density polyethylene. After evaporation, the remaining salts are removed for treatment or recovery.

In cases where batteries cannot be fully discharged to remove all the lithium, for example, because of loss of electrolyte, a conductive graphite slurry is added to the cell head space to induce many low-level shorts within the cell. After one month, all available $SOCl_2$ is consumed, and the cell has no current capability. It can then be safely transported to the disposal site and deactivated in the same manner as the lithium-limited units.

REFERENCES

1. R. F. Bis, J. A. Barnes, W. V. Zajac, P. B. Davis and R. M. Murphy, Safety Characteristics of Lithium Primary and Secondary Battery Systems, Report NSWC-TR 86-296, Naval Surface Weapons Center, White Oak, Maryland (1986).
2. W. N. C. Garrard, An Introduction to Lithium Batteries, MRL-GD-0018, Department of Defence, Materials Research Laboratory, Melbourne, Australia (1988).
3. Department of the Navy, Technical Manual for Batteries, Navy Lithium Safety Program Responsibilities and Procedures, S9310-AQ-SAF-010, Naval Sea Systems Command, Washington D.C. (1988).
4. Code of Federal Regulations, 49 CFR, 173.206f, Office of the Federal Register, National Archives and Records Administration (1992).
5. S. C. Levy, Transportation container for Li/SO$_2$ batteries on passenger aircraft, in *Proc. Lithium '87,* Waste Resource Associates, Niagara Falls, New York (1987).
6. R. F. Bis and W. V. Zajac, U.S. Navy Lithium Battery Disposal, Report ESD 88-179, Naval Weapons Support Center, Crane, Indiana (1989).
7. P. Bro and S. C. Levy, Batteries and the environment, in *Environmental Oriented Electrochemistry* (C. A. C. Sequeira, ed.), Elsevier, Amsterdam (1993).
8. A. Wilde, The proper disposal of lithium batteries, in *Proceedings of the 2nd Annual Battery Conference on Applications and Advances,* California State University, Long Beach (1987).
9. W. V. Zajac, Jr., H. G. Kautz, D. J. Kautz, A. J. Bossert, and S. Cohen, Method and apparatus for neutralizing the reactive materials in lithium batteries, U.S. Patent 4,637,928 (January 20, 1987).
10. R. C. McDonald, F. Goebel, J. S. Shambaugh, and M. A. Slavin, Deactivation and disposal of large 10,000 Ah Li/SOCl$_2$ batteries, *Electrochemical Society Extended Abstracts,* Vol. 81-2, pp. 48–49 (1981).
11. T. B. Haskins, Innovative, large 10K amp-hour battery treatment and disposal, *Proc. Lithium '87,* Waste Resource Associates, Niagara Falls, New York (1987).

Definitions and Abbreviations

All the letter symbols employed in the mathematical equations are defined in the text where used. Standard chemical symbols are used in the chemical reaction equations.

COMMONLY USED TERMS

Acceptor	A substance that accepts electrons
Anode	Negative electrode of a primary cell
bar	A pressure equal to 0.987 atm.
Bar	A pressure of one million bars, a megabar
Battery	A single cell or several cells connected in series or parallel
Brisance	The time rate of pressure increase in an explosion
Capacity	Electrical charge in ampere-hours or coulombs
Cathode	Positive electrode of a primary cell
Cell	A single self-contained electrochemical system capable of delivering electrical power and comprising an anode, a cathode, electrolyte, separator, and containing structures
Conflagration	Rapidly expanding fire
Deflagration	Explosive process in which the combustion wave progresses at subsonic velocity
Depolarizer	Active electrode material of positive electrode
Detonation	Explosive process in which the combustion wave progresses at supersonic velocity
Donor	A substance that delivers electrons
Mach number	The actual velocity of an object or of an advancing reaction front divided by the sound velocity in the same medium

Negative	Electrode undergoing reduction on charge and/or oxidation on discharge
Overcharge	Electrical charge or process in which more charge is supplied to a battery than is required to convert the electrodes to the fully charged state
Overvoltage	Difference between the operating voltage of a cell on charge or discharge and the open-circuit voltage of the cell, both at the same state of charge
Oxidation	Chemical process in which a substance donates electrons
Polarization	Departure of the voltage of an electrode or of a battery from its equilibrium value
Positive	Electrode undergoing oxidation on charge and/or reduction on discharge
Primary	Nonrechargeable battery
Reduction	Chemical process in which a substance accepts electrons
Reversal	Process in which discharge current is forced through a cell after its capacity has been exhausted
Secondary	Rechargeable battery
Separator	Porous, nonconductive material interposed between the electrodes of a cell to prevent internal short circuits

ABBREVIATIONS

AA cell	Standard cell size: 14 mm diameter, 50 mm high
AIT	Autoignition temperature
ARC	Accelerating rate calorimetry
BCI	Battery Council International
CFR	Code of Federal Regulations (U.S.)
CRS	Cold-rolled steel
D Cell	Standard cell size: 33 mm diameter, 60 mm high
DOD	Depth of discharge
DOT	Department of Transportation (U.S.)
DSC	Differential scanning calorimetry
DTA	Differential thermal analysis
EOL	End of life, generally to a specific voltage
EPA	Environmental Protection Agency (U.S.)
FHSA	Federal Hazardous Substances Act (U.S.)
FTC	Federal Trade Commission (U.S.)
GTM	Glass-to-metal seal
IATA	International Air Transport Association
IBMA	Independent Battery Manufacturers' Association

ICAO	International Civil Aviation Organization
IR	Infrared
LC_{50}	Lethal concentration, 50% kill probability
LD_{50}	Lethal dose, 50% kill probability
LED	Light-emitting diode
LEL	Lower explosion limit, a concentration
MCL	Maximum contaminant level, a concentration
OSHA	Occupational Safety and Health Administration (U.S.)
ppm	Parts per million, a concentration
PPPA	Poison Prevention Packaging Act (U.S.)
psig	Gauge pressure, pressure in excess of atmospheric pressure
PTC	Positive temperature coefficient
PTCR	Positive temperature coefficient of resistance
RCRA	Resource Conservation and Recovery Act (U.S.)
SOC	State of charge
SS	Stainless steel
STP	Standard temperature and pressure, 0°C, 1 atm
TLV	Threshold limit value, a concentration
TWA	Time-weighted average concentration of a substance
UEL	Upper explosion limit, a concentration
UL	Underwriters Laboratory

SOME CHEMICAL DESIGNATIONS

BCX	Bromine complex
CSC	Chlorine in sulfuryl chloride
DME	Dimethoxyethane
LAN	Lithium anode
PC	Propylene carbonate
PTFE	Polytetrafluoroethylene (Teflon)
TAA	Tetraazoannulene
THF	Tetrahydrofuran
2-Me THF	2-Methyltetrahydrofuran

Appendix
Specialty Batteries

A number of battery systems have been developed for special applications and have seen limited use. Included are both primary and secondary, aqueous and nonaqueous chemistries. In this chapter we will review the safety considerations of some of the more common of these systems, although in some cases the data are rather sparse.

1. NICKEL/HYDROGEN CELLS

The nickel/hydrogen cell is very similar to the nickel/cadmium cell except the Cd anode is replaced with a H_2 electrode. This electrode consists of a Teflon-bonded platinum black catalyst layer, an expanded nickel current collector, and a microporous Teflon backing. The catalyst loading is 3 mg of Pt/ cm^2.[1] A potassium titanate composite is used as the separator. The nickel oxyhydroxide cathode and potassium hydroxide electrolyte remain the same as in the nickel/cadmium cell. The overall appearance of the cell is quite different, since the hydrogen gas is stored under pressure. Typically, 34 atm (500 psi) is the storage pressure for a fully charged cell.[1] The cell case is a stainless steel pressure vessel capable of safely holding this pressure.

$Ni(OH)_2$ is a poison by the subcutaneous route, the lowest published toxic dose for rats being 480 mg/kg.[2] It is also a carcinogen and tumorigen. Hydrogen gas is nontoxic but is a highly dangerous fire and explosive hazard when exposed to heat, flame, or oxidizers.[2]

Hazardous Chemical Reactions. Cell reversal in a hydrogen-limited design will result in oxygen generation at the positive electrode[1,3]:

$$2NiOOH + H_2O \rightarrow 2Ni(OH)_2 + \tfrac{1}{2}O_2 \qquad (A.1)$$

This will result in the buildup of pressure, but well within the limits of the pressure vessel. The oxygen will then react exothermically with hydrogen at the catalytic surface of the anode to form water. This recombination usually maintains the partial oxygen pressure to $\sim 1\%$ of the total.[3]

Cells are normally designed to be cathode limited by adding additional hydrogen gas during fabrication, which also imparts protection against cell reversal. The nickel oxyhydroxide electrode, when completely discharged, generates hydrogen gas as current passes through it. An equivalent amount of hydrogen is oxidized at the anode, maintaining pressure equilibrium. Therefore, cells can be operated indefinitely in reversal with no hazardous buildup of pressure.[3]

Nickel/hydrogen cells also contain a built-in overcharge protection mechanism. Oxygen generated at the nickel oxyhydroxide electrode reacts immediately with hydrogen at the catalytic platinum surface to form water.

2. SILVER/HYDROGEN CELLS

The silver/hydrogen system is similar to the silver/zinc battery, with the zinc anode replaced by a hydrogen anode. The design of silver/hydrogen cells closely resembles that of nickel/hydrogen cells described above.

Silver oxide, Ag_2O, is moderately toxic by ingestion, the LD_{50} in rats being 2820 mg/kg.[2] It is a strong oxidizing agent which may be flammable through chemical reactions.

Hazardous Chemical Reactions. No safety-related reactions have been reported in the literature, but silver oxide is quite soluble in alkaline electrolyte. Unless an argentistatic membrane is used as separator, for example, as in silver/zinc cells, dissolved silver oxide will migrate to the hydrogen anode, which can result in internal shorting.[4]

3. REDOX FLOW BATTERIES

Redox flow batteries utilize completely soluble reactants and products at all stages of charge. The reactants are stored in external tanks as acidified aqueous solutions. The reactants are continually circulated from the storage tanks through the active cells and back to the storage tanks. The anodic and cathodic solutions are separated within the cell by an anion-exchange membrane. Each electrode is a porous carbon or graphite felt pad, which is compressed between the membrane and a terminal plate.[5]

3.1. Iron–Chromium Redox System

The positive reactant in the iron–chromium redox system is a solution of the ferric–ferrous redox couple. The negative reactant is a solution of the chromous–chromic couple. The carbon or graphite felt used for the chromium electrode is lightly coated with a gold–lead catalyst. Chromium compounds in general are known carcinogens. The chromous (Cr^{2+}) ion is a powerful reducing agent.[2]

Use of the gold–lead catalyst inhibits the formation of hydrogen gas during charging. However, if an oxidizing environment (e.g., air or ferric ion) contacts the catalyzed electrode after some period of normal use, the ability of the catalyst to inhibit hydrogen formation could be lost.[5]

3.2. Zinc/Ferricyanide System

The zinc/ferricyanide system is a hybrid redox battery, since zinc metal is the negative electrode in a fully charged system. Sodium ferricyanide solution is the positive, and a sodium hydroxide solution serves as electrolyte. On discharge, zinc is oxidized to ZnO, and ferricyanide is reduced to ferrocyanide. Materials are circulated between the active electrodes and storage tanks, the negative material in the form of sodium zincate.

Zinc oxide is a skin and eye irritant and an experimental teratogen. It is poisonous by the intraperitoneal route. The lowest published toxic concentration for inhalation by humans is 60 mg/m^3.[2] Ferrocyanides exhibit low toxicity but, when mixed with hot, concentrated acid, decompose to form highly toxic hydrogen cyanide (HCN). Upon strong irradiation, acidic, basic, and neutral solutions liberate HCN.[2]

The main concern with zinc/ferricyanide batteries, from a safety point, is leakage or spilling of the $2N$ NaOH electrolyte. Contact with the skin can cause chemical burns.[6]

4. LIQUID AMMONIA BATTERY

The liquid ammonia battery is a reserve system that was developed for ordnance applications. The negative electrode, or anode, is magnesium. The anolyte is a solution of potassium thiocyanate (KSCN) in liquid ammonia.

The separator can be either cellulose or glass fiber. The catholyte consists of ammonium thiocyanate (NH_4SCN) in liquid ammonia. The cathode is either 1,3-dinitrobenzene[7] or mercuric sulfate.[8] The battery consists of two sections: the battery compartment, which contains the electrode assembly, and the activator section, which contains an ammonia reservoir and a gas generator assembly. Activation of the battery is accomplished by firing the gas generator, which then forces electrolyte into the battery compartment by rupturing a diaphragm separating the two sections. The vapor pressure of NH_3 at $+74°C$ is 520 psi, while the battery case is designed to withstand a pressure of 3000 psi.[7]

Ammonia is a human poison, as well as an eye, mucous membrane, and systemic irritant by inhalation. The lowest published toxic concentration for inhalation by humans is 20 ppm, while the lowest published lethal concentration is 30,000 ppm for 5 min. Ammonia is an explosion hazard when exposed to flame or in a fire. Potassium thiocyanate is a human poison. Large doses can cause skin eruptions, psychosis, and collapse. The lowest published lethal dose by ingestion for humans is 80 mg/kg. Ammonium thiocyanate is a poison by ingestion. Human systemic effects include hallucinations and distorted perception, nausea or vomiting, and gastrointestinal effects. The lowest published toxic dose by human ingestion is 430 mg/kg. 1,3-Dinitrobenzene is a human poison by ingestion and skin contact. It can result in cyanosis and motor activity changes. The lowest published lethal dose orally for humans is 28 mg/kg. Mercuric sulfate is also a poison by ingestion and is moderately toxic by skin contact. The oral LD_{50} in rats if 57 mg/kg, while for skin contact in rats the LD_{50} is 625 mg/kg. When mercuric sulfate is heated to decomposition, toxic fumes of mercury and SO_x are emitted.[2]

Several thousand destructive tests were performed during the development of a specifically designed liquid ammonia reserve battery for one fuze application.[9] Additional batteries were subjected to simulated environments, resulting in this system being accepted as a power source for electronic fuzing.

5. ALUMINUM–AIR SYSTEM

The aluminum–air system can be considered as a hybrid fuel cell and battery. The negative electrode is aluminum, and the active component of the positive electrode is air (oxygen).

The aluminum–air system has been developed as a power source for electric vehicles. It consists of an aluminum alloy anode and an air cathode. The air cathode comprises a semiporous mixture of carbon, wetproofing agents, and catalyst on a metal screen current collector. An aqueous sodium or potassium hydroxide solution serves as the electrolyte. Both flowing electrolyte and flowing air streams transport reactants and control the cell temperature.[10]

Sodium hydroxide, both as a solid and in solution, is extremely corrosive to all body tissue, causing burns and, frequently, deep ulceration. Ingestion of sodium hydroxide causes very serious damage to the mucous membranes and to any other tissue with which it comes in contact.[2] Aluminum metal in bulk is not a human poison; however, in powder form it can be hazardous.

The net cell reaction is the oxidation of aluminum in sodium hydroxide solution to form sodium aluminate:

$$Al + NaOH + \tfrac{3}{2}H_2O + \tfrac{3}{4}O_2 \rightarrow NaAl(OH)_4 \qquad (A.2)$$

In a crystallizer unit, the aluminate is broken down to sodium and aluminum hydroxides:

$$NaAl(OH)_4 \rightarrow NaOH + Al(OH)_3 \qquad (A.3)$$

Thus, the net reaction of the battery is

$$Al + \tfrac{3}{2}H_2O + \tfrac{3}{4}O_2 \rightarrow Al(OH)_3 \qquad (A.4)$$

A parasitic corrosion reaction occurs at the anode which produces hydrogen gas:

$$Al + 3H_2O + NaOH \rightarrow NaAl(OH)_4 + \tfrac{3}{2}H_2 \qquad (A.5)$$

Thus, during extended periods of shutdown, care must be taken when draining the caustic electrolyte and in safely releasing the hydrogen gas.[10]

6. ZINC/BROMINE BATTERY

The zinc/bromine battery has been developed for stationary energy storage (load leveling for electric utilities) and electric vehicle propulsion. It is a flowing electrolyte system in which the active materials are stored externally and pumped into the power conversion stack as required during charge and discharge. The battery stack is composed of carbon-filled polyethylene electrodes, separators, and two current collectors within a high-density polyethylene frame.[11] Metallic zinc, plated on the negative electrode, serves as the anode, and complexed bromine is the positive reactant. Depending on the application and manufacturer, the electrolyte may be an aqueous solution of zinc bromide, with either ammonium chloride or zinc chloride added, hydrobromic acid, a quaternary ammonium salt (methyl- or ethylpyrrolidinium bromide, methyl- or ethylmorpholinium bromide), and a small amount of bromine. The cell discharge reaction is[12]

$$Zn + Br_2 \rightarrow ZnBr_2 \qquad (A.6)$$

Bromine is a poison by ingestion and moderately toxic by inhalation. It is an irritant to the mucous membranes, eyes, and upper respiratory tract. Severe exposure can result in pulmonary edema. The lowest published lethal dose for humans orally is 14 mg/kg; by inhalation the lowest published lethal concentration is 1000 ppm.[2] Bromine is a corrosive material and flammable by spontaneous chemical reaction with reducing agents. Complexing of the bromine by a quatenary ammonium salt reduces its vapor pressure, thus lowering the risk associated with this system in the event of a spill or vent.

7. ZINC/CHLORINE BATTERY

The zinc/chlorine system, commonly referred to as zinc–chloride, has been developed for utility load leveling. It consists of a cell compartment and a separate container, known as a store, which is filled with water. The negative electrode comprises a graphite substrate onto which zinc plates, and the positive electrode is fabricated from either porous graphite or ruthenia-catalyzed porous titania. The electrolyte is aqueous $ZnCl_2$, ranging in concentration from $0.5M$ in the fully charged state to $2M$ when the battery is fully discharged. During charge, zinc is plated onto the negative electrode, and chlorine is evolved at the positive. The chlorine passes into the store, where it reacts with the water, which is cooled to $<10°C$, to form chlorine hydrate, $Cl_2 \cdot xH_2O$. This is an exothermic reaction having a heat of formation of approximately -18 kcal/mol, necessitating cooling of the store during charge.[13]

Discharge is initiated by heating the chlorine hydrate via heat exchange with the warmer electrolyte. This breaks down the hydrate, allowing chlorine gas to pass back into the cell. The cell reaction is exothermic, and this heat is transferred to the store to continue the release of chlorine. Chlorine gas is moderately toxic to humans by inhalation. A concentration of 15 ppm causes irritation of the throat, 50 ppm is dangerous for short exposure, and 1000 ppm may be fatal even for brief exposure.[2]

Safety Issues. The major hazard of the zinc–chloride system is the release of gaseous chlorine and its spread in the atmosphere. This can only occur if the battery case ruptures. A number of features have been incorporated into these units to prevent such ruptures[13]:

(a) Operating temperature and pressure are near ambient.
(b) Safety interlocks are located at the module level to monitor failure and deactivate the module in case of malfunction.
(c) A nonflammable coolant is used.
(d) The unit contains a hydrogen–chlorine reactor to maintain low concentrations of hydrogen within the module.

In the event of a rupture, the rate of chlorine release is suppressed by the following features[13]:

(a) Chlorine is stored in the form of a solid hydrate.
(b) If the fracture is minor, the hydrate will remain inside the module, and the chlorine release rate will be small.
(c) A modular concept is used so, in the event of a significant rupture, the amount of hydrate that can exit the unit is limited to ~90 pounds. Tests have demonstrated that dangerous levels of chlorine from a single module rupture are unlikely to extend beyond 50 feet from the accident.

REFERENCES

1. M. Klein and M. George, Nickel–hydrogen secondary batteries, in *Proceedings of the 26th Power Sources Symposium*, pp. 18–20, PSC Publications Committee, Red Bank, New Jersey (1974).
2. N. I. Sax and R. J. Lewis, Sr., *Dangerous Properties of Industrial Materials*, Van Nostrand Reinhold, New York (1989).
3. J. Dunlop, J. Stockel and G. van Ommering, Sealed metal oxide–hydrogen secondary cells, *Power Sources 5* (D. H. Collins, ed.), pp. 315–328, Academic Press, London (1975).
4. P. O'D. Offenhartz and G. L. Holleck, *J. Electrochem. Soc.* **127**(6), 1213 (1980).
5. N. H. Hagedorn, NASA Redox Storage System Development Project, Report DOE/NASA/ 12726-24, NASA TM-83677 (1984).
6. R. P. Hollandsworth, J. Zegarski and J. R. Selman, Zinc/Ferricyanide Battery Development, Phase IV, Report SAND85-7195, Sandia National Laboratories, Albuquerque, New Mexico (1985).
7. L. J. Minnick, Ammonia batteries Part I, Reserve liquid ammonia fuze battery, in *Proceedings of the 17th Annual Power Sources Conference*, pp. 128–131, PSC Publications Committee, Red Bank, New Jersey (1963).
8. D. J. Doan and L. R. Wood, Liquid NH_3 battery systems, in *Proceedings of the 16th Annual Power Sources Conference*, pp. 141–144, PSC Publications Committee, Red Bank, New Jersey (1962).
9. H. R. Smith and B. C. Tierney, Ammonia batteries, in *Proceedings of the 19th Annual Power Sources Conference*, pp. 97–100, PSC Publications Committee, Red Bank, New Jersey (1965).
10. E. Behrin, R. L. Wood, J. D. Salisbury, D. J. Whisler and C. L. Hudson, Design Analysis of an Aluminum–Air Battery for Vehicle Operations, Report UCRL-53382, Lawrence Livermore National Laboratory (1983).

11. L. Richards, W. Vanschalwijk, G. Albert, M. Tarjanyi, A. Leo and S. Lott, Zinc–Bromine Battery Development, Sandia Contract 48-8838, Final Report, SAND90-7016, Sandia National Laboratories, Albuquerque, New Mexico (1990).
12. G. Clerici, M. de Rossi and M. Marchetto, Zinc–bromine storage battery for electric vehicles, in *Power Sources 5* (D. H. Collins, ed.), p. 167, Academic Press, London (1975).
13. D. L. Douglas, Development of the Zinc–Chloride Battery for Utility Applications, Report EM-1417, Energy Development Associates, Madison Heights, Michigan (1980).

Index

Abuse of batteries
  electrical, 72
  mechanical, 71
  thermal, 75
Abuse test protocols
  lithium batteries
    U.S. Navy, 291
    Underwriters Laboratory, 291–293
  nonlithium batteries
    Underwriters Laboratory, 293, 294
Abuse tests
  ammonia batteries, 340
  electrical, 287–289
  environmental, 289, 290
  lithium ion/cobalt oxide cells, 269
  lithium/bromine chloride (BCX) cells, 250, 251
  lithium/chromium oxide cells, 202
  lithium/cobalt oxide cells, 266
  lithium/copper oxide cells, 196, 197
  lithium/copper sulfide cells, 200
  lithium/CSC cells, 253, 254
  lithium/iodine cells, 205
  lithium/iron disulfide batteries, 281
  lithium/iron sulfide batteries, 279, 280
  lithium/iron sulfide cells, 203
  lithium/manganese dioxide cells, 192, 193, 263
  lithium/molybdenum disulfide cells, 258, 259
  lithium/niobium triselenide cells, 267

Abuse tests (cont.)
  lithium/poly(carbonmonofluoride) cells, 194, 195
  lithium/silver vanadium oxide cells, 204
  lithium/sulfur dioxide cells, 241, 242
  lithium/sulfur dioxide cells, rechargeable, 264, 265
  lithium/sulfuryl chloride cells, 252
  lithium/thionyl chloride cells, 224–226
  lithium/titanium disulfide cells, 261, 262
  lithium/vanadium oxide cells, 268
  lithium/vanadium pentoxide cells, 201
  sodium/sulfur batteries, 276–278
Accident reports, 36–38
Acetonitrile toxicity, 233
Age distribution of injured persons, 38
Air electrodes, 176
Air standards to exposure, 34
Alkaline electrolyte
  exposure
    eye, 154
    skin, 154
  ingestion, 154
  inhalation, 154
  toxicity, 153
Aluminum toxicity, 184
Ammonia toxicity, 340
Ammoniacal solution toxicity, 152
Ammonium chloride, 136, 138
Ammonium thiocyanate toxicity, 340
Anodes, 4

Antimony, additive in lead/acid battery, 161
Arsine toxicity, 178, 179
Autoignition temperature, 105

Batteries
    lithium, 3, 4, 6
        configuration, 7
    multicell
        primary, 19, 20
        rechargeable, 20, 21
    primary, 4
    secondary, 4
    with zinc electrodes, explosions, 101
Battery; see also Cell
    alkaline manganese, 4, 6, 11 139
    aluminum/air, 340, 341
    cadmium/mercuric oxide, 10, 144, 145
    iron/chromium, 339
    lead/acid, 4, 5, 11, 12, 18, 19, 158
        construction, 159
    liquid ammonia, 339, 340
    lithium ion/cobalt oxide, 269
    lithium/bismuth oxide, 206
    lithium/bismuth oxychromate, 206
    lithium/bismuth–lead iodide, 206
    lithium/bromine complex (BCX), 249–
        251
    lithium/carbon monofluoride, 4
    lithium/chlorine in sulfuryl chloride
        (CSC), 252, 253
    lithium/chromium oxide, 201, 202, 268
    lithium/cobalt oxide, 265, 266
    lithium/cobalt polysulfide, 206
    lithium/copper oxide, 4, 195–197
    lithium/copper oxyphosphate, 197, 198
    lithium/copper sulfide, 4, 198–200
    lithium/iodine, 204, 205
    lithium/iron disulfide (thermal battery),
        280, 281
    lithium/iron sulfide, 202, 203
    lithium/iron sulfide (lithium monosul-
        fide), 278–280
    lithium/lead iodide, 206
    lithium/manganese, 11
    lithium/manganese dioxide, 189–193
        rechargeable, 262, 263
    lithium/manganese oxide, 4
    lithium/molybdenum disulfide, 257–259
    lithium/niobium pentoxide, 268

Battery
    lithium/niobium triselenide, 266, 267
    lithium/poly(carbonmonofluoride), 193–
        195
    lithium/silver chromate, 205
    lithium/silver vanadium oxide, 203, 204
    lithium/silver–bismuth chromate, 205
    lithium/sulfur dioxide, 4
        rechargeable, 263–265
    lithium/sulfuryl chloride, 251, 252
    lithium/thionyl chloride, 4
    lithium/titanium disulfide, 259–262
    lithium/vanadium oxide, 4, 267, 268
    lithium/vanadium pentoxide, 200, 201
    manganese/zinc, rechargeable, 173
    mercuric oxide, 142–144
    metal/air, 174
    nickel/cadmium, 4, 5, 12, 162
    nickel/hydrogen, 5
    nickel/metal hydride, 12, 165
    nickel/zinc, 168
    redox flow, 338, 339
    silver oxide, 140–142
    silver/zinc, 10, 169–173
    sodium/sulfur, 273–278
    zinc/air, 4, 10, 11, 174–177
    zinc/bromine, 341, 342
    zinc/carbon (Leclanché), 4, 8, 9
    zinc/chloride, 138
    zinc/chlorine, 342, 343
    zinc/ferricyanide, 339
    zinc/mercuric oxide, 9, 10
    zinc/silver oxide, 4, 5
Battery accidents and charging, 303
Battery case corrosion, 69
Battery closures and leakage, 60
Battery fires
    ignition, 27
    lithium batteries, 27
    solvents, 27
Body injuries, 37
Bromine toxicity, 342
Bromine chloride toxicity, 249
γ-Butyrolactone toxicity, 194

Cadmium
    anode, 144
    as corrosion inhibitor, 136
    equivalent weight, 164

Cadmium (*cont.*)
  maximum allowable concentration
    in air, 34
    in drinking water, 33
  toxicity, 148
Calcium, additive in lead/acid battery, 161
Capacity imbalance of cells in rechargeable
    batteries, 20
Catalytic recombination of hydrogen, 19
Cathodes, 4
Cell; *see also* Battery
  bobbin, 137
  cadmium/mercuric oxide, 13
  lithium sulfur dioxide, shock sensitivity,
    234
  lithium/iodine, 17
  lithium/thionyl chloride, thermal mass,
    222
  nickel/cadmium, 17, 18
  nickel/hydrogen, 337, 338
  silver/cadmium, 172
  silver/hydrogen, 338
  zinc/air, 13, 145–147
  zinc/carbon, 136–138
  zinc/manganese dioxide, 13
  zinc/mercuric oxide, 13
  zinc/silver oxide, 13
Cell case deformation, 88
Cell pressure, 7
  effects of gas leakage, 57
  estimation from gassing rates, 55, 56
  gas evolution, 7, 14
  tolerance of D cells, 91
Cell reversal, 48, 49
Cell voltages, 4, 5
Cerium toxicity, 184
Charging
  of lead acid batteries
    controlled voltage, 53–55
    float, 53
  Li/SO$_2$ cells, analysis, 126–131
  primary cells, hazards of, 14, 74, 75
Chemical burns, 29; *see also*, Injuries by
    electrolyte exposure
Chemical hazards and injuries, 28–31
Chemical health hazards
  alkaline manganese cells, 139, 140
  cadmium/mercuric oxide cells, 144,
    145

Chemical health hazards (*cont.*)
  lead/acid batteries, 159–161
  nickel/cadmium batteries, 163, 164
  nickel/metal hydride batteries, 166,
    167
  nickel/zinc cells, 168, 169
  silver/zinc cells, 141, 142, 170–173
  zinc/air batteries, 146, 147, 175–177
  zinc/carbon cells, 137, 138
  zinc/chloride batteries, 138, 139
  zinc/manganese, 173, 174
  zinc/mercuric oxide cells, 143, 144
Chlorine toxicity, 342
Chromium oxide toxicity, 201
Chromium compounds toxicity, 339
Cobalt
  oxide toxicity, 265
  phthalocyanine catalyst, 223
  tetraazoannulene catalyst, 223
  toxicity, 179, 180
Cold-rolled steel
  tensile strength, 90
  yield strength, 90
Composition of batteries: *see* Chemical
    health hazards
Conflagration, 100
Copper oxyphosphate toxicity, 198
Copper II oxide toxicity, 196
Copper II sulfide toxicity, 198
Corrosion and leakage, 69–71
Creepage, electrochemical, 59
Crimp seals and leakage, 61–65
Cylindrical cell cases
  hoop stresses, 89
  longitudinal stresses, 89

Deflagration, 100, 106, 107
Detonation, 100, 106, 107
Diethyl carbonate toxicity, 269
Dimethoxyethane toxicity, 189, 198
Dimethyl sulfite toxicity, 194
1,3-Dinitrobenzene toxicity, 340
Diode
  blocking, 302, 317, 319
  bypass, 301
  reverse-bias, 318
  shunt, 319
Dioxolane toxicity, 196
Drinking water standards, 33

Electrochemical creepage, 59
Electrolyte
    leakage, 28
        electrical abuse, 72–75
        mechanical abuse, 71, 72
        thermal abuse, 75, 76
    spills, 29, 30, 37
Electron acceptors, 3
Electron donors, 3
Energy density
    alkaline primary cells, 25
    lead/acid batteries, 25
    lithium/carbon monofluoride cells, 193
    lithium/chromium oxide cells, 201
    lithium/cobalt oxide, 265, 269
    lithium/copper oxide cells, 195
    lithium/manganese oxide cells, 189
    lithium/manganese oxide rechargeable, 262
    lithium/molybdenum disulfide, 257
    lithium/niobium triselenide, 266
    lithium/silver bismuth oxide, 205
    lithium/silver chromate, 205
    lithium/silver vanadium oxide, 203
    lithium/sulfur dioxide cells, 26
    lithium/thionyl chloride cells, 211
    lithium/titanium disulfide, 259
    lithium/vanadium oxide, 200
    tables of, 4, 5
Entropic heat, 115, 116
Equipment damage and corrosion protection
    of circuits, 33
Ethylene carbonate toxicity, 257
Exothermic reactions in lithium batteries, 102
Explosion of lead/acid batteries, 36, 101

Fault tree analysis of lithium sulfur dioxide
    cells, 241
Finite element methods, 89
Float charging of lead/acid batteries, 53
Fragmentation patterns in battery explosions,
    107
Fusible separators, 301

Gas evolution
    in aqueous electrolyte cells, 20
    charging of aqueous electrolyte batteries,
        49–55
    charging of multicell batteries, 21
    charging of series connected cells, 54

Gas evolution (*cont.*)
    estimating amount, 49
    in lead/acid batteries, 19, 29, 44, 49–
        55
        reduction, 55
    in nickel/cadmium cells, 18
    by zinc corrosion, 13, 14, 44
Gas recombination, 7
Gas scavenging, 304
Gassing inhibition on zinc electrodes, 13
Glass-to-metal seals, integrity of, 66–69

Hazard reduction by
    chemical means, 304
    correct charging methods, 303, 307
    electrical fusing, 301
    electrochemical means, 303
    improved separators, 301, 305
    mechanical design, 299
    safe usage and disposal, 307–311
    thermal control, 306
    use of diodes, 301
Health hazards of batteries, 133; *see also*
        Chemical health hazards
Heat generation
    and cell safety, 123
    in lithium/sulfur dioxide cells on charge,
        127
    in normal discharge, 115–117
    processes and thermal runaway, 115
    rate, 113
    in short circuits, 118
Heat transfer
    conduction, 123–125
    convective, 128
    lithium/sulfur dioxide cells on charge,
        127–131
    overall coefficients, 121, 125
    radiative, 128
    rate, 113
Hermetic seals, leakage paths, 65–69
Hydrogen
    explosions, 54
        spark ignition, 105
        thermal ignition, 105
    hazards, 337
    lower explosion limit, 106
    solubility in KOH, 46
    upper explosion limit, 106

Hydrogen diffusion through porous membranes, 57
Hydrogen evolution at zinc electrodes, 45
Hydrogen generation, maximum rate on charging, 54

Incineration, 23, 35, 312
Inhalation of electrolyte mists, 28, 29
Inhalation of solvent vapors, 28
Injuries by electrolyte exposure
    clothing, 29
    eye contact, 29
    first-aid, 29–31
    ingestion, 29–30
Iron sulfide toxicity, 202

Jump-starting cars, proper procedure, 306, 307

Knudsen diffusion, 81
Knudsen flow, 84

Labeling of batteries, 310
Lanthanum toxicity, 184
Lead, maximum allowable concentration
    in air, 34
    in drinking water, 33
    toxicity, 180
Lead/acid batteries, sealed, relief valves, 95
Lead/acid battery
    environmental issues, 34
    explosions, 26, 28, 29, 101
Lithium alloy anodes in thionyl chloride cells, 221
Lithium batteries
    safety vents, 95
    thermal runaway, 15, 16
    explosions, 102
Lithium battery types, 14, 65
Lithium dithionite in lithium/sulfur dioxide cells, 233, 234
Lithium fires and extinguishers, 27, 28
Lithium hexafluoroarsenate, 16
Lithium hexafluoroarsenate toxicity, 194
Lithium hexafluorophosphate toxicity, 260
Lithium iodide toxicity, 205
Lithium perchlorate

as oxidant, 16
    toxicity, 189
Lithium tetrafluoroborate toxicity, 194
Lithium thiocyanate toxicity, 260
Lithium trifluoromethanesulfonate (triflate) toxicity, 189
Lower explosion limit, 105

Manganese dioxide
    equivalent weight, 137, 140
    toxicity, 189, 190
Manganese toxicity, 149
Mercuric oxide, 143, 144
Mercuric sulfate toxicity, 340
Mercury
    maximum allowable concentration in air, 34
    maximum allowable concentration in drinking water, 33
    toxicity, 150
Methane in lithium/sulfur dioxide cells, 234
Methyl acetate toxicity, 265
4-Methyl-1,3-dioxolane toxicity, 260
Methyl formate toxicity, 200
3-Methyl-2-oxazolidone toxicity, 202
2-Methyl tetrahydrofuran toxicity, 260
Multicell batteries, 6

Negative electrodes, 5
Nickel hydroxide toxicity, 337
Nickel oxyhydroxide equivalent weight, 164
Nickel toxicity, 181, 182
Nickel/cadmium batteries, environmental issues, 34
Nickel/metal hydride batteries, alternative to nickel/cadmium batteries, 34
Niobium triselenide toxicity, 266

Overcharge protection in lithium/titanium disulfide cells, 261
Overvoltage, 115, 116
Oxygen evolution
    lead/acid batteries, 45
    nickel/cadmium batteries, 45
Oxygen generation, maximum rate on charging, 54

Physical damage of explosions, 110
Physical hazards of batteries, 25–28
Pollution by batteries, environmental aspects, 33–35
Poly(carbonmonofluoride) toxicity, 194
Positive electrodes, 5
Potassium hydroxide equivalent weight, 164
Potassium thiocyanate toxicity, 340
Power, 4
Prepressurization and explosive violence, 108
Pressure increase in explosions, 107, 108
Pressure reduction by gas diffusion elements, 57
Pressure relief valves, 7, 95, 109
Pressure waves in flame fronts, 107
Propylene carbonate toxicity, 189
Protective film on lithium, 16, 103

Rare earth metals toxicity, 184
Reaction wave in fuel–air explosions, 106
Recycling, 34
Redox reactions, 3
Redox shuttle, 320
Rubber gloves, 29, 30
Rupture disks, 7

Safety goggles, 26, 29–30
Safety vents in lithium batteries, 95
Salt deposits on batteries, 28, 61
Shock wave, Mach number, 110
Short circuit prevention, in lithium/manganese dioxide cells, 191
Short circuits
    and heat generation, 118
    in lithium batteries, 15
Silver
    equivalent weight, 171
    toxicity, 150
Silver oxide
    equivalent weight, 142
    toxicity, 338
Sodium hydroxide toxicity, 341
Solvent–air explosions, 16, 25, 26
    thermal ignition, 105
Solvent oxidation in lithium batteries, 15
Stability of organic solvents in lithium cells, 16
Stainless steel
    tensile strength, 90

Stainless steel (cont.)
    yield strength, 90
Stibine
    in lead/acid batteries, 161
    toxicity, 178
Sulfur dioxide toxicity, 233
Sulfuric acid
    eye exposure, 185
    ingestion, 185
    inhalation, 185
    skin exposure, 185
    toxicity, 185
Sulfuryl chloride toxicity, 251

Tafel equation, 116
Taguchi analysis, lithium/thionyl chloride cell safety, 218
Temperature rise of cells
    with heat transfer, 121
    with no heat transfer, 118
Tetraglyme toxicity, 197, 198
Tetrahydrofuran (THF) toxicity, 260
Thermal coupling between cells and their environment, 125
Thermal expansion and void space in cells, 58, 91
Thermal runaway, 103, 104, 113
    due to charging of primary batteries, 126
Thionyl chloride toxicity, 211
Titanium disulfide toxicity, 260
Titanium toxicity, 183
Toxicity, general considerations, 31–33
Triglyme toxicity, 260

Upper explosion limit, 106

Vanadium oxide toxicity, 267
Vanadium pentoxide toxicity, 200
Vanadium toxicity, 183, 184
Vent designs, in lithium cells, 314
Ventilation of explosive gases, 19
Venting of cells, 14, 18, 19
Violence of explosions, 107, 108
Voltage, gas-generating reactions, 44
Voltage equivalence of chemical reactions, 18
Voltage reversal, 20
    and gas generation, 15, 20, 48, 49

Water decomposition
   in lead/acid battery, 19, 51, 161
   in nickel/cadmium battery, 18
Welding and battery leakage, 70, 71

Zinc corrosion, 13
Zinc dendrites, 169, 172, 174, 177

Zinc equivalent weight, 137, 142
Zinc hydroxide, 140
Zinc oxide toxicity, 339
Zinc toxicity, 151
Zinc chloride electrolyte toxicity, 155
Zirconium toxicity, 184